Riemann's Zeta Function

Riemann's Zeta Function

H. M. Edwards
New York University
New York, New York

ACADEMIC PRESS, INC.
Harcourt Brace Jovanovich, Publishers
Boston San Diego New York
London Sydney Tokyo Toronto

This book is printed on acid-free paper. ♾

ACADEMIC PRESS, INC.
1250 Sixth Avenue, San Diego, CA 92101-4311

United Kingdom Edition published by
ACADEMIC PRESS LIMITED
24-28 Oval Road, London NW1 7DX

Library of Congress Cataloging in Publication Data

AMS(MOS) 1970 Subject Classifications: 10-02, 10-03, 10H05, 10H15

Edwards, Harold M
Riemann's zeta function.
(Pure and applied mathematics; a series of monographs and textbooks)
1. Numbers, Theory of. 2. Functions, Zeta. I. Title. II. Series.
QA3.P8 [QA241] 510'.8s [512'.73] 73-794
ISBN 0-12-232750-0

PRINTED IN THE UNITED STATES OF AMERICA
92 93 94 9 8 7 6 5

Contents

PREFACE ix
ACKNOWLEDGMENTS xiii

Chapter 1 **Riemann's Paper**

1.1 The Historical Context of the Paper 1
1.2 The Euler Product Formula 6
1.3 The Factorial Function 7
1.4 The Function $\zeta(s)$ 9
1.5 Values of $\zeta(s)$ 11
1.6 First Proof of the Functional Equation 12
1.7 Second Proof of the Functional Equation 15
1.8 The Function $\xi(s)$ 16
1.9 The Roots ρ of ξ 18
1.10 The Product Representation of $\xi(s)$ 20
1.11 The Connection between $\zeta(s)$ and Primes 22
1.12 Fourier Inversion 23
1.13 Method for Deriving the Formula for $J(x)$ 25
1.14 The Principal Term of $J(x)$ 26
1.15 The Term Involving the Roots ρ 29
1.16 The Remaining Terms 31
1.17 The Formula for $\pi(x)$ 33
1.18 The Density dJ 36
1.19 Questions Unresolved by Riemann 37

Chapter 2 **The Product Formula for ξ**

2.1 Introduction 39
2.2 Jensen's Theorem 40
2.3 A Simple Estimate of $|\xi(s)|$ 41
2.4 The Resulting Estimate of the Roots ρ 42
2.5 Convergence of the Product 42
2.6 Rate of Growth of the Quotient 43
2.7 Rate of Growth of Even Entire Functions 45
2.8 The Product Formula for ξ 46

Chapter 3 **Riemann's Main Formula**

3.1 Introduction 48
3.2 Derivation of von Mangoldt's Formula for $\psi(x)$ 50
3.3 The Basic Integral Formula 54
3.4 The Density of the Roots 56
3.5 Proof of von Mangoldt's Formula for $\psi(x)$ 58
3.6 Riemann's Main Formula 61
3.7 Von Mangoldt's Proof of Riemann's Main Formula 62
3.8 Numerical Evaluation of the Constant 66

Chapter 4 **The Prime Number Theorem**

4.1 Introduction 68
4.2 Hadamard's Proof That Re $\rho < 1$ for All ρ 70
4.3 Proof That $\psi(x) \sim x$ 72
4.4 Proof of the Prime Number Theorem 76

Chapter 5 **De la Vallée Poussin's Theorem**

5.1 Introduction 78
5.2 An Improvement of Re $\rho < 1$ 79
5.3 De la Vallée Poussin's Estimate of the Error 81
5.4 Other Formulas for $\pi(x)$ 84
5.5 Error Estimates and the Riemann Hypothesis 88
5.6 A Postscript to de la Vallée Poussin's Proof 91

Chapter 6 **Numerical Analysis of the Roots by Euler–Maclaurin Summation**

6.1 Introduction 96
6.2 Euler–Maclaurin Summation 98
6.3 Evaluation of Π by Euler–Maclaurin Summation. Stirling's Series 106
6.4 Evaluation of ζ by Euler–Maclaurin Summation 114
6.5 Techniques for Locating Roots on the Line 119
6.6 Techniques for Computing the Number of Roots in a Given Range 127
6.7 Backlund's Estimate of $N(T)$ 132
6.8 Alternative Evaluation of $\zeta'(0)/\zeta(0)$ 134

Chapter 7 **The Riemann–Siegel Formula**

7.1 Introduction 136
7.2 Basic Derivation of the Formula 137
7.3 Estimation of the Integral away from the Saddle Point 141
7.4 First Approximation to the Main Integral 145
7.5 Higher Order Approximations 148
7.6 Sample Computations 155
7.7 Error Estimates 162
7.8 Speculations on the Genesis of the Riemann Hypothesis 164
7.9 The Riemann–Siegel Integral Formula 166

Chapter 8 **Large-Scale Computations**

8.1 Introduction 171
8.2 Turing's Method 172
8.3 Lehmer's Phenomenon 175
8.4 Computations of Rosser, Yohe, and Schoenfeld 179

Chapter 9 **The Growth of Zeta as $t \to \infty$ and the Location of Its Zeros**

9.1 Introduction 182
9.2 Lindelöf's Estimates and His Hypothesis 183
9.3 The Three Circles Theorem 187
9.4 Backlund's Reformulation of the Lindelöf Hypothesis 188
9.5 The Average Value of $S(t)$ Is Zero 190
9.6 The Bohr–Landau Theorem 193
9.7 The Average of $|\zeta(s)|^2$ 195
9.8 Further Results. Landau's Notation o, O 199

Chapter 10 **Fourier Analysis**

10.1 Invariant Operators on R^+ and Their Transforms 203
10.2 Adjoints and Their Transforms 205
10.3 A Self-Adjoint Operator with Transform $\xi(s)$ 206
10.4 The Functional Equation 209
10.5 $2\xi(s)/s(s-1)$ as a Transform 212
10.6 Fourier Inversion 213
10.7 Parseval's Equation 215
10.8 The Values of $\zeta(-n)$ 216
10.9 Möbius Inversion 217
10.10 Ramanujan's Formula 218

Chapter 11 **Zeros on the Line**

11.1 Hardy's Theorem 226
11.2 There Are at Least KT Zeros on the Line 229
11.3 There Are at Least $KT \log T$ Zeros on the Line 237
11.4 Proof of a Lemma 246

Chapter 12 **Miscellany**

12.1 The Riemann Hypothesis and the Growth of $M(x)$ 260
12.2 The Riemann Hypothesis and Farey Series 263
12.3 Denjoy's Probabilistic Interpretation of the Riemann Hypothesis 268
12.4 An Interesting False Conjecture 269
12.5 Transforms with Zeros on the Line 269
12.6 Alternative Proof of the Integral Formula 273
12.7 Tauberian Theorems 278
12.8 Chebyshev's Identity 281

12.9 Selberg's Inequality 284
12.10 Elementary Proof of the Prime Number Theorem 288
12.11 Other Zeta Functions. Weil's Theorem 298

APPENDIX On the Number of Primes Less Than a Given Magnitude (By Bernhard Riemann) 299

REFERENCES 306
INDEX 311

Preface

My primary objective in this book is to make a point, not about analytic number theory, but about the way in which mathematics is and ought to be studied. Briefly put, I have tried to say to students of mathematics that they should *read the classics* and beware of secondary sources.

This is a point which Eric Temple Bell makes repeatedly in his biographies of great mathematicians in *Men of Mathematics*. In case after case, Bell points out that the men of whom he writes learned their mathematics not by studying in school or by reading textbooks, but by going straight to the sources and reading the best works of the masters who preceded them. It is a point which in most fields of scholarship at most times in history would have gone without saying.

No mathematical work is more clearly a classic than Riemann's memoir *Ueber die Anzahl der Primzahlen unter einer gegebenen Grösse,* published in 1859. Much of the work of many of the great mathematicians since Riemann—men like Hadamard, von Mangoldt, de la Vallée Poussin, Landau, Hardy, Littlewood, Siegel, Polya, Jensen, Lindelöf, Bohr, Selberg, Artin, Hecke, to name just a few of the most important—has stemmed directly from the ideas contained in this eight-page paper. According to legend, the person who acquired the copy of Riemann's collected works from the library of Adolph Hurwitz after Hurwitz's death found that the book would automatically fall open to the page on which the Riemann hypothesis was stated.

Yet it is safe to say both that the dictum "read the classics" is not much heard among contemporary mathematicians and that few students read *Ueber die Anzahl* . . . today. On the contrary, the mathematics of previous generations is generally considered to be unrigorous and naïve, stated in obscure terms which can be vastly simplified by modern terminology, and full of false starts and misstatements which a student would be best advised to avoid. Riemann in particular is avoided because of his reputation for lack of rigor (his "Dirichlet principle" is remembered more for the fact that Weierstrass pointed out that its proof was inadequate than it is for the fact that it was after all correct and that with it Riemann revolutionized the study of Abelian functions), because of his difficult style, and because of a general impression that the valuable parts of his work have all been gleaned and incorporated into subsequent more rigorous and more readable works.

These objections are all valid. When Riemann makes an assertion, it may be something which the reader can verify himself, it may be something which Riemann has proved or intends to prove, it may be something which was not proved rigorously until years later, it may be something which is still unproved, and, alas, it may be something which is not true unless the hypotheses are strengthened. This is especially distressing for a modern reader who is trained to digest each statement before going on to the next. Moreover, Riemann's style is extremely difficult. His tragically brief life was too occupied with mathematical creativity for him to devote himself to elegant exposition or to the polished presentation of all of his results. His writing is extremely condensed and *Ueber die anzahl* . . . in particular is simply a resumé of very extensive researches which he never found the time to expound upon at greater length; it is the only work he ever published on number theory, although Siegel found much valuable new material on number theory in Riemann's private papers. Finally, it is certainly true that most of Riemann's best ideas have been incorporated in later, more readable works.

Nonetheless, it is just as true that one should read the classics in this case as in any other. No secondary source can duplicate Riemann's insight. Riemann was so far ahead of his time that it was 30 years before anyone else began really to grasp his ideas—much less to have their own ideas of comparable value. In fact, Riemann was so far ahead of his time that the results which Siegel found in the private papers were a major contribution to the field when they were published in 1932, seventy years after Riemann discovered them. Any simplification, paraphrasing, or reworking of Riemann's ideas runs a grave risk of missing an important idea, of obscuring a point of view which was a source of Riemann's insight, or of introducing new technicalities or side issues which are not of real concern. There is no mathematician since Riemann whom I would trust to revise his work.

The perceptive reader will of course have noted the paradox here of a secondary source denouncing secondary sources. I might seem to be saying, "Do not read this book." But he will also have seen the answer to the paradox. What I am saying is: Read the classics, not just Riemann, but all the major contributions to analytic number theory that I discuss in this book. The purpose of a secondary source is to make the primary sources accessible to you. If you can read and understand the primary sources without reading this book, more power to you. If you read this book without reading the primary sources you are like a man who carries a sack lunch to a banquet.

Acknowledgments

It is a pleasure to acknowledge my indebtedness to many individuals and institutions that have aided me in the writing of this book. The two institutions which have supported me during this time are the Washington Square College of New York University and the School of General Studies of the Australian National University; I have been proud to be a part of these particular schools because of their affirmation of the importance of undergraduate teaching in an age when the pressures toward narrow over-specialization are so great.

I am grateful to many libraries for the richness of the resources which they make available to us all. Of the many I have used during the course of the preparation of this book, I think particularly of the following institutions which provided access to relatively rare documents: The New York Public Library, the Courant Institute of Mathematical Sciences, the University of Illinois, Sydney University, the Royal Society of Adelaide, the Linnean Society of New South Wales, the Public Library of New South Wales, and the Australian National University. I am especially grateful to the University Library in Göttingen for giving me access to Riemann's *Nachlass* and for permitting me to photocopy the portions of it relevant to this book.

Among the individuals I would like to thank for their comments on the manuscript are Gabriel Stolzenberg of Northeastern University, David Lubell of Adelphi, Bruce Chandler of New York, Ian Richards of Minnesota, Robert Spira of Michigan State, and Andrew Coppel of the Australian National University. J. Barkley Rosser and D. H. Lehmer were very helpful in providing information on their researches. Carl Ludwig Siegel was very hospitable and generous with his time during my brief visit to Göttingen. And, finally, I am deeply grateful to Wilhelm Magnus for his understanding of my objectives and for his encouragement, which sustained me through many long days when it seemed that the work would never be done.

Chapter 1

Riemann's Paper

1.1 THE HISTORICAL CONTEXT OF THE PAPER

This book is a study of Bernhard Riemann's epoch-making 8-page paper "On the Number of Primes Less Than a Given Magnitude,"† and of the subsequent developments in the theory which this paper inaugurated. This first chapter is an examination and an amplification of the paper itself, and the remaining 11 chapters are devoted to some of the work which has been done since 1859 on the questions which Riemann left unanswered.

The theory of which Riemann's paper is a part had its beginnings in Euler's theorem, proved in 1737, that the sum of the reciprocals of the prime numbers

$$\frac{1}{2}+\frac{1}{3}+\frac{1}{5}+\frac{1}{7}+\frac{1}{11}+\frac{1}{13}+\frac{1}{17}+\cdots \tag{1}$$

is a divergent series. This theorem goes beyond Euclid's ancient theorem that there are infinitely many primes [E2] and shows that the primes are rather *dense* in the set of all integers—denser than the squares, for example, in that the sum of the reciprocals of the square numbers converges.

Euler in fact goes beyond the mere statement of the divergence of (1) to say that since $1+\frac{1}{2}+\frac{1}{3}+\frac{1}{4}+\cdots$ diverges like the logarithm and since the series (1) diverges like‡ the logarithm of $1+\frac{1}{2}+\frac{1}{3}+\frac{1}{4}\cdots$, the series (1)

†The German title is *Ueber die Anzahl der Primzahlen unter einer gegebenen Grösse.* An English translation of the paper is given in the Appendix.

‡This is true by dint of the Euler product formula which gives $\sum(1/n)=\prod(1-p^{-1})^{-1}$ (see Section 1.2); hence $\log\sum(1/n)=-\sum\log(1-p^{-1})=\sum(p^{-1}+\frac{1}{2}p^{-2}+\frac{1}{3}p^{-3}+\cdots)=\sum(1/p)+$ convergent.

must diverge like the log of the log, which Euler writes [E4] as

$$\frac{1}{2} + \frac{1}{3} + \frac{1}{5} + \frac{1}{7} + \frac{1}{11} + \cdots = \log(\log \infty). \tag{2}$$

It is not clear exactly what Euler understood this equation to mean—if indeed he understood it as anything other than a mnemonic—but an obvious interpretation of it would be

$$\sum_{p<x} 1/p \sim \log(\log x) \qquad (x \to \infty), \tag{2'}$$

where the left side denotes the sum of $1/p$ over all primes p less than x and where the sign $\sim$ means that the relative error is arbitrarily small for x sufficiently large or, what is the same, that the ratio of the two sides approaches one as $x \to \infty$. Now

$$\log(\log x) = \int_1^{\log x} \frac{du}{u} = \int_e^x \frac{dv}{v \log v}$$

so (2′) says that the integral of $1/v$ relative to the measure $dv/\log v$ diverges in the same way as the integral of $1/v$ relative to the point measure which assigns weight 1 to primes and weight 0 to all other points. In this sense (2′) can be regarded as saying that the density of primes is roughly $1/\log v$. However, there is no evidence that Euler thought about the density of primes, and his methods were not adequate to prove the formulation (2′) of his statement (2).

Gauss states† in a letter [G2] written in 1849 that he had observed as early as 1792 or 1793 that the density of prime numbers appears on the average to be $1/\log x$ and he says that each new tabulation of primes which was published in the ensuing years had tended to confirm his belief in the accuracy of this approximation. However, he does not mention Euler's formula (2) and he gives no analytical basis for the approximation, which he presents solely as an empirical observation. He gives, in particular, Table I.

TABLE I[a]

x	Count of primes $< x$	$\int \frac{dn}{\log n}$	Difference
500,000	41,556	41,606.4	50.4
1,000,000	78,501	78,627.5	126.5
1,500,000	114,112	114,263.1	151.1
2,000,000	148,883	149,054.8	171.8
2,500,000	183,016	183,245.0	229.0
3,000,000	216,745	216,970.6	225.6

[a]From Gauss [G2].

†For some corroboration of Gauss's claim see his collected works [G3].

Gauss does not say exactly what he means by the symbol $\int(dn/\log n)$, but the data given in Table II, taken from D.N. Lehmer [L9], would indicate that he means n to be a continuous variable integrated from 2 to x, that is, $\int_2^x (dt/\log t)$. Note that Lehmer's count† of primes, which can safely be assumed to be accurate, differs from Gauss's information and that the difference is in *favor* of Gauss's estimate for the larger values of x.

TABLE II[a]

x	Count of primes $< x$	$\int_2^x \frac{dt}{\log t}$	Difference
500,000	41,538	41,606	68
1,000,000	78,498	78,628	130
1,500,000	114,155	114,263	108
2,000,000	148,933	149,055	122
2,500,000	183,072	183,245	173
3,000,000	216,816	216,971	155

[a]Data from Lehmer [L9].

Around 1800 Legendre published in his *Theorie des Nombres* [L11] an empirical formula for the number of primes less than a given value which amounted more or less to the same statement, namely, that the density of primes is $1/\log x$. Although Legendre made some slight attempt to prove his formula, his argument amounts to nothing more than the statement that if the density of primes is assumed to have the form

$$1/(A_1x^{m_1} + A_2x^{m_2} + \cdots)$$

where $m_1 > m_2 > \ldots$, then m_1 cannot be positive [because then the sum (1) would converge]; hence m_1 must be "infinitely small" and the density must be of the form

$$1/(A \log x + B).$$

He then determines A and B empirically. Legendre's formula was well known in the mathematical world and was mentioned prominently by Abel [A2], Dirichlet [D3], and Chebyshev [C2] during the period 1800–1850.

The first significant results beyond Euler's were obtained by Chebyshev around 1850. Chebyshev proved that the relative error in the approximation

$$\pi(x) \sim \int_2^x \frac{dt}{\log t}, \tag{3}$$

†Lehmer insists on counting 1 as a prime. To conform to common usage his counts have therefore been reduced by one in Table II.

where $\pi(x)$ denotes the number of primes less than x, is less than 11% for all sufficiently large x; that is, he proved† that

$$(0.89) \int_2^x \frac{dt}{\log t} < \pi(x) < (1.11) \int_2^x \frac{dt}{\log t}$$

for all sufficiently large x. He proved, moreover, that no approximation of Legendre's form

$$\pi(x) \sim x/(\mathrm{A} \log x + B)$$

can be better than the approximation (3) and that if the ratio of $\pi(x)$ to $\int_2^x (dt/\log t)$ approaches a limit as $x \longrightarrow \infty$, then this limit must be 1. It is clear that Chebyshev was attempting to prove that the relative error in the approximation (3) approaches zero as $x \longrightarrow \infty$, but it was not until almost 50 years later that this theorem, which is known as the "prime number theorem," was proved. Although Chebyshev's work was published in France well before Riemann's paper, Riemann does not refer to Chebyshev in his paper. He does refer to Dirichlet, however, and Dirichlet, who was acquainted with Chebyshev (see Chebyshev's report on his trip to Western Europe [C5, Vol. 5, p. 245 and pp. 254–255]) would probably have made Riemann aware of Chebyshev's work. Riemann's unpublished papers do contain several of Chebyshev's formulas, indicating that he had studied Chebyshev's work, and contain at least one direct reference to Chebyshev (see Fig. 1).

The real contribution of Riemann's 1859 paper lay not in its results but in its methods. The principal result was a formula‡ for $\pi(x)$ as the sum of an infinite series in which $\int_2^x (dt/\log t)$ is by far the largest term. However, Riemann's proof of this formula was inadequate; in particular, it is by no means clear from Riemann's arguments that the infinite series for $\pi(x)$ even *converges*, much less that its largest term $\int_2^x (dt/\log t)$ dominates it for large x. On the other hand, Riemann's methods, which included the study of the function $\zeta(s)$ as a function of a complex variable, the study of the complex zeros of $\zeta(s)$, Fourier inversion, Möbius inversion, and the representation of functions such as $\pi(x)$ by "explicit formulas" such as his infinite series, all have had important parts in the subsequent development of the theory.

For the first 30 years after Riemann's paper was published, there was

†Chebyshev did not state his result in this form. This form can be obtained from his estimate of the number of primes between l and L (see Chebyshev [C3, Section 6]) by fixing l, letting $L \longrightarrow \infty$, and using $\int_2^L (\log t)^{-1}\, dt \sim L/\log L$.

‡See Section 1.17. Note that $Li(x) = \int_2^x (dt/\log t) + \text{const.}$

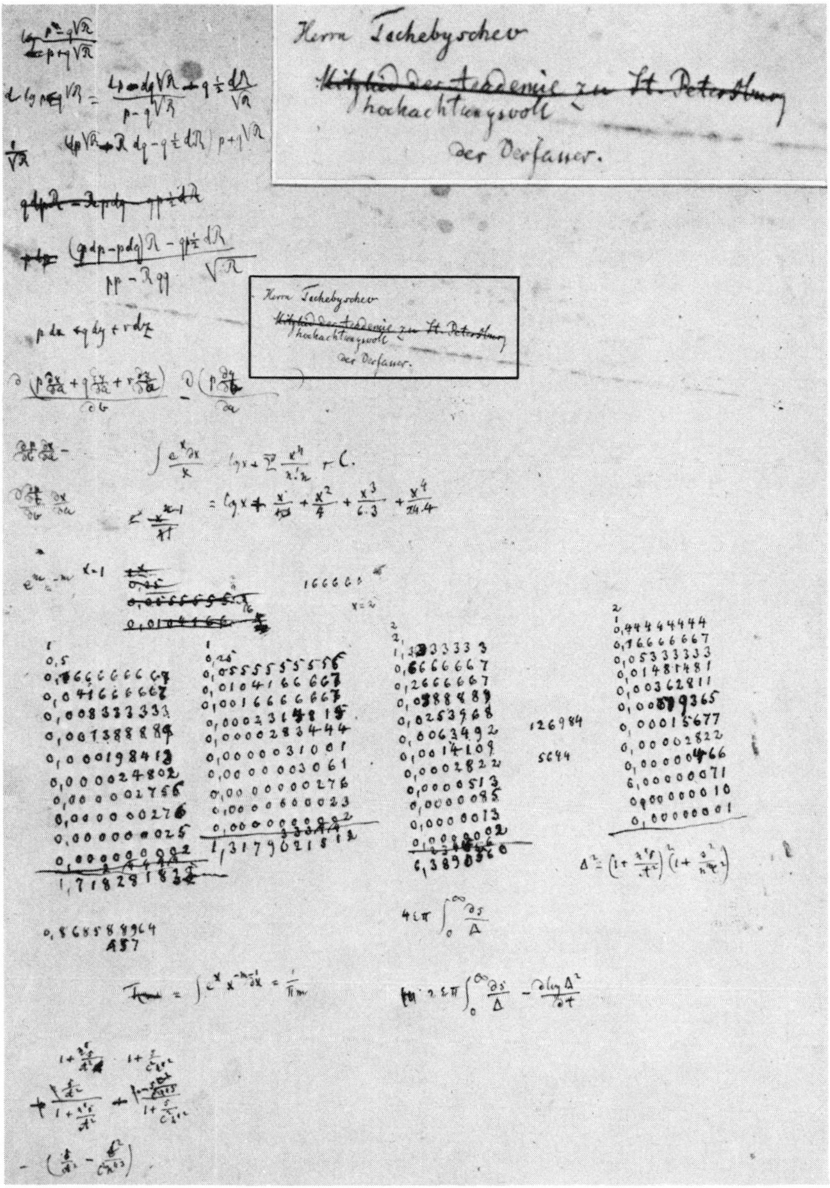

Fig. 1. A scrap sheet used to hold some other loose sheets in Riemann's papers. The note seems to prove that Riemann was aware of Chebyshev's work and intended to send him an offprint of his own paper. In all likelihood Riemann was practicing his penmanship in forming Roman, rather than German, letters to write a dedication to Chebyshev. (Reproduced with the permission of the Niedersächsische Staats- und Universitätsbibliothek, Handschriftenabteilung, Göttingen.)

virtually no progress† in the field. It was as if it took the mathematical world that much time to digest Riemann's ideas. Then, in a space of less than 10 years, Hadamard, von Mangoldt, and de la Vallée Poussin succeeded in proving both Riemann's main formula for $\pi(x)$ and the prime number theorem (3), as well as a number of other related theorems. In all these proofs Riemann's ideas were crucial. Since that time there has been no shortage of new problems and no shortage of progress in analytic number theory, and much of this progress has been inspired by Riemann's ideas.

Finally, no discussion of the historical context of Riemann's paper would be complete without a mention of the Riemann hypothesis. In the course of the paper, Riemann says that he considers it "very likely" that the complex zeros of $\zeta(s)$ all have real part equal to $\frac{1}{2}$, but that he has been unable to prove that this is true. This statement, that the zeros have real part $\frac{1}{2}$, is now known as the "Riemann hypothesis." The experience of Riemann's successors with the Riemann hypothesis has been the same as Riemann's—they also consider its truth "very likely" and they also have been unable to prove it. Hilbert included the problem of proving the Riemann hypothesis in his list [H9] of the most important unsolved problems which confronted mathematics in 1900, and the attempt to solve this problem has occupied the best efforts of many of the best mathematicians of the twentieth century. It is now unquestionably the most celebrated problem in mathematics and it continues to attract the attention of the best mathematicians, not only because it has gone unsolved for so long but also because it appears tantalizingly vulnerable and because its solution would probably bring to light new techniques of far-reaching importance.

1.2 THE EULER PRODUCT FORMULA

Riemann takes as his starting point the formula

$$\sum_n \frac{1}{n^s} = \prod_p \frac{1}{\left(1 - \dfrac{1}{p^s}\right)} \tag{1}$$

of Euler. Here n ranges over all positive integers ($n = 1, 2, 3, \ldots$) and p ranges over all primes ($p = 2, 3, 5, 7, 11, \ldots$). This formula, which is now known as the "Euler product formula," results from expanding each of the

†A major exception to this statement was Mertens's theorem [M5] of 1874 stating that (2′) is true in the strong sense that the difference of the two sides approaches a limit as $x \longrightarrow \infty$, namely, Euler's constant plus $\sum_p [\log(1 - p^{-1}) + p^{-1}]$. Another perhaps more natural statement of Mertens's theorem is

$$\lim_{x\to\infty} \log x \prod_{p<x} (1 - p^{-1}) = e^{-\gamma},$$

where γ is Euler's constant. See, for example, Hardy and Wright [H7].

factors on the right

$$\frac{1}{\left(1-\frac{1}{p^s}\right)} = 1 + \frac{1}{p^s} + \frac{1}{(p^2)^s} + \frac{1}{(p^3)^s} + \cdots$$

and observing that their product is therefore a sum of terms of the form

$$\frac{1}{(p_1^{n_1} p_2^{n_2} \cdots p_r^{n_r})^s},$$

where $p_1, \ldots, p_r$ are distinct primes and $n_1, n_2, \ldots, n_r$ are natural numbers, and then using the fundamental theorem of arithmetic (every integer can be written in essentially only one way as a product of primes) to conclude that this sum is simply $\sum(1/n)^s$. Euler used this formula principally as a formal identity and principally for integer values of s (see, for example, Euler [E5]).

Dirichlet also based his work† in this field on the Euler product formula. Since Dirichlet was one of Riemann's teachers and since Riemann refers to Dirichlet's work in the first paragraph of his paper, it seems certain that Riemann's use of the Euler product formula was influenced by Dirichlet. Dirichlet, unlike Euler, used the formula (1) with s as a real variable and, also unlike Euler, he proved‡ rigorously that (1) is true for all real $s > 1$.

Riemann, as one of the founders of the theory of functions of a complex variable, would naturally be expected to consider s as a *complex* variable. It is easy to show that both sides of the Euler product formula converge for complex s in the halfplane Re $s > 1$, but Riemann goes much further and shows that even though both sides of (1) diverge for other values of s, the function they define is meaningful for *all* values of s except for a pole at $s = 1$. This extension of the range of s requires a few facts about the factorial function which will be covered in the next section.

1.3 THE FACTORIAL FUNCTION

Euler extended the factorial function $n! = n(n-1)(n-2)\cdots 3\cdot 2\cdot 1$ from the natural numbers n to all real numbers greater than -1 by observing that¶

(1) $$n! = \int_0^\infty e^{-x}x^n\,dx \qquad (n = 1, 2, 3, \ldots)$$

†Dirichlet's major contribution to the theory was his proof that if m is relatively prime to n, then the congruence $p \equiv m \pmod{n}$ has infinitely many prime solutions p. He was also interested in questions concerning the density of the distribution of primes, but he did not have significant success with these questions.

‡Dirichlet [D3]. Since the terms p^{-s} are all positive, there is nothing subtle or difficult about this proof—it is essentially a reordering of absolutely convergent series—but it has the important effect of transforming (1) from a formal identity true for various values of s to an analytical formula true for all real $s > 1$.

¶However Euler wrote the integral in terms of $y = e^{-x}$ as $n! = \int_0^1 (\log 1/y)^n\,dy$ (see Euler [E3]).

(integration by parts) and by observing that the integral on the right converges for noninteger values of n, provided only that $n > -1$. Gauss [G1] introduced the notation†

$$\Pi(s) = \int_0^\infty e^{-x} x^s \, dx \qquad (s > -1) \tag{2}$$

for Euler's integral on the right side of (1). Thus $\Pi(s)$ is defined for all real numbers s greater than -1, in fact for all complex numbers s in the halfplane $\operatorname{Re} s > -1$, and $\Pi(s) = s!$ whenever s is a natural number. There is another representation of $\Pi(s)$ which was also known‡ to Euler, namely,

$$\Pi(s) = \lim_{N\to\infty} \frac{1\cdot 2 \cdots N}{(s+1)(s+2)\cdots(s+N)}(N+1)^s. \tag{3}$$

This formula is valid for all s for which (2) defines $\Pi(s)$, that is, for all s in the halfplane $\operatorname{Re} s > -1$. On the other hand, it is not difficult to show [use formula (4) below] that the limit (3) exists for *all* values of s, real or complex, provided only that the denominator is not zero, that is, provided only that s is not a negative integer. In short, formula (3) extends the definition of $\Pi(s)$ to all values of s other than $s = -1, -2, -3, \ldots$.

In addition to the fact that the two definitions (2) and (3) of $\Pi(s)$ coincide for real $s > -1$, the following facts will be used without proof:

$$\Pi(s) = \prod_{n=1}^\infty \frac{n^{1-s}(n+1)^s}{s+n} = \prod_{n=1}^\infty \left(1 + \frac{s}{n}\right)^{-1}\left(1 + \frac{1}{n}\right)^s, \tag{4}$$

$$\Pi(s) = s\Pi(s-1), \tag{5}$$

$$\frac{\pi s}{\Pi(s)\Pi(-s)} = \sin \pi s, \tag{6}$$

$$\Pi(s) = 2^s \Pi\left(\frac{s}{2}\right)\Pi\left(\frac{s-1}{2}\right)\pi^{-1/2}. \tag{7}$$

For the proofs of these facts the reader is referred to any book which deals with factorial function or the "Γ-function," for example, Edwards [E1, pp. 421–425]. Identity (4) is a simple reformulation of formula (3). Using it one can prove that $\Pi(s)$ *is an analytic function of the complex variable* s *which has simple poles at* $s = -1, -2, -3, \ldots$. *It has no zeros.* Identity (5) is

†Unfortunately, Legendre subsequently introduced the notation $\Gamma(s)$ for $\Pi(s-1)$. Legendre's reasons for considering $(n-1)!$ instead of $n!$ are obscure (perhaps he felt it was more natural to have the first pole occur at $s = 0$ rather than at $s = -1$) but, whatever the reason, this notation prevailed in France and, by the end of the nineteenth century, in the rest of the world as well. Gauss's original notation appears to me to be much more natural and Riemann's use of it gives me a welcome opportunity to reintroduce it.

‡See Euler [E3, E8].

called the "functional equation of the factorial function"; together with $\Pi(0) = 1$ [from (4)] it gives $\Pi(n) = n!$ immediately. Identity (6) is essentially the product formula for the sine; when $s = \frac{1}{2}$ it combines with (5) to give the important value $\Pi(-\frac{1}{2}) = \pi^{1/2}$. Identity (7) is known as the *Legendre relation.* It is the case $n = 2$ of a more general identity

$$\frac{\Pi(s)}{n^s\Pi\left(\frac{s}{n}\right)\Pi\left(\frac{s-1}{n}\right)\cdots\Pi\left(\frac{s-n+1}{n}\right)} = \left[\frac{2\pi n}{(2\pi)^n}\right]^{1/2}$$

which will not be needed.

1.4 THE FUNCTION $\zeta(s)$

It is interesting to note that Riemann does not speak of the "analytic continuation" of the function $\sum n^{-s}$ beyond the halfplane Re $s > 1$, but speaks rather of finding a formula for it which "remains valid for all s." This indicates that he viewed the problem in terms more analogous to the extension of the factorial function by formula (3) of the preceding section than to a piece-by-piece extension of the function in the manner that analytic continuation is customarily taught today. The view of analytic continuation in terms of chains of disks and power series convergent in each disk descends from Weierstrass and is quite antithetical to Riemann's basic philosophy that analytic functions should be dealt with *globally,* not locally in terms of power series.

Riemann derives his formula for $\sum n^{-s}$ which "remains valid for all s" as follows. Substitution of nx for x in Euler's integral for $\Pi(s-1)$ gives

$$\int_0^\infty e^{-nx}x^{s-1}\,dx = \frac{\Pi(s-1)}{n^s}$$

($s > 0$, $n = 1, 2, 3, \ldots$). Riemann sums this over n and uses $\sum_{n=1}^\infty r^{-n} = (r-1)^{-1}$ to obtain†

(1) $$\int_0^\infty \frac{x^{s-1}}{e^x - 1}\,dx = \Pi(s-1)\sum_{n=1}^\infty \frac{1}{n^s}$$

($s > 1$). Convergence of the improper integral on the left and the validity of

†This formula, with $s = 2n$, occurs in a paper [A1] of Abel which was included in the 1839 edition of Abel's collected works. It seems very likely that Riemann would have been aware of this. A very similar formula

$$\int_0^\infty (e^x - 1)^{-1}e^{-x}x^\rho\,dx = \Pi(\rho)\sum_{n=2}^\infty n^{-1-\rho}$$

is the point of departure of Chebyshev's 1848 paper [C2].

the interchange of summation and integration are not difficult to establish.

Next he considers the contour integral

$$\int_{+\infty}^{+\infty} \frac{(-x)^s}{e^x - 1} \frac{dx}{x}.$$

The limits of integration are intended to indicate a path of integration which begins at $+\infty$, moves to the left down the positive real axis, circles the origin once in the positive (counterclockwise) direction, and returns up the positive real axis to $+\infty$. The definition of $(-x)^s$ is $(-x)^s = \exp[s \log(-x)]$, where the definition of $\log(-x)$ conforms to the usual definition of $\log z$ for z not on the negative real axis as the branch which is real for positive real z; thus $(-x)^s$ is not defined on the positive real axis and, strictly speaking, the path of integration must be taken to be slightly above the real axis as it descends from $+\infty$ to 0 and slightly below the real axis as it goes from 0 back to $+\infty$. When this integral is written in the form

$$\int_{+\infty}^{\delta} \frac{(-x)^s\, dx}{(e^x - 1)x} + \int_{|x|=\delta} \frac{(-x)^s\, dx}{(e^x - 1)x} + \int_{\delta}^{+\infty} \frac{(-x)^s\, dx}{(e^x - 1)x},$$

the middle term is $2\pi i$ times the average value of $(-x)^s(e^x - 1)^{-1}$ on the circle $|x| = \delta$ [because on this circle $i\, d\theta = (dx/x)$]. Thus the middle term approaches zero as $\delta \longrightarrow 0$ provided $s > 1$ [because $x(e^x - 1)^{-1}$ is nonsingular near $x = 0$]. The other two terms can then be combined to give

$$\begin{aligned}\int_{+\infty}^{+\infty} \frac{(-x)^s}{e^x - 1} \cdot \frac{dx}{x} &= \lim_{\delta \to 0} \left\{ \int_{+\infty}^{\delta} \frac{\exp[s(\log x - i\pi)]\, dx}{(e^x - 1)x} \right. \\ &\qquad \left. + \int_{\delta}^{+\infty} \frac{\exp[s(\log x + i\pi)]\, dx}{(e^x - 1)x} \right\} \\ &= (e^{i\pi s} - e^{-i\pi s}) \int_0^{\infty} \frac{x^{s-1}\, dx}{e^x - 1}\end{aligned}$$

which combines with the previous formula (1) to give

$$\int_{+\infty}^{+\infty} \frac{(-x)^s}{e^x - 1} \cdot \frac{dx}{x} = 2i \sin(\pi s) \Pi(s - 1) \sum_{n=1}^{\infty} \frac{1}{n^s}$$

or, finally, when both sides are multiplied by $\Pi(-s)s/2\pi is$ and identity (6) of the preceding section is used,

$$\frac{\Pi(-s)}{2\pi i} \int_{+\infty}^{+\infty} \frac{(-x)^s}{e^x - 1} \cdot \frac{dx}{x} = \sum_{n=1}^{\infty} \frac{1}{n^s}. \tag{2}$$

In other words, if $\zeta(s)$ is defined by the formula†

$$\zeta(s) = \frac{\Pi(-s)}{2\pi i} \int_{+\infty}^{+\infty} \frac{(-x)^s}{e^x - 1} \cdot \frac{dx}{x}, \tag{3}$$

†This formula is misstated by the editors of Riemann's works in the notes; they put the factor π on the wrong side of their equation.

then, for real values of s greater than one, $\zeta(s)$ is equal to Dirichlet's function

(4) $$\zeta(s) = \sum_{n=1}^{\infty} \frac{1}{n^s}.$$

However, formula (3) for $\zeta(s)$ "remains valid for all s." In fact, since the integral in (3) clearly converges for all values of s, real or complex (because e^x grows much faster than x^s as $x \longrightarrow \infty$), and since the function it defines is complex analytic (because convergence is uniform on compact domains), the function $\zeta(s)$ of (3) is defined and analytic at all points with the possible exception of the points $s = 1, 2, 3, \ldots$, where $\Pi(-s)$ has poles. Now at $s = 2, 3, 4, \ldots$, formula (4) shows that $\zeta(s)$ has no pole [hence the integral in (3) must have a zero which cancels the pole of $\Pi(-s)$ at these points, a fact which also follows immediately from Cauchy's theorem], and at $s = 1$ formula (4) shows that $\lim \zeta(s) = \infty$ as $s \downarrow 1$, hence that $\zeta(s)$ has a simple [because the pole of $\Pi(-s)$ is simple] pole at $s = 1$. Thus *formula (3) defines a function $\zeta(s)$ which is analytic at all points of the complex s-plane except for a simple pole at $s = 1$.* This function coincides with $\sum n^{-s}$ for real values of $s > 1$ and in fact, by analytic continuation, throughout the halfplane $\text{Re } s > 1$.

The function $\zeta(s)$ is known as the Riemann zeta function.

1.5 VALUES OF $\zeta(s)$

The function $x(e^x - 1)^{-1}$ is analytic near $x = 0$; therefore it can be expanded as a power series

(1) $$\frac{x}{e^x - 1} = \sum_{n=0}^{\infty} \frac{B_n x^n}{n!}$$

valid near zero [in fact valid in the disk $|x| < 2\pi$ which extends to the nearest singularities $x = \pm 2\pi i$ of $x(e^x - 1)^{-1}$]. The coefficients B_n of this expansion are by definition the *Bernoulli numbers;* the first few are easily determined to be

$$\begin{aligned} B_0 &= 1, & B_1 &= -\tfrac{1}{2}, \\ B_2 &= \tfrac{1}{6}, & B_3 &= 0, \\ B_4 &= -\tfrac{1}{30}, & B_5 &= 0, \\ B_6 &= \tfrac{1}{42}, & B_7 &= 0, \\ B_8 &= -\tfrac{1}{30}, & B_9 &= 0. \end{aligned}$$

The odd Bernoulli numbers B_{2n+1} are all zero† after the first, and the even Bernoulli numbers B_{2n} can be determined successively, but there is no simple

†This can be proved directly by noting that $(-t)(e^{-t} - 1)^{-1} + (-t/2) = (-te^t + t - t)(1 - e^t)^{-1} - (t/2) = t(e^t - 1)^{-1} + (t/2)$, that is, $t(e^t - 1)^{-1} + (t/2)$ is an even function. For alternative proofs see the note of Section 1.6 and formula (10) of Section 6.2.

computational formula for them. (See Euler [E6] for a list of the values of $(-1)^{n-1} B_{2n}$ up to B_{30}.)

When $s = -n$ $(n = 0, 1, 2, \ldots)$, this expansion (1) can be used in the defining equation of $\zeta(s)$ to obtain

$$\begin{aligned}\zeta(-n) &= \frac{\Pi(n)}{2\pi i}\int_{+\infty}^{\infty} \frac{(-x)^{-n}}{e^x - 1}\cdot\frac{dx}{x} \\ &= \frac{\Pi(n)}{2\pi i}\int_{|x|=\delta}\left(\sum_m \frac{B_m x^m}{m!}\right)\frac{(-x)^{-n}}{x}\cdot\frac{dx}{x} \\ &= \sum_m \Pi(n)\frac{B_m}{m!}(-1)^n \cdot \frac{1}{2\pi}\int_0^{2\pi} x^{m-n-1}\, d\theta \\ &= n!\frac{B_{n+1}}{(n+1)!}(-1)^n = (-1)^n \frac{B_{n+1}}{n+1}.\end{aligned}$$

Riemann does not give this formula for $\zeta(-n)$, but he does state the particular consequence $\zeta(-2) = \zeta(-4) = \zeta(-6) = \cdots = 0$. He was surely aware, however, not only of the values†

$$\zeta(0) = -1/2, \qquad \zeta(-1) = -1/12, \qquad \zeta(-3) = 1/120,$$

etc., which it implies, but also of the values

$$\zeta(2) = \pi^2/6,\ \zeta(4) = \pi^4/90, \ldots,$$

and, in general,

$$\zeta(2n) = \frac{(2\pi)^{2n}(-1)^{n+1}B_{2n}}{2\cdot(2n)!} \tag{2}$$

which had been found by Euler [E6]. There is no easy way to deduce this famous formula of Euler's from Riemann's integral formula for $\zeta(s)$ [(3) of Section 1.4] and it may well have been this problem of deriving (2) anew which led Riemann to the discovery‡ of the functional equation of the zeta function which is the subject of the next section.

1.6 FIRST PROOF OF THE FUNCTIONAL EQUATION

For negative real values of s, Riemann evaluates the integral

$$\zeta(s) = \frac{\Pi(-s)}{2\pi i}\int_{+\infty}^{+\infty} \frac{(-x)^s}{e^x - 1}\cdot\frac{dx}{x} \tag{1}$$

†The editors of Riemann's collected works give the erroneous value $\zeta(0) = \frac{1}{2}$.

‡Actually the functional equation occurs in Euler's works [E7] in a slightly different form, and it is entirely possible that Riemann found it there. (See also Hardy [H5, pp. 23–26].) In any case, Euler had nothing but an empirical (!) proof of the functional equation and Riemann, in a reversal of his usual role, gave the first rigorous proof of a statement which had been made, but not adequately proved, by someone else.

as follows. Let D denote the domain in the s-plane which consists of all points other than those which lie within ϵ of the positive real axis or within ϵ of one of the singularities $x = \pm 2\pi in$ of the integrand of (1). Let ∂D be the boundary of D oriented in the usual way. Then, ignoring for the moment the fact that D is not compact, Cauchy's theorem gives

(2) $$\frac{\Pi(-s)}{2\pi i}\int_{\partial D}\frac{(-x)^s}{e^x - 1}\cdot\frac{dx}{x} = 0.$$

Now one component of this integral is the integral (1) with the orientation reversed, whereas the others are integrals over the circles $|x \pm 2\pi in| = \epsilon$ oriented clockwise. Thus when the circles are oriented in the usual counterclockwise sense, (2) becomes

(3) $$-\zeta(s) - \sum\frac{\Pi(-s)}{2\pi i}\int_{|x\pm 2\pi in|=\epsilon}\frac{(-x)^s}{e^x - 1}\cdot\frac{dx}{x} = 0.$$

The integrals over the circles can be evaluated by setting $x = 2\pi in + y$ for $|y| = \epsilon$ to find

$$\begin{aligned}&\frac{\Pi(-s)}{2\pi i}\int_{|y|=\epsilon}\frac{(-2\pi in - y)^s}{e^{2\pi in+y} - 1}\frac{dy}{2\pi in + y}\\ &= -\frac{\Pi(-s)}{2\pi i}\int_{|y|=\epsilon}(-2\pi in - y)^{s-1}\cdot\frac{y}{e^y - 1}\cdot\frac{dy}{y}\\ &= -\Pi(-s)(-2\pi in)^{s-1}\end{aligned}$$

by the Cauchy integral formula. Summing over all integers n other than $n = 0$ and using (3) then gives

$$\begin{aligned}\zeta(s) &= \sum_{n=1}^{\infty}\Pi(-s)[(-2\pi in)^{s-1} + (2\pi in)^{s-1}]\\ &= \Pi(-s)(2\pi)^{s-1}[i^{s-1} + (-i)^{s-1}]\sum_{n=1}^{\infty}n^{s-1}.\end{aligned}$$

Finally, using the simplification

$$\begin{aligned}i^{s-1} + (-i)^{s-1} &= \frac{1}{i}[e^{s\log i} - e^{s\log(-i)}]\\ &= \frac{1}{i}[e^{s\pi i/2} - e^{-s\pi i/2}] = 2\sin\frac{s\pi}{2},\end{aligned}$$

one obtains the desired formula

(4) $$\zeta(s) = \Pi(-s)(2\pi)^{s-1}\,2\sin(s\pi/2)\zeta(1 - s).$$

This relationship between $\zeta(s)$ and $\zeta(1 - s)$ is known as the *functional equation of the zeta function.*

In order to prove rigorously that (4) holds for $s < 0$, it suffices to modify the above argument by letting D_n be the intersection of D with the disk $|s| \leq (2n + 1)\pi$ and letting $n \to \infty$; then the integral (2) splits into two parts, one being an integral over the circle $|s| = (2n + 1)\pi$ with the points within ϵ of the positive real axis deleted, and the other being an integral whose limit as $n \longrightarrow \infty$ is the left side of (3). The first of these two parts approaches zero because the length of the path of integration is less than $2\pi(2n + 1)\pi$, because the factor $(e^x - 1)^{-1}$ is bounded on the circle $|s| = (2n + 1)\pi$, and because the modulus of $(-x)^s/x$ on this circle is $|x|^{s-1} \leq [(2n + 1)\pi]^{-\delta-1}$ for $s \leq -\delta < 0$. Thus the second part, which by Cauchy's theorem is the negative of the first part, also approaches zero, which implies (3) and hence (4).

This completes the proof of the functional equation (4) in the case $s < 0$. However, both sides of (4) are analytic functions of s, so this suffices to prove (4) for all values of s [except for $s = 0, 1, 2, \ldots$, where† one or more of the terms of (4) have poles].

For $s = 1 - 2n$ the functional equation plus the identity

$$\zeta(-(2n - 1)) = (-1)^{2n-1}\frac{B_{2n}}{2n}$$

of the previous section gives

$$(-1)^{2n-1}\frac{B_{2n}}{2n} = \Pi(2n - 1)(2\pi)^{-2n}2(-1)^n\zeta(2n)$$

and hence Euler's famous formula for $\zeta(2n)$ [(2) of Section 1.5].

Riemann uses two of the basic identities of the factorial function [(6) and (7) of Section 1.3] to rewrite the functional equation (4) in the form

$$\zeta(s) = \pi^{-1/2}2^{-s}\Pi\left(-\frac{s}{2}\right)\Pi\left(-\frac{s+1}{2}\right)2^s\pi^{s-1}\frac{\pi s/2}{\Pi\left(\frac{s}{2}\right)\Pi\left(-\frac{s}{2}\right)}\zeta(1 - s)$$

and hence in the form

$$\Pi\left(\frac{s}{2} - 1\right)\pi^{-s/2}\zeta(s) = \Pi\left(\frac{1-s}{2} - 1\right)\pi^{-(1-s)/2}\zeta(1 - s). \tag{5}$$

In words, then, *the function on the left side of (5) is unchanged by the substitution $s = 1 - s$.*

Riemann appears to consider this symmetrical statement (5) as the natural statement of the functional equation, because he gives‡ an alternative proof

†When $s = 2n + 1$, the fact that $\zeta(s)$ has no pole at $2n + 1$ implies, since Π has a pole at $-2n - 1$ and $\sin(s\pi/2)$ has no zero at $2n + 1$, that $\zeta(-2n) = 0$ and hence, by the formula for $\zeta(-2n)$ of the preceding section, that the odd Bernoulli numbers $B_3, B_5, B_7, \ldots$ are all zero.

‡Since the second proof renders the first proof wholly unnecessary, one may ask why Riemann included the first proof at all. Perhaps the first proof shows the argument by which he originally discovered the functional equation or perhaps it exhibits some properties which were important in his understanding of ζ.

which exhibits this symmetry in a more satisfactory way. This second proof is given in the next section.

1.7 SECOND PROOF OF THE FUNCTIONAL EQUATION

Riemann first observes that the change of variable $x = n^2\pi x$ in Euler's integral for $\Pi(s/2 - 1)$ gives

$$\frac{1}{n^s}\pi^{-s/2}\Pi\left(\frac{s}{2}-1\right) = \int_0^\infty e^{-n^2\pi x}x^{s/2}\frac{dx}{x} \qquad (\operatorname{Re} s > 1).$$

Thus summation over n gives

$$\Pi\left(\frac{s}{2}-1\right)\pi^{-s/2}\zeta(s) = \int_0^\infty \psi(x)x^{s/2}\cdot\frac{dx}{x} \qquad (\operatorname{Re} s > 1), \tag{1}$$

where† $\psi(x) = \sum_{n=1}^\infty \exp(-n^2\pi x)$. The symmetrical form of the functional equation is the statement that the function (1) is unchanged by the substitution $s = 1 - s$. To prove directly that the integral on the right side of (1) is unchanged by this substitution Riemann uses the *functional equation of the theta function* in a form taken from Jacobi,‡ namely, in the form

$$\frac{1+2\psi(x)}{1+2\psi(1/x)} = \frac{1}{\sqrt{x}}. \tag{2}$$

[Since $\psi(x)$ approaches zero very rapidly as $x \to \infty$, this shows in particular that $\psi(x)$ is like $\frac{1}{2}(x^{-1/2} - 1)$ for x near zero and hence that the integral on the right side of (1) is convergent for $s > 1$. Once this has been established, the validity of (1) for $s > 1$ can be proved by an elementary argument using absolute convergence to justify the interchange of summation and integration.] Using (2), Riemann reformulates the integral on the right side of (1) as

$$\begin{aligned}\int_0^\infty \psi(x)x^{s/2}\frac{dx}{x} &= \int_1^\infty \psi(x)x^{s/2}\cdot\frac{dx}{x} - \int_\infty^1 \psi\left(\frac{1}{x}\right)x^{-s/2}\cdot\frac{dx}{x}\\ &= \int_1^\infty \psi(x)x^{s/2}\cdot\frac{dx}{x} + \int_1^\infty\left[x^{1/2}\psi(x) + \frac{x^{1/2}}{2} - \frac{1}{2}\right]x^{-s/2}\frac{dx}{x}\\ &= \int_1^\infty \psi(x)[x^{s/2} + x^{(1-s)/2}]\frac{dx}{x}\\ &\quad + \frac{1}{2}\int_1^\infty [x^{-(s-1)/2} - x^{-s/2}]\frac{dx}{x}.\end{aligned}$$

†This function $\psi(x)$ has nothing whatsoever to do with the function $\psi(x)$ which appears in Chapter 3.

‡Riemann refers to Section 65 of Jacobi's treatise "Fundmenta Nova Theoriae Functionum Ellipticarum." Although the needed formula is not given explicitly there, Jacobi in another place [J1] shows how the needed formula follows from formula (6) of Section 65. Jacobi attributes the formula to Poisson. For a proof of the formula see Section 10.4.

Now $\int_1^\infty x^{-a}\,(dx/x) = 1/a$ for $a > 0$ so the second integral is

$$\frac{1}{2}\left[\frac{1}{(s-1)/2} - \frac{1}{s/2}\right] = \frac{1}{s(s-1)}$$

for $s > 1$. Thus for $s > 1$ the formula

(3) $$\Pi\left(\frac{s}{2} - 1\right)\pi^{-s/2}\zeta(s) = \int_1^\infty \psi(x)[x^{s/2} + x^{(1-s)/2}]\frac{dx}{x} - \frac{1}{s(1-s)}$$

holds. But, because $\psi(x)$ decreases more rapidly than any power of x as $x \longrightarrow \infty$, the integral in this formula converges for all† s. Since both sides are analytic, the same equation holds for all s. Because the right side is obviously unchanged by the substitution $s = 1 - s$, this proves the functional equation of the zeta function.

1.8 THE FUNCTION $\xi(s)$

The function $\Pi((s/2) - 1)\pi^{-s/2}\zeta(s)$, which occurs in the symmetrical form of the functional equation, has poles at $s = 0$ and $s = 1$. [This follows immediately from (3) of the preceding section.] Riemann multiplies it by $s(s-1)/2$ and defines‡

(1) $$\xi(s) = \Pi(s/2)(s-1)\pi^{-s/2}\zeta(s).$$

Then $\xi(s)$ is an entire function—that is, an analytic function of s which is defined for all values of s—and the functional equation of the zeta function is equivalent to $\xi(s) = \xi(1-s)$.

Riemann next derives the following representation of $\xi(s)$. Equation (3) of the preceding section gives

$$\begin{aligned}\xi(s) &= \frac{1}{2} - \frac{s(1-s)}{2}\int_1^\infty \psi(s)(x^{s/2} + x^{(1-s)/2})\frac{dx}{x} \\ &= \frac{1}{2} - \frac{s(1-s)}{2}\int_1^\infty \frac{d}{dx}\left\{\psi(x)\left[\frac{x^{s/2}}{s/2} + \frac{x^{(1-s)/2}}{(1-s)/2}\right]\right\}dx \\ &\quad + \frac{s(1-s)}{2}\int_1^\infty \psi'(x)\left[\frac{x^{s/2}}{s/2} + \frac{x^{(1-s)/2}}{(1-s)/2}\right]dx\end{aligned}$$

†Note that this gives, therefore, another formula for $\zeta(s)$ which is "valid for all s" other than $s = 0, 1$; that is, it gives an alternative proof of the fact that $\zeta(s)$ can be analytically continued.

‡Actually Riemann uses the letter ξ to denote the function which it is now customary to denote by Ξ, namely, the function $\Xi(t) = \xi(\frac{1}{2} + it)$, where ξ is defined as above. I follow Landau, and almost all subsequent writers, in rejecting Riemann's change of variable $s = \frac{1}{2} + it$ in formula (1) as being confusing. In fact, there is reason to believe that Riemann himself was confused by it [see remarks concerning $\xi(0)$ in Section 1.16].

$$= \frac{1}{2} + \frac{s(1-s)}{2}\psi(1)\left[\frac{2}{s} + \frac{2}{1-s}\right]$$

$$+ \int_1^\infty \psi'(x)[(1-s)x^{s/2} + sx^{(1-s)/2}]\, dx$$

$$= \frac{1}{2} + \psi(1) + \int_1^\infty x^{3/2}\psi'(x)[(1-s)x^{[(s-1)/2]-1} + sx^{-(s/2)-1}]\, dx$$

$$= \frac{1}{2} + \psi(1) + \int_1^\infty \frac{d}{dx}[x^{3/2}\psi'(x)(-2x^{(s-1)/2} - 2x^{-s/2})]\, dx$$

$$- \int_1^\infty \frac{d}{dx}[x^{3/2}\psi'(x)]\,[-2x^{(s-1)/2} - 2x^{-s/2}]\, dx$$

$$= \frac{1}{2} + \psi(1) - \psi'(1)[-2-2]$$

$$+ \int_1^\infty \frac{d}{dx}[x^{3/2}\psi'(x)]\,(2x^{(s-1)/2} + 2x^{-s/2})\, dx.$$

Now differentiation of

$$2\psi(x) + 1 = x^{-1/2}[2\psi(1/x) + 1]$$

easily gives

$$\tfrac{1}{2} + \psi(1) + 4\psi'(1) = 0,$$

and using this puts the formula in the final form

$$\xi(s) = 4\int_1^\infty \frac{d[x^{3/2}\psi'(x)]}{dx}\, x^{-1/4}\cosh\left[\frac{1}{2}\left(s - \frac{1}{2}\right)\log x\right] dx \tag{2}$$

or, as Riemann writes it,

$$\xi\left(\frac{1}{2} + it\right) = 4\int_1^\infty \frac{d[x^{3/2}\psi'(x)]}{dx}\, x^{-1/4}\cos\left(\frac{t}{2}\log x\right) dx.$$

If $\cosh[\frac{1}{2}(s-\frac{1}{2})\log x]$ is expanded in the usual power series $\cosh y = \frac{1}{2}(e^y + e^{-y}) = \sum y^{2n}/(2n)!$, formula (2) shows that

$$\xi(s) = \sum_{n=0}^{\infty} a_{2n}(s - \tfrac{1}{2})^{2n} \tag{3}$$

where

$$a_{2n} = 4\int_1^\infty \frac{d[x^{3/2}\psi'(x)]}{dx}\, x^{-1/4}\frac{(\frac{1}{2}\log x)^{2n}}{(2n)!}\, dx.$$

Riemann states that this series representation of $\xi(s)$ as an even function of $s - \frac{1}{2}$ "converges very rapidly," but he gives no explicit estimates and he does not say what role this series plays in the assertions which he makes next.

The two paragraphs which follow the formula (2) for $\xi(s)$ are the most difficult portion of Riemann's paper. Their goal is essentially to prove that

$\xi(s)$ can be expanded as an infinite product

$$\xi(s) = \xi(0) \prod_{\rho} \left(1 - \frac{s}{\rho}\right), \tag{4}$$

where ρ ranges† over the roots of the equation $\xi(\rho) = 0$. Now any *polynomial* $p(s)$ can be expanded as a finite product $p(s) = p(0) \prod_{\rho} [1 - (s/\rho)]$, where ρ ranges over the roots of the equation $p(\rho) = 0$ [except that the product formula for $p(s)$ is slightly different if $p(0) = 0$]; hence the product formula (4) states that $\xi(s)$ is *like a polynomial of infinite degree*. (Similarly, Euler thought of $\sin x$ as a "polynomial of infinite degree" when he conjectured, and finally proved, the formula $\sin \pi x = \pi x \prod_{n=1}^{\infty} [1 - (x/n)^2]$.) On the other hand, the statement that the series (3) converges "very rapidly" is also a statement that $\xi(s)$ is like a polynomial of infinite degree—a finite number of terms gives a very good approximation in any finite part of the plane. Thus there is some relationship between the series (3) and the product formula (4)—in fact it is *precisely* the rapid decrease of the coefficients a_n which Hadamard (in 1893) proved was necessary and sufficient for the validity of the product formula—but the steps of the argument by which Riemann went from the one to the other are obscure, to say the very least.

The next section contains a discussion of the distribution of the roots ρ of $\xi(\rho) = 0$, and the following section returns to the discussion of the product formula for $\xi(s)$.

1.9 THE ROOTS ρ OF ξ

In order to prove the convergence of the product $\xi(s) = \xi(0) \prod_{\rho} [1 - (s/\rho)]$, Riemann needed, of course, to investigate the distribution of the roots ρ of $\xi(\rho) = 0$. He begins by observing that the Euler product formula

$$\zeta(s) = \prod_{p} (1 - p^{-s})^{-1} \qquad (\text{Re}\, s > 1)$$

shows immediately that $\zeta(s)$ has no zeros in the halfplane $\text{Re}\, s > 1$ (because a convergent infinite product can be zero only if one of its factors is zero). Since $\xi(s) = \Pi(s/2)(s - 1)\pi^{-s/2}\zeta(s)$ and since the factors other than $\zeta(s)$ have only the simple zero at $s = 1$, it follows that none of the roots ρ of $\xi(\rho) = 0$ lie in the halfplane $\text{Re}\, s > 1$. Since $1 - \rho$ is a root if and only if ρ is, this implies that none of the roots lie in the halfplane $\text{Re}\, s < 0$ either, and hence that *all the roots ρ of $\xi(\rho) = 0$ lie in the strip* $0 \leq \text{Re}\, \rho \leq 1$.

He then goes on to say that the number of roots ρ whose imaginary parts

†Here, and in the many formulas in the remainder of the book which involve sums or products over the roots ρ, it is understood that multiple roots—if there are any—are to be counted with multiplicities.

lie between 0 and T is approximately

$$\frac{T}{2\pi}\log\frac{T}{2\pi}-\frac{T}{2\pi} \tag{1}$$

and that the relative† error in this approximation is of the order of magnitude $1/T$. His "proof" of this is simply to say that the number of roots in this region is equal to the integral of $\xi'(s)\,ds/2\pi i\xi(s)$ around the boundary of the rectangle $\{0 \leq \operatorname{Re} s \leq 1,\ 0 \leq \operatorname{Im} s \leq T\}$ and that this integral is equal to (1) with a relative error T^{-1}. Unfortunately he gives no hint whatsoever of the method he used to estimate the integral. He himself was a master at evaluating and estimating definite integrals (see, for example, Section 1.14 or 7.4) and it is quite possible that he assumed that his readers would be able to carry out their own estimation of this integral, but if so he was wrong; it was not until 1905 that von Mangoldt succeeded in proving that Riemann's estimate was correct (see Section 6.7).

Riemann's next statement is even more baffling. He states that the number of roots *on the line* $\operatorname{Re} s = \frac{1}{2}$ is also "about" (1). He does not make precise the sense in which this approximation is true, but it is generally assumed that he meant that the relative error in the approximation of the number of zeros of $\xi(\frac{1}{2}+it)$ for $0 \leq t \leq T$ by (1) approaches zero as $T \to \infty$. He gives no indication of a proof at all, and no one since Riemann has been able to prove (or disprove) this statement. It was proved in 1914 that $\xi(\frac{1}{2}+it)$ has infinitely many real roots (Hardy [H3]), in 1921 that the number of real roots between 0 and T is at least KT for some positive constant K and all sufficiently large T (Hardy and Littlewood [H6]), in 1942 that this number is in fact at least $KT \log T$ for some positive K and all large T (Selberg, [S1]), and in 1914 that the number of complex roots t of $\xi(\frac{1}{2}+it) = 0$ in the range $\{0 \leq \operatorname{Re} t \leq T,\ -\epsilon \leq \operatorname{Im} t \leq \epsilon\}$ is equal, for any $\epsilon > 0$, to (1) with a relative error which approaches zero as $T \to \infty$ (Bohr and Landau, [B8]). However, these partial results are still far from Riemann's statement. We can only guess what lay behind this statement (see Siegel [S4 p. 67], Titchmarsh [T8, pp. 213–214], or Section 7.8 of this book), but we do know that it led Riemann to conjecture an even stronger statement, namely, that *all* the roots lie on $\operatorname{Re} s = \frac{1}{2}$.

This is of course the famous "Riemann hypothesis." He says he considers it "very likely" that the roots all do lie on $\operatorname{Re} s = \frac{1}{2}$, but says that he was not able to prove it (which would seem to imply, incidentally, that he did feel he had rigorous proofs of the preceding two statements). Since it is not necessary for his main goal, which is the proof of his formula for the number of primes less than a given magnitude, he simply leaves the matter there—where it has remained ever since—and goes on to the product formula for $\xi(s)$.

†Titchmarsh, in an unfortunate lapse which he did not catch in the 21 years between the publication of his two books on the zeta function, failed to realize that Riemann meant the *relative* error and believed that Riemann had made a mistake at this point. See Titchmarsh [T8, p. 213].

1.10 THE PRODUCT REPRESENTATION OF $\xi(s)$

A recurrent theme in Riemann's work is the *global characterization of analytic functions by their singularities*.† Since the function $\log \xi(s)$ has logarithmic singularities at the roots ρ of $\xi(s)$ and no other singularities, it has the same singularities as the formal sum

$$\sum_{\rho} \log\left(1 - \frac{s}{\rho}\right). \tag{1}$$

Thus if this sum converges and if the function it defines is in some sense as well behaved near ∞ as $\log \xi(s)$ is, then it should follow that the sum (1) differs from $\log \xi(s)$ by at most an additive constant; setting $s = 0$ gives the value $\log \xi(0)$ for this constant, and hence exponentiation gives

$$\xi(s) = \xi(0) \prod_{\rho} \left(1 - \frac{s}{\rho}\right) \tag{2}$$

as desired. This is essentially the proof of the product formula (2) which Riemann sketches.

There are two problems associated with the sum (1). The first is the determination of the imaginary parts of the logarithms it contains. Riemann passes over this point without comment and, indeed, it is not a very serious problem. For any fixed s the ambiguity in the imaginary part of $\log[1 - (s/\rho)]$ disappears for large ρ; hence the sum (1) is defined except for a (finite) multiple of $2\pi i$ which drops out when one exponentiates (2). Furthermore, one can ignore the imaginary parts altogether; the real parts of the terms of (1) are unambiguously defined and their sum is a harmonic function which differs from Re $\log \xi(s)$ by a harmonic function without singularities, and if this difference function can be shown to be constant, it will follow that its harmonic conjugate is constant also.

The second problem associated with the sum (1) is its convergence. It is in fact a conditionally convergent sum, and the *order* of the series must be specified in order for the sum to be well determined. Roughly speaking the natural order for the terms would be the order of increasing $|\rho|$, or perhaps of increasing $|\rho - \frac{1}{2}|$, but specifically it suffices merely to stipulate that each

†See, for example, the Inauguraldissertation, especially article 20 (*Werke*, pp. 37–39) or part 3 of the introduction to the article "Theorie der Abel'schen Functionen," which is entitled "Determination of a function of a complex variable by boundary values and singularities [R1]." See also Riemann's introduction to Paper XI of the collected works, where he writes ". . . our method, which is based on the determination of functions by means of their singularities (*Unstetigkeiten und Unendlichwerden*) . . . [R1]." Finally, see Ahlfors [A3], the section at the end entitled "Riemann's point of view."

term be paired with its "twin" $\rho \leftrightarrow 1 - \rho$,

(3) $$\sum_{\operatorname{Im}\rho>0}\left[\log\left(1-\frac{s}{\rho}\right)+\log\left(1-\frac{s}{1-\rho}\right)\right],$$

because this sum converges absolutely. The proof of the absolute convergence of (3) is roughly as follows.

To prove the absolute convergence of

$$\sum_{\operatorname{Im}\rho>0}\log\left[\left(1-\frac{s}{\rho}\right)\left(1-\frac{s}{1-\rho}\right)\right]=\sum_{\operatorname{Im}\rho>0}\log\left[1-\frac{s(1-s)}{\rho(1-\rho)}\right],$$

it suffices to prove the absolute convergence of

(4) $$\sum_{\operatorname{Im}\rho>0}\frac{1}{\rho(1-\rho)}.$$

(In other words, to prove the absolute convergence of a product $\prod(1 + a_i)$, it suffices to prove the absolute convergence of the sum $\sum a_i$.) But the estimate of the distribution of the roots ρ given in the preceding section indicates that their density is roughly

$$d\left(\frac{T}{2\pi}\log\frac{T}{2\pi}-\frac{T}{2\pi}\right)=\frac{1}{2\pi}\log\frac{T}{2\pi}\,dT.$$

Hence

$$\sum_{\operatorname{Im}\rho>0}\frac{1}{\rho(1-\rho)}\sim\int^{\infty}\frac{1}{T^2}\frac{1}{2\pi}\log\frac{T}{2\pi}\,dT<\infty,$$

or, in short, the terms are like T^{-2} and their density is like $\log T$ so their sum converges. As will be seen in Chapter 2, the only serious difficulty in making this into a rigorous proof of the absolute convergence of (3) is the proof that the vertical density of the roots ρ is in some sense a constant times $\log(T/2\pi)$. Riemann merely states this fact without proof.

Riemann then goes on to say that the function defined by (3) grows only as fast as $s \log s$ for large s; hence, because it differs from $\log \xi(s)$ by an even function of $s - \frac{1}{2}$ [and because $\log \xi(s)$ also grows like $s \log s$ for large s], this difference must be constant because it can contain no terms in $(s - \frac{1}{2})^2$, $(s - \frac{1}{2})^4, \ldots$. It will be shown in Chapter 2 that the steps in this argument can all be filled in more or less as Riemann indicates, but it must be admitted that Riemann's sketch is so abbreviated as to make it virtually useless in constructing a proof of (2).

The first proof of the product representation (2) of $\xi(s)$ was published by Hadamard [H1] in 1893.

1.11 THE CONNECTION BETWEEN $\zeta(s)$ AND PRIMES

The essence of the relationship between $\zeta(s)$ and prime numbers is the Euler product formula

$$\zeta(s) = \prod_p \frac{1}{1 - p^{-s}} \qquad (\text{Re } s > 1) \tag{1}$$

in which the product on the right is over all prime numbers p. Taking the log of both sides and using the series $\log(1 - x) = -x - \frac{1}{2}x^2 - \frac{1}{3}x^3 - \cdots$ puts this in the form

$$\log \zeta(s) = \sum_p \left[\sum_n (1/n)p^{-ns} \right] \qquad (\text{Re } s > 1).$$

Since the double series on the right is absolutely convergent for $\text{Re } s > 1$, the order of summation is unimportant and the sum can be written simply

$$\log \zeta(s) = \sum_p \sum_n (1/n)p^{-ns} \qquad (\text{Re } s > 1). \tag{2}$$

It will be convenient in what follows to write this sum as a Stieltjes integral

$$\log \zeta(s) = \int_0^\infty x^{-s}\, dJ(x) \qquad (\text{Re } s > 1) \tag{3}$$

where $J(x)$ is† the function which begins at 0 for $x = 0$ and increases by a jump of 1 at primes p, by a jump of $\frac{1}{2}$ at prime squares p^2, by a jump of $\frac{1}{3}$ at prime cubes, etc. As is usual in the theory of Stieltjes integrals, the value of $J(x)$ at each jump is defined to be halfway between its new value and its old value. Thus $J(x)$ is zero for $0 \le x < 2$, is $\frac{1}{2}$ for $x = 2$, is 1 for $2 < x < 3$, is $1\frac{1}{2}$ for $x = 3$, is 2 for $3 < x < 4$, is $2\frac{1}{4}$ for $x = 4$, is $2\frac{1}{2}$ for $4 < x < 5$, is 3 for $x = 5$, is $3\frac{1}{2}$ for $5 < x < 7$, etc. A formula for $J(x)$ is

$$J(x) = \frac{1}{2}\left[\sum_{p^n < x} \frac{1}{n} + \sum_{p^n \le x} \frac{1}{n} \right].$$

Riemann did not, of course, have the vocabulary of Stieltjes integration available to him, and he stated (3) in the slightly different form

$$\log \zeta(s) = s \int_0^\infty J(x) x^{-s-1}\, dx \qquad (\text{Re } s > 1) \tag{4}$$

which can be obtained from (3) by integration by parts. [As $x \downarrow 0$, clearly $x^{-s}J(x) = 0$ because $J(x) \equiv 0$ for $x < 2$. On the other hand, $J(x) < x$ for all x, so $x^{-s}J(x) \longrightarrow 0$ as $x \longrightarrow \infty$ for $\text{Re } s > 1$.] The integral in (4) can be con-

†Riemann denotes this function $f(x)$, and most other writers denote it $\Pi(x)$. Since $f(x)$ now is commonly used to denote a generic function and since $\Pi(x)$ in this book denotes the factorial function, I have taken the liberty of introducing a new notation $J(x)$ for this function.

sidered to be an ordinary Riemann integral and the formula itself can be derived without using Stieltjes integration by setting

$$p^{-ns} = s\int_{p^n}^{\infty} x^{-s-1}\,dx \qquad (\operatorname{Re} s > 1)$$

in (2), which is Riemann's derivation of (4).

Formulas (2)–(4) should all be thought of as minor variations of the Euler product formula (1) which is the basic idea connecting $\zeta(s)$ and primes.

1.12 FOURIER INVERSION

Riemann was a master of Fourier analysis and his work in developing this theory must certainly be counted among his greatest contributions to mathematics. It is not surprising, therefore, that he immediately applies Fourier inversion to the formula

(1) $$\frac{\log \zeta(s)}{s} = \int_0^{\infty} J(x)x^{-s-1}\,dx \qquad (\operatorname{Re} s > 1)$$

to conclude

(2) $$J(x) = \frac{1}{2\pi i}\int_{a-i\infty}^{a+i\infty} \log \zeta(s) x^s \frac{ds}{s} \qquad (a > 1).$$

Then using an alternative formula for $\log \zeta(s)$, he obtains an alternative formula for $J(x)$ which is the main result of the paper.

[The improper integral in (2) is only conditionally convergent and an "order of summation" must be specified. Here it is understood that the integral in (2) means the limit as $T \longrightarrow \infty$ of the integral over the vertical line segment from $a - iT$ to $a + iT$. More generally, conditionally convergent integrals and series are very common in Fourier analysis, and it is always understood that such integrals and series are summed in their "natural order"; for example,

$$\sum_{n=-\infty}^{\infty} c_n e^{inx} \quad \text{means} \quad \lim_{N\to\infty} \sum_{n=-N}^{N} c_n e^{inx},$$

$$\int_{-\infty}^{\infty} f(y)e^{iyx}\,dy \quad \text{means} \quad \lim_{T\to\infty} \int_{-T}^{T} f(y)e^{iyx}\,dy,$$

etc. This is analogous to the convention that discontinuous functions such as $J(x)$ assume the middle value $J(x) = \frac{1}{2}[J(x-\epsilon) + J(x+\epsilon)]$ at any jump x, that divergent integrals such as $\operatorname{Li}(x)$ (see Section 1.14 below) are taken to mean the Cauchy principal value, and that the product $\prod [1 - (s/\rho)]$ is ordered in such a way as to pair ρ with $1 - \rho$, or, later on, ordered by $|\operatorname{Im} \rho|$.]

In deriving (2) from (1) Riemann makes use of "Fourier's theorem," by which he means† the Fourier inversion formula

$$\phi(x) = \frac{1}{2\pi}\int_{-\infty}^{\infty}\left[\int_{-\infty}^{\infty}\phi(\lambda)e^{i(x-\lambda)\mu}\,d\lambda\right]d\mu. \tag{3}$$

Otherwise stated, "Fourier's theorem" is the statement that in order to write a given function $\phi(x)$ as a superposition of exponentials

$$\phi(x) = \int_{-\infty}^{\infty}\Phi(\mu)e^{i\mu x}\,d\mu,$$

it is necessary and sufficient (under suitable conditions) that the "coefficients" $\Phi(\mu)$ of the expansion be defined by

$$\Phi(\mu) = \frac{1}{2\pi}\int_{-\infty}^{\infty}\phi(\lambda)e^{-i\lambda\mu}\,d\lambda.$$

This statement of Fourier's theorem brings out the analogy with Fourier series

$$f(x) = \sum_{-\infty}^{\infty} a_n e^{inx} \Longleftrightarrow a_n = \frac{1}{2\pi}\int_0^{2\pi} f(\lambda)e^{-in\lambda}\,d\lambda,$$

and in fact theorem (3) for Fourier integrals follows formally from a passage to the limit in the theorem for Fourier series.

To derive (2) from (1), let $s = a + i\mu$, where a is a constant $a > 1$ and μ is a real variable, let $\lambda = \log x$, and let $\phi(x) = 2\pi J(e^x)e^{-ax}$. Then (1) becomes

$$\begin{aligned}\frac{\log\zeta(a+i\mu)}{a+i\mu} &= \int_{-\infty}^{\infty} J(e^\lambda)e^{-(a+i\mu)\lambda}\,d\lambda \\ &= \frac{1}{2\pi}\int_{-\infty}^{\infty}\phi(\lambda)e^{-i\mu\lambda}\,d\lambda \qquad (a > 1),\end{aligned}$$

and when this function is taken to be $\Phi(\mu)$, Fourier's theorem gives

$$\begin{aligned}2\pi J(e^x)e^{-ax} &= \int_{-\infty}^{\infty}\frac{\log\zeta(a+i\mu)}{a+i\mu}e^{i\mu x}\,d\mu, \\ J(y) &= \frac{1}{2\pi}\int_{-\infty}^{\infty}\frac{\log\zeta(a+i\mu)}{a+i\mu}y^{a+i\mu}\,d\mu,\end{aligned}$$

from which (2) follows immediately.

Riemann completely ignores the question of the applicability of Fourier's theorem to the function $J(e^x)e^{-ax}$ and states simply that (2) holds in "complete generality." However, $J(e^x)e^{-ax}$ is a very well-behaved function—it has simple well-behaved jumps, it is identically zero for $x < 0$, and it goes to zero

†See Riemann [R2, p. 86].

faster than $e^{-(a-1)x}$ as $x \longrightarrow \infty$—and the very simplest theorems† on Fourier integrals suffice to prove rigorously Riemann's statement that (2) holds in complete generality.

1.13 METHOD FOR DERIVING THE FORMULA FOR $J(x)$

The two formulas for $\xi(s)$, namely,

$$\xi(s) = \Pi\left(\frac{s}{2}\right)\pi^{-s/2}(s-1)\zeta(s) \qquad \text{and} \qquad \xi(s) = \xi(0)\prod_{\rho}\left(1 - \frac{s}{\rho}\right),$$

combine to give

$$\begin{aligned}\log \zeta(s) &= \log \xi(s) - \log \Pi\left(\frac{s}{2}\right) + \frac{s}{2}\log \pi - \log(s-1)\\ &= \log \xi(0) + \sum_{\rho} \log\left(1 - \frac{s}{\rho}\right) - \log \Pi\left(\frac{s}{2}\right)\\ &\quad + \frac{s}{2}\log \pi - \log(s-1).\end{aligned}$$

Riemann's formula for $J(x)$, which is the main result of his paper, is obtained essentially by substituting this formula for $\log \zeta(s)$ in the formula

$$J(x) = \frac{1}{2\pi i}\int_{a-i\infty}^{a+i\infty} \log \zeta(s) x^s \frac{ds}{s} \qquad (a > 1)$$

of the preceding section and integrating termwise. However, because a direct substitution leads to divergent integrals [the term $(s/2)\log \pi$, for example, leads to an integral which is a constant times $(i)^{-1}\int x^s\, ds = e^a \int e^{iu \log x}\, du$ which oscillates and does not converge even conditionally], Riemann first integrates by parts to obtain

$$(1) \qquad J(x) = -\frac{1}{2\pi i}\cdot\frac{1}{\log x}\int_{a-i\infty}^{a+i\infty} \frac{d}{ds}\left[\frac{\log \zeta(s)}{s}\right] x^s\, ds \qquad (a > 1)$$

before substituting the above expression for $\log \zeta(s)$. The validity of the integration by parts by which (1) is obtained depends merely on showing that

$$(2) \qquad \lim_{T\to\infty} \frac{\log \zeta(a \pm iT)}{a \pm iT} x^{a \pm iT} = 0,$$

†See, for example, Taylor [T2]. The particular form of Fourier inversion that Riemann uses here—which is essentially Fourier analysis on the multiplicative group of positive reals rather than on the additive group of all real numbers—is often called Mellin inversion. Riemann's work precedes that of Mellin by 40 years.

which follows easily from the inequality

$$\begin{aligned} |\log \zeta(a \pm iT)| &= \left| \sum_n \sum_p (1/n) p^{-n(a \pm iT)} \right| \\ &\leq \sum_n \sum_p (1/n) p^{-na} = \log \zeta(a) = \text{const} \end{aligned} \tag{3}$$

because this shows that the numerator in (2) is bounded while the denominator goes to infinity.

The substitution of

$$\log \zeta(s) = \log \xi(0) + \sum_\rho \log\left(1 - \frac{s}{\rho}\right) - \log \Pi\left(\frac{s}{2}\right) + \frac{s}{2} \log \pi - \log(s-1)$$

into (1) expresses $J(x)$ as a sum of five terms (the integral of a finite sum is always the sum of the integrals provided the latter converge) and the derivation of Riemann's formula for $J(x)$ depends now on the evaluation of these five definite integrals.

It should be noted that for any fixed s there is some ambiguity in the definition of $\log[1 - (s/\rho)]$ for those roots ρ which are not large relative to s. In order to remove this ambiguity in $\operatorname{Re} s > 1$ let $\log[1 - (s/\rho)]$ be defined to be $\log(s - \rho) - \log(-\rho)$; this is meaningful because none† of the ρ's are real and greater than or equal to 0. In this way $\log[1 - (s/\rho)]$ is unambiguously defined throughout $\operatorname{Re} s > 1$ and, in particular, on the path of integration $\operatorname{Re} s = a > 1$.

1.14 THE PRINCIPAL TERM OF $J(x)$

It will be seen below that the principal term in the formula for $J(x)$ is the term corresponding to the term $-\log(s - 1)$ of the expansion of $\log \zeta(s)$. This term is

$$\frac{1}{2\pi i} \frac{1}{\log x} \int_{a-i\infty}^{a+i\infty} \frac{d}{ds}\left[\frac{\log(s-1)}{s}\right] x^s \, ds \qquad (a > 1).$$

Riemann shows that for $x > 1$ the value of this definite integral is the logarithmic integral

$$\operatorname{Li}(x) = \lim_{\epsilon \downarrow 0} \left[\int_0^{1-\epsilon} \frac{dt}{\log t} + \int_{1+\epsilon}^{x} \frac{dt}{\log t}\right];$$

†See Section 2.3, or observe that the series $1 - 2^{-s} + 3^{-s} - 4^{-s} + 5^{-s} - \cdots$ converges to a positive number for $s > 0$ and that this number is

$$\zeta(s) - 2 \cdot 2^{-s}(1 + 2^{-s} + 3^{-s} + \cdots) = (1 - 2^{1-s})\zeta(s).$$

that is, it is the Cauchy principal value of the divergent integral $\int_0^x (dt/\log t)$. His argument is as follows:

Fix $x > 1$ and consider the function of β defined by

$$F(\beta) = \frac{1}{2\pi i}\frac{1}{\log x}\int_{a-i\infty}^{a+i\infty} \frac{d}{ds}\left\{\frac{\log[(s/\beta) - 1]}{s}\right\} x^s \, ds$$

so that the desired number is $F(1)$. The definition of $F(\beta)$ can be extended to all real or complex numbers β other than real numbers $\beta \leq 0$ by taking $a >$ Re β and defining $\log[(s/\beta) - 1]$ to be $\log(s - \beta) - \log \beta$, where, as usual, $\log z$ is defined for all z other than real $z \leq 0$ by the condition that it be real for real $z > 0$. The integral $F(\beta)$ converges absolutely because

$$\left|\frac{d}{ds}\left\{\frac{\log[(s/\beta) - 1]}{s}\right\}\right| \leq \frac{|\log[(s/\beta) - 1]|}{|s|^2} + \frac{1}{|s(s - \beta)|}$$

is integrable while x^s oscillates on the line of integration. Because

$$\frac{d}{d\beta}\left\{\frac{\log[(s/\beta) - 1]}{s}\right\} = \frac{1}{(\beta - s)\beta},$$

differentiation under the integral sign and integration by parts give

$$\begin{aligned} F'(\beta) &= \frac{1}{2\pi i}\frac{1}{\log x}\int_{a-i\infty}^{a+i\infty} \frac{d}{ds}\left[\frac{1}{(\beta - s)\beta}\right] x^s \, ds \\ &= -\frac{1}{2\pi i}\int_{a-i\infty}^{a+i\infty} \frac{x^s}{(\beta - s)\beta}\, ds. \end{aligned}$$

This last integral can be evaluated by applying Fourier inversion to the formula

$$\frac{1}{s - \beta} = \int_1^\infty x^{-s} x^{\beta - 1}\, dx \qquad [\mathrm{Re}(s - \beta) > 0],$$

$$\frac{1}{a + i\mu - \beta} = \int_0^\infty e^{-i\lambda\mu} e^{\lambda(\beta - a)}\, d\lambda \qquad [a > \mathrm{Re}\,\beta],$$

to obtain

(1) $$\int_{-\infty}^{\infty} \frac{1}{a + i\mu - \beta} e^{i\mu x}\, d\mu = \begin{cases} 2\pi e^{x(\beta - a)} & \text{if } x > 0, \\ 0 & \text{if } x < 0. \end{cases}$$

from which it follows that

(2) $$\frac{1}{2\pi i}\int_{a-i\infty}^{a+i\infty} \frac{1}{s - \beta} y^s \, ds = \begin{cases} y^\beta & \text{if } y > 1, \\ 0 & \text{if } y < 1, \end{cases}$$

provided $a >$ Re β. Since $x > 1$ by assumption, this gives $F'(\beta) = x^\beta/\beta$.

Now let C^+ denote the contour in the complex t-plane which consists of the line segment from 0 to $1 - \epsilon$ (where ϵ is a small positive number), followed by the semicircle in the upper halfplane Im $t \geq 0$ from $1 - \epsilon$ to $1 + \epsilon$, followed by the line segment from $1 + \epsilon$ to x, and let

$$G(\beta) = \int_{C+} \frac{t^{\beta-1}}{\log t}\,dt.$$

Then

$$G'(\beta) = \int_{C+} t^{\beta-1}\,dt = \frac{t^\beta}{\beta}\bigg|_0^x = F'(\beta).$$

Now $G(\beta)$ is defined and analytic for $\operatorname{Re}\beta > 0$ (if $\operatorname{Re}\beta < 0$, then the integral which defines G diverges at $t = 0$) as is $F(\beta)$; hence they differ by a constant (which might depend on x) throughout $\operatorname{Re}\beta > 0$. Riemann states that this constant can be evaluated by holding $\operatorname{Re}\beta$ fixed and letting $\operatorname{Im}\beta \to +\infty$ in both $F(\beta)$ and $G(\beta)$, but he does not carry out this evaluation.

To evaluate the limit of $G(\beta)$, set $\beta = \sigma + i\tau$, where σ is fixed and $\tau \to \infty$. The change of variable $t = e^u$, $u = \log t$ puts $G(\beta)$ in the form

$$\int_{i\delta-\infty}^{i\delta+\log x} \frac{e^{\beta u}}{u}\,du + \int_{i\delta+\log x}^{\log x} \frac{e^{\beta u}}{u}\,du,$$

where the path of integration has been altered slightly using Cauchy's theorem. The changes of variable $u = i\delta + v$ in the first integral and $u = \log x + iw$ in the second put this in the form

$$G(\beta) = e^{i\delta\sigma}e^{-\delta\tau}\int_{-\infty}^{\log x} \frac{e^{\sigma v}}{i\delta + v}e^{i\tau v}\,dv - ix^\beta \int_0^\delta \frac{e^{-\tau w}e^{\sigma i w}}{\log x + iw}\,dw.$$

Both integrals in this expression approach zero as $\tau \to \infty$, the first because $e^{-\delta\tau} \to 0$ and the second because $e^{-\tau w} \to 0$ except at $w = 0$. Thus *the limit of $G(\beta)$ as $\tau \to \infty$ is zero.* (Note, however, that this argument would not be valid if C^+ were changed to follow the lower semicircle because then $e^{-\delta\tau}$ would be replaced by $e^{\delta\tau}$ and $e^{-\tau w}$ would be replaced by $e^{\tau w}$.)

To evaluate the limit of $F(\beta)$ let

$$H(\beta) = \frac{1}{2\pi i}\frac{1}{\log x}\int_{a-i\infty}^{a+i\infty} \frac{d}{ds}\left\{\frac{\log[1-(s/\beta)]}{s}\right\}x^s\,ds$$

where $a > \operatorname{Re}\beta$ and where $\log[1-(s/\beta)]$ is defined for all complex numbers β other than real numbers $\beta \geq 0$ to be $\log(s-\beta) - \log(-\beta)$. The difference $H(\beta) - F(\beta)$ is defined for all complex numbers β other than the real axis, and in the upper halfplane $\operatorname{Im}\beta > 0$ it is equal to

$$\begin{aligned} H(\beta) - F(\beta) &= \frac{1}{2\pi i}\frac{1}{\log x}\int_{a-i\infty}^{a+i\infty} \frac{d}{ds}\left[\frac{\log\beta - \log(-\beta)}{s}\right]x^s\,ds, \\ &= \frac{1}{2\pi i}\frac{1}{\log x}\int_{a-i\infty}^{a+i\infty} \frac{d}{ds}\left[\frac{i\pi}{s}\right]x^s\,ds \\ &= -\frac{1}{2\pi i}\int_{a-i\infty}^{a+i\infty} \frac{i\pi}{s}x^s\,ds = -i\pi \end{aligned}$$

by the case $\beta = 0$ of (2). Thus $F(\beta) = H(\beta) + i\pi$ throughout the upper halfplane, and it will suffice to evaluate the limit of $H(\beta)$ as $\tau \to \infty$ ($\beta = \sigma + i\tau$). Now $1 - (s/\beta) \to 1$; hence its log goes to zero and it appears plausible therefore that $H(\beta)$ also goes to zero. This can be proved by carrying out the differentiation

$$\frac{d}{ds}\left\{\frac{\log[1-(s/\beta)]}{s}\right\} = -\frac{\log[1-(s/\beta)]}{s^2} + \frac{1}{s(s-\beta)}$$
$$= -\frac{\log[1-(s/\beta)]}{s^2} + \frac{1}{\beta(s-\beta)} - \frac{1}{\beta s},$$

multiplying by $x^s\,ds/2\pi i$, and integrating from $a - i\infty$ to $a + i\infty$ (in the usual sense, namely, the limit as $T \to \infty$ of the integral from $a - iT$ to $a + iT$). Because of the s^2 in the denominator of the first integral, it is not difficult to show, using the Lebesgue bounded convergence theorem (see Edwards [E1]), that the limit of this integral as $\tau \to \infty$ is the integral of the limit, namely, zero. The remaining two integrals can be evaluated using (2) to find they are $x^\beta/\beta - x^0/\beta = (x^\beta - 1)/\beta$. Since the numerator is bounded and $|\beta| \to \infty$, this approaches zero; hence $H(\beta)$ approaches zero and $F(\beta)$ therefore approaches $i\pi$. Hence $F(\beta) = G(\beta) + i\pi$ in the halfplane $\operatorname{Re}\beta > 0$. Thus the desired number $F(1)$ is

$$F(1) = \int_0^{1-\epsilon} \frac{dt}{\log t} + \int_{1-\epsilon}^{1+\epsilon} \frac{(t-1)}{\log t}\cdot\frac{dt}{t-1} + \int_{1+\epsilon}^{x} \frac{dt}{\log t} + i\pi,$$

where the second integral is over the semicircle in the upper halfplane; as $\epsilon \downarrow 0$, the quotient $(t-1)/\log t$ approaches 1 along this semicircle, and hence the integral approaches $\int_{1-\epsilon}^{1+\epsilon} dt/(t-1) = -i\pi$. Thus the limit as $\epsilon \downarrow 0$ of the above formula is

$$F(1) = \operatorname{Li}(x)$$

as was to be shown.

1.15 THE TERM INVOLVING THE ROOTS ρ

Consider next the term in the formula for $J(x)$ arising from the term $\sum \log[1 - (s/\rho)]$ in the formula for $\log \zeta(s)$, namely,

(1) $$-\frac{1}{2\pi i}\frac{1}{\log x}\int_{a-i\infty}^{a+i\infty} \frac{d}{ds}\left\{\frac{\sum \log[1-(s/\rho)]}{s}\right\} x^s\,ds.$$

If the operation of summation over ρ can be interchanged with the differentiation and the integration, *then* this is equal to $-\Sigma H(\rho)$, where $H(\rho)$ is defined as in the preceding section. Now it was shown that $H(\rho) \equiv G(\rho)$ for ρ in the first quadrant ($\operatorname{Re}\rho > 0$, $\operatorname{Im}\rho > 0$) and in exactly the same way it can be shown that for ρ in the fourth quadrant (Re $\rho > 0$, Im $\rho < 0$) the value of $H(\rho)$ is equal to the integral $G(\rho)$ except that the integral must be over the contour C^- which goes over the lower semicircle from $1-\epsilon$ to $1+\epsilon$ rather than over the upper semicircle as C^+ did. Thus, pairing terms of the sum over ρ in the usual way, the integral (1) would be

$$-\sum_{\operatorname{Im}\rho>0}\left(\int_{C^+}\frac{t^{\rho-1}}{\log t}\,dt+\int_{C^-}\frac{t^{-\rho}}{\log t}\,dt\right) \tag{2}$$

if it could be evaluated termwise. Now if β is real and positive, then the change of variable $u = t^\beta$, which implies $\log t = \log u/\beta$, $dt/t = du/u\beta$, gives

$$\int_{C^+}\frac{t^{\beta-1}}{\log t}\,dt=\int_0^{x^\beta}\frac{du}{\log u}=\operatorname{Li}(x^\beta)-i\pi,$$

where the second integral is over a path which passes above the singularity at $u = 1$. Since the integral on the left converges throughout the halfplane Re $\beta > 0$, this formula gives the analytic continuation of $\operatorname{Li}(x^\beta)$ to this half-plane (when x is, as always, a fixed number $x > 1$). In the same way

$$\int_{C^-}\frac{t^{\beta-1}}{\log t}\,dt=\operatorname{Li}(x^\beta)+i\pi,$$

and (2) becomes

$$-\sum_{\operatorname{Im}\rho>0}[\operatorname{Li}(x^\rho)+\operatorname{Li}(x^{1-\rho})]. \tag{3}$$

Thus, if termwise evaluation is valid, the desired integral (1) is equal to (3).

Riemann states that termwise evaluation is valid and that (3) is indeed the desired value (1) but that the series (3) is only conditionally convergent—even though the terms ρ, $1-\rho$ are paired—and that it must be summed in the order of increasing† Im ρ. He concedes that the validity of this termwise evaluation of (1) requires "a more exact discussion of the function ξ," but says that this is "easy" and passes on to the next point.

One other small remark about the sum (3) is necessary. The computations above assume Re $\rho > 0$, but it has not been shown that this is true for all roots ρ. Although Hadamard later proved that there are no roots ρ on the line Re $\rho = 0$ (see Section 4.2), Riemann has not excluded this possibility and he is therefore not justified in ignoring the point as he does.

†It is interesting to note that Riemann writes $\rho = \frac{1}{2} + i\alpha$ and says first that the sum (3) is over all *positive* values of α in order of size before then adding parenthetically that it is over all α's with Re(α) > 0 in order of size. Thus he admits, albeit parenthetically, the possibility that the Riemann hypothesis is false.

1.16 THE REMAINING TERMS

One of the three remaining terms in the formula for $J(x)$, namely, the term arising from $(s/2)\log\pi$, drops out when it is divided by s and differentiated with respect to s. The term arising from the constant $\log\xi(0)$ is

$$-\frac{1}{2\pi i}\frac{1}{\log x}\int_{a-i\infty}^{a+i\infty}\frac{d}{ds}\left(\frac{\log\xi(0)}{s}\right)x^s\,ds$$
$$=\frac{1}{2\pi i}\int_{a-i\infty}^{a+i\infty}\frac{\log\xi(0)}{s}x^s\,ds=\log\xi(0)$$

using (2) of Section 1.14 in the case $\beta = 0$. Now $\xi(0) = \Pi(0)\pi^{-0}(0-1)\zeta(0) = -\zeta(0) = \frac{1}{2}$ so $\log\xi(0) = -\log 2$ is the numerical value of this term.

Riemann writes $\log\xi(0)$ instead of $-\log 2$, but since he uses ξ to denote a different function—namely, the function $\xi(\frac{1}{2}+it)$ of t—his $\xi(0)$ denotes $\xi(\frac{1}{2}) \neq \frac{1}{2}$ and thus his formula is in error. It is hard to guess what the source of this trivial error might be, other than to say that it arises from some confusion between the product formula

$$\xi(s)=\xi(0)\prod_{\rho}[1-(s/\rho)]$$

in the form it is given above and the product formula

$$\xi\left(\frac{1}{2}+it\right)=\xi(0)\prod\left(1-\frac{\frac{1}{2}+it}{\frac{1}{2}+i\alpha}\right)$$
$$=\xi(0)\prod\left(\frac{i\alpha-it}{\frac{1}{2}+i\alpha}\right)$$
$$=\xi(0)\prod\left(\frac{i\alpha}{\frac{1}{2}+i\alpha}\right)\prod\left(1-\frac{it}{i\alpha}\right)$$
$$=\xi(0)\prod\left(1-\frac{\frac{1}{2}}{\frac{1}{2}+i\alpha}\right)\prod\left(1-\frac{t}{\alpha}\right)$$
$$=\xi\left(\frac{1}{2}\right)\prod_{\operatorname{Re}\alpha>0}\left(1-\frac{t^2}{\alpha^2}\right)$$

in the form given by Riemann, and a concomitant confusion of the integral

$$\frac{1}{2\pi i}\frac{1}{\log x}\int_{a-i\infty}^{a+i\infty}\frac{d}{ds}\left\{\frac{\log[1-(s/\rho)]}{s}\right\}x^s\,ds$$

which he evaluates, with the integral

$$\frac{1}{2\pi i}\frac{1}{\log x}\int_{a-i\infty}^{a+i\infty}\frac{d}{ds}\left\{\frac{\log[1-(s-\frac{1}{2})/i\alpha]}{s}\right\}x^s\,ds$$

which differs from it by a constant. Whatever the source of the error, Riemann makes the same error in the letter quoted by the editors in the notes which follow the paper in the collected works, and his unpublished papers [R1a] include a computation of $\log\xi(\frac{1}{2})$ to several decimal places, so it was definitely not a typographical error as the editors of the collected works suppose. The error was noticed by Genocchi [G4] during Riemann's lifetime.

This leaves only one term

$$(1)\qquad \frac{1}{2\pi i}\frac{1}{\log x}\int_{a-i\infty}^{a+i\infty}\frac{d}{ds}\left[\frac{\log\Pi(s/2)}{s}\right]x^s\,ds$$

to be evaluated. Now by formula (4) of Section 1.3

$$\log\Pi\left(\frac{s}{2}\right)=\sum_{n=1}^{\infty}\left[-\log\left(1+\frac{s}{2n}\right)+\frac{s}{2}\log\left(1+\frac{1}{n}\right)\right].$$

Using this formula in (1) and assuming that termwise integration is valid puts (1) in the form

$$-\sum_{n=1}^{\infty}\frac{1}{2\pi i}\frac{1}{\log x}\int_{a-i\infty}^{a+i\infty}\frac{d}{ds}\left\{\frac{\log[1+(s/2n)]}{s}\right\}x^s\,ds=-\sum_{n=1}^{\infty}H(-2n),$$

where H is as in Section 1.14. The previous formulas for $H(\beta)$ apply only in the halfplane $\operatorname{Re}\beta>0$. To obtain a formula for H in $\operatorname{Re}\beta<0$ set

$$E(\beta)=-\int_x^{\infty}\frac{t^{\beta-1}}{\log t}\,dt.$$

Then $E(\beta)$ converges for $\operatorname{Re}\beta<0$ and satisfies

$$E'(\beta)=-\int_x^{\infty}t^{\beta-1}\,dt=\frac{x^{\beta}}{\beta}=F'(\beta)=H'(\beta)$$

so $E(\beta)$ differs from $H(\beta)$ by a constant throughout $\operatorname{Re}\beta<0$. Since both E and H approach zero as $\beta\to-\infty$, the constant is zero and $E\equiv H$. Thus (1) becomes

$$\sum_{n=1}^{\infty}\int_x^{\infty}\frac{t^{-2n-1}}{\log t}\,dt=\int_x^{\infty}\frac{1}{t\log t}\left(\sum t^{-2n}\right)dt=\int_x^{\infty}\frac{dt}{t(t^2-1)\log t}$$

provided termwise integration is valid. The proof that termwise integration is valid, which Riemann (tacitly) leaves to the reader, can be given as follows.

Note first that the series

$$\frac{d}{ds}\left[\frac{\log\Pi(s/2)}{s}\right]=-\sum_{n=1}^{\infty}\frac{d}{ds}\left\{\frac{\log[1+(s/2n)]}{s}\right\}$$

converges uniformly in any disk $|s|\le K$. [For large n the series expansion $\log(1+x)=x-\frac{1}{2}x^2+\frac{1}{3}x^3-\cdots$ can be used, and the summand on the right contains only terms in $n^{-2}, n^{-3},\ldots$.] This justifies the termwise differentiation and also justifies termwise integration over any finite interval

$$(2)\qquad \frac{1}{2\pi i}\frac{1}{\log x}\int_{a-iT}^{a+iT}\frac{d}{ds}\left[\frac{\log\Pi(s/2)}{s}\right]x^s\,ds$$

$$=-\sum_{n=1}^{\infty}\frac{1}{2\pi i}\frac{1}{\log x}\int_{a-iT}^{a+iT}\frac{d}{ds}\left\{\frac{\log[1+(s/2n)]}{s}\right\}x^s\,ds.$$

To estimate the nth term of the sum on the right set $v=(s-a)/2n$, $b=a/2n$,

$c = T/2n$, so $s = 2n(v + b)$ and the nth term is minus

$$\frac{1}{2\pi i}\frac{1}{\log x}\int_{-ic}^{ic}\frac{d}{2n\,dv}\left[\frac{\log(1+v+b)}{2n(v+b)}\right]x^{2nv+a}2n\,dv$$
$$= \frac{1}{2\pi i}\frac{x^a}{2n\log x}\int_{-ic}^{ic}\frac{d}{dv}\left[\frac{\log(1+v+b)}{v+b}\right]x^{2nv}\,dv.$$

Integration by parts puts this in the form

$$= \frac{1}{2\pi i}\frac{x^a}{2n\log x}\cdot\frac{1}{2n\log x}\left(\left\{\frac{d}{dv}\left[\frac{\log(1+v+b)}{v+b}\right]x^{2nv}\right|_{v=-ic}^{v=ic}\right.$$
$$\left.-\int_{-ic}^{ic}\frac{d^2}{dv^2}\left[\frac{\log(1+v+b)}{v+b}\right]x^{2nv}\,dv\right).$$

Now b is a real number $0 \le b \le a$, the function

$$\frac{d}{dv}\left[\frac{\log(1+v+b)}{v+b}\right] = \frac{1}{(v+b)(v+b+1)} - \frac{\log(1+v+b)}{(v+b)^2}$$

is bounded on the imaginary axis, and its derivative is absolutely integrable over $(-i\infty, i\infty)$, from which it follows that the modulus of the nth term of the series on the right side of (2) is at most a constant times n^{-2} *for all* T. Thus the series converges uniformly in T and one can pass to the limit $T \to \infty$ termwise, as was to be shown.

This completes the evaluation of the terms in the formula for $J(x)$. Combining them gives the final result

$$(3)\qquad J(x) = \mathrm{Li}(x) - \sum_{\mathrm{Im}\,\rho>0}[\mathrm{Li}(x^\rho) + \mathrm{Li}(x^{1-\rho})] + \int_x^\infty \frac{dt}{t(t^2-1)\log t} + \log\xi(0) \qquad (x > 1)$$

which is Riemann's formula [except that, as noted above, $\log\xi(0)$ equals $\log(\frac{1}{2})$ and not $\log\xi(\frac{1}{2})$ as in Riemann's notation it should]. This analytic formula for $J(x)$ is the principal result of the paper.

1.17 THE FORMULA FOR $\pi(x)$

Of course Riemann's goal was to obtain a formula not for $J(x)$ but for the function $\pi(x)$, that is, for the number of primes less than any given magnitude x. Since the number of prime squares less than x is obviously equal to the number of primes less than $x^{1/2}$, that is, equal to $\pi(x^{1/2})$, and since in the same way the number of prime nth powers p^n less than x is $\pi(x^{1/n})$, it follows that J and π are related by the formula

$$(1)\qquad J(x) = \pi(x) + \frac{1}{2}\pi(x^{1/2}) + \frac{1}{3}\pi(x^{1/3}) + \cdots + \frac{1}{n}\pi(x^{1/n}) + \cdots.$$

The series in this formula is finite for any given x because $x^{1/n} < 2$ for n sufficiently large, which implies $\pi(x^{1/n}) = 0$. Riemann inverts this relationship by means of the Möbius inversion formula† (see Section 10.9) to obtain

$$(2) \qquad \pi(x) = J(x) - \frac{1}{2}J(x^{1/2}) - \frac{1}{3}J(x^{1/3}) - \frac{1}{5}J(x^{1/5}) + \frac{1}{6}J(x^{1/6}) + \cdots + \frac{\mu(n)}{n}J(x^{1/n}) + \cdots,$$

where $\mu(n)$ is 0 if n is divisible by a prime square, 1 if n is a product of an even number of distinct primes, and -1 if n is a product of an odd number of distinct primes. The series (2) is a finite series for any fixed x and when combined with the analytical formula for $J(x)$

$$(3) \qquad J(x) = \text{Li}(x) - \sum_{\rho} \text{Li}(x^{\rho}) - \log 2 + \int_x^{\infty} \frac{dt}{t(t^2 - 1)\log t} \qquad (x > 1),$$

it gives an analytical formula for $\pi(x)$ as desired.

The formula for $\pi(x)$ which results from substituting (3) in the (finite) series (2) consists of three kinds of terms, namely, those which do not grow as x grows [arising from the last two terms of (3)], those which grow as x grows but which oscillate in sign [the terms arising from $\text{Li}(x^{\rho})$ which Riemann calls "periodic"], and those which grow steadily as x grows [the terms arising from $\text{Li}(x)$]. If all but the last type are ignored, the terms in the formula for $\pi(x)$ are just

$$\text{Li}(x) - \tfrac{1}{2}\text{Li}(x^{1/2}) - \tfrac{1}{3}\text{Li}(x^{1/3}) - \tfrac{1}{5}\text{Li}(x^{1/5}) + \tfrac{1}{6}\text{Li}(x^{1/6}) - \tfrac{1}{7}\text{Li}(x^{1/7}) + \cdots.$$

Now *empirically* this is found to be a good approximation to $\pi(x)$. In fact, the first term alone is essentially Gauss's approximation

$$\pi(x) \sim \int_2^x \frac{dt}{\log t} = \text{Li}(x) - \text{Li}(2)$$

[$\text{Li}(2) = 1.04\ldots$] and the first two terms indicate that

$$\pi(x) \sim \text{Li}(x) - \tfrac{1}{2}\text{Li}(x^{1/2})$$

†Very simply this inversion is effected by performing successively for each prime $p = 2, 3, 5, 7, 11, \ldots$ the operation of replacing the functions $f(x)$ on each side of the equation with the functions $f(x) - (1/p)f(x^{1/p})$. This gives successively

$$J(x) - \tfrac{1}{2}J(x^{1/2}) = \pi(x) + \tfrac{1}{3}\pi(x^{1/3}) + \tfrac{1}{5}\pi(x^{1/5}) + \cdots,$$

$$J(x) - \tfrac{1}{2}J(x^{1/2}) - \tfrac{1}{3}J(x^{1/3}) + \tfrac{1}{6}J(x^{1/6}) = \pi(x) + \tfrac{1}{5}\pi(x^{1/5}) + \tfrac{1}{7}\pi(x^{1/7}) + \cdots,$$

etc., where at each step the sum on the left consists of those terms of the right side of (2) for which the factors of n contain *only* the primes already covered and the sum on the right consists of those terms of the right side of (1) for which the factors of n contain *none* of the primes already covered. Once p is sufficiently large, the latter are all zero except for $\pi(x)$.

which gives, for example,

$$\pi(10^6) \sim 78{,}628 - \tfrac{1}{2}\cdot 178 = 78{,}539$$

which is better than Gauss's approximation and which becomes still better if the third term is used. The extent to which Riemann's suggested approximation

$$\pi(x) \sim \mathrm{Li}(x) + \sum_{n=2}^{\infty} \frac{\mu(n)}{n} \mathrm{Li}(x^{1/n}) \tag{4}$$

is better than $\pi(x) \sim \mathrm{Li}(x)$ is stunningly illustrated by one of Lehmer's tables [L9], an extract of which is given in Table III.

TABLE III[a]

x	Riemann's error	Gauss's error
1,000,000	30	130
2,000,000	−9	122
3,000,000	0	155
4,000,000	33	206
5,000,000	−64	125
6,000,000	24	228
7,000,000	−38	179
8,000,000	−6	223
9,000,000	−53	187
10,000,000	88	339

[a]From Lehmer [L9].

Of course Riemann did not have such extensive empirical data at his disposal, but he seems well aware of the fact that (4) is a better approximation, as well as a more natural approximation, to $\pi(x)$.

Riemann was also well aware, however, of the defects of the approximation (4) and of his analysis of it. Although he has succeeded in giving an exact analytical formula for the error

$$\pi(x) - \sum_{n=1}^{N} \frac{\mu(n)}{n} \mathrm{Li}(x^{1/n}) = \sum_{n=1}^{N} \sum_{\rho} \mathrm{Li}(x^{\rho/n}) + \text{lesser terms}$$

(where N is large enough that $x^{1/(N+1)} < 2$) he has no estimate at all of the size of these "periodic" terms $\sum \sum \mathrm{Li}(x^{\rho/n})$. Actually, the empirical fact that they are as small as Lehmer found them to be is somewhat surprising in view of the fact that the series $\sum [\mathrm{Li}(x^{\rho}) + \mathrm{Li}(x^{1-\rho})]$ is only conditionally convergent—hence the smallness of its sum for any x depends on wholesale cancellation of signs among the terms—and in view of the fact that the individual terms $\operatorname{Li}(x^\rho)$ grow in magnitude like $|x^\rho/\log x^\rho| = x^{\operatorname{Re}\rho}/|\rho|\log x$ (see Section 5.5) so that many of them grow at least as fast as $x^{1/2}/\log x \sim 2\operatorname{Li}(x^{1/2}) > \operatorname{Li}(x^{1/3})$ and would therefore be expected to be as significant for large x as the term $-\frac{1}{2}\operatorname{Li}(x^{1/2})$ and more significant than any of the following terms of (4). On these subjects Riemann restricts himself to the statement that it would be interesting in later counts of primes to study the effect of the particular "periodic" terms on their distribution.

In short, although formulas (2) and (3) combine to give an analytical formula for $\pi(x)$, the validity of the new approximation (4) to $\pi(x)$ to which it leads is based, like that of the old approximation $\pi(x) \sim \operatorname{Li}(x)$, solely on empirical evidence.

1.18 THE DENSITY *dJ*

A simple formulation of the main result

(1)
$$J(x) = \operatorname{Li}(x) - \sum_\rho \operatorname{Li}(x^\rho) - \log 2 + \int_x^\infty \frac{dt}{t(t^2-1)\log t}$$

can be obtained by differentiating to find

(2)
$$dJ = \left[\frac{1}{\log x} - \sum_{\operatorname{Re}\alpha>0} \frac{2\cos(\alpha\log x)}{x^{1/2}\log x} - \frac{1}{x(x^2-1)\log x}\right] dx \qquad (x > 1),$$

where α ranges over all values such that $\rho = \frac{1}{2} + i\alpha$—in other words $\alpha = -i(\rho - \frac{1}{2})$, where ρ ranges over the roots—so that

$$x^{\rho-1} + x^{-\rho} = x^{-1/2}[x^{i\alpha} + x^{-i\alpha}] = 2x^{-1/2}\cos(\alpha\log x).$$

[The Riemann hypothesis is that the α's are all real. In writing formula (2) in this form Riemann is clearly thinking of the α's as being real since otherwise the natural form would be $x^{\rho-1} + x^{\bar\rho-1} = 2x^{\beta-1}\cos(\gamma\log x)$, where $\rho = \beta + i\gamma$.]

By the definition of J, the measure dJ is dx times the density of primes plus $\frac{1}{2}$ the density of prime squares, plus $\frac{1}{3}$ the density of prime cubes plus, etc. Thus $1/\log x$ should be considered to be an approximation not to the density of primes as Gauss suggested but rather to dJ, that is, to the density of primes *plus* $\frac{1}{2}$ the density of prime squares, plus, etc.

Given two large numbers $a < b$ the approximation obtained by taking a finite number of the α's

(3)
$$J(b) - J(a) \sim \int_a^b \frac{dt}{\log t} - 2\sum \int_a^b \frac{\cos(\alpha\log t)\,dt}{t^{1/2}\log t}$$

should be a fairly good approximation because the omitted term $\int dx/x(x^2-1)\log x$

is entirely negligible and because the integrals involving the large α's oscillate very rapidly for large x and therefore should make very small contributions. In fact, the basic formula (1) implies immediately that the error in (3) approaches the negligible omitted term as more and more of the α's are included in the sum.

It is in the sense of investigating the number of α's which are significant in (3) that Riemann meant to investigate empirically the influence of the "periodic" terms on the distribution of primes. So far as I know, no such investigation has ever been carried out.

1.19 QUESTIONS UNRESOLVED BY RIEMANN

Riemann himself, in a letter quoted in the notes which follow this paper in his collected works, singles out two statements of the paper as not having been fully proved as yet, namely, the statement that the equation $\xi(\frac{1}{2} + i\alpha) = 0$ has approximately $(T/2\pi)\log(T/2\pi)$ *real* roots α in the range $0 < \alpha < T$ and the statement that the integral of Section 1.15 can be evaluated termwise. He expresses no doubt about the truth of these statements, however, and says that they follow from a new representation of the function ξ which he has not yet simplified sufficiently to publish. Nonetheless, as was stated in Section 1.9, the first of these two statements—at least if it is understood to mean that the relative error in the approximation approaches zero as $T \longrightarrow \infty$—has never been proved. The second was proved by von Mangoldt in 1895, but by a method completely different from that suggested by Riemann, namely by proving first that Riemann's formula for $J(x)$ is valid and by concluding from this that the termwise value of the integral in Section 1.15 must be correct.

Riemann evidently believed that he had given a proof of the product formula for $\xi(s)$, but, at least from the reading of the paper given above, one cannot consider his proof to be complete, and, in particular, one must question Riemann's estimate of the number of roots ρ in the range $\{0 \leq \operatorname{Im} \rho \leq T\}$ on which this proof is based. It was not until 1893 that Hadamard proved the product formula, and not until 1905 that von Mangoldt proved the estimate of the number of roots in $\{0 \leq \operatorname{Im} \rho \leq T\}$.

Next, the original question of the validity of the approximation $\pi(x) \sim \int_2^x (dt/\log t)$ remained entirely unresolved by Riemann's paper. It can be shown that the relative error of this approximation approaches zero as $x \longrightarrow \infty$ if and only if the same is true of the relative error in Riemann's approximation $J(x) \sim \operatorname{Li}(x)$, so the original question is equivalent to the question of whether $\sum \operatorname{Li}(x^\rho)/\operatorname{Li}(x) \longrightarrow 0$, but this unfortunately does not bring the problem any nearer to a solution. It was not until 1896 that Hadamard and, independently, de la Vallée Poussin proved the prime number theorem to the effect that the relative error in $\pi(x) \sim \int_2^x (dt/\log t)$ does approach zero as $x \longrightarrow \infty$.

Finally, the paper raised a question much greater than any question it answered, the question of the truth or falsity of the Riemann hypothesis.

The remainder of this book is devoted to the subsequent history of these six questions. In summary, they are as follows:

(a) Is Riemann's estimate of the number of roots ρ on the line segment from $\frac{1}{2}$ to $\frac{1}{2} + iT$ correct as $T \longrightarrow \infty$? (Unknown.)

(b) Is termwise evaluation of the integral of Section 1.15 valid? (Yes, von Mangoldt, 1895.)

(c) Is the product formula for $\xi(s)$ valid? (Yes, Hadamard, 1893.)

(d) Is Riemann's estimate of the number of roots ρ in the strip $\{0 \leq \operatorname{Im} \rho \leq T\}$ correct? (Yes, von Mangoldt, 1905.)

(e) Is the prime number theorem true? [Yes, Hadamard and de la Vallée Poussin (independently), 1896.]

(f) Is the Riemann hypothesis true? (Unknown.)

Chapter 2

The Product Formula for ξ

2.1 INTRODUCTION

In 1893 Hadamard published a paper [H1] in which he studied entire functions (functions of a complex variable which are defined and analytic at all points of the complex plane) and their representations as infinite products. One consequence of the general theory which he developed in this paper is the fact that the product formula

$$\xi(s) = \xi(0) \prod_{\rho} \left(1 - \frac{s}{\rho}\right) \tag{1}$$

is valid; here ξ is the entire function defined in Section 1.8, ρ ranges over all roots ρ of $\xi(\rho) = 0$, and the infinite product is understood to be taken in an order which pairs each root ρ with the corresponding root $1 - \rho$. Hadamard's proof of the product formula for ξ was called by von Mangoldt [M1] "the first real progress in the field in 34 years," that is, the first since Riemann's paper.

This chapter is devoted to the proof of the product formula (1). Since only the specific function ξ is of interest here, Hadamard's methods for general entire functions can, of course, be considerably specialized and simplified† for this case, and in the end the proof which results is closer to the one outlined by Riemann than to Hadamard's proof. The first step of the proof is to make an estimate of the distribution of the roots ρ. This estimate, which is that the number of roots ρ in the disk $|\rho - \frac{1}{2}| < R$ is less than a constant times $R \log R$ as $R \to \infty$, is based on Jensen's theorem and is much less exact than Riemann's estimate that the number of roots in the strip $\{0 < \text{Im}$

†A major simplification is the use of Jensen's theorem, which was not known at the time Hadamard was writing.

$\rho < T\}$ is $(T/2\pi)\log(T/2\pi) - (T/2\pi)$ with a relative error which is of the order of magnitude of T^{-1}. It is exact enough, however, to prove the convergence of the product (1). Once it has been shown that this product converges, the rest of the proof can be carried out more or less as Riemann suggests.

2.2 JENSEN'S THEOREM

Theorem Let $f(z)$ be a function which is defined and analytic throughout a disk $\{|z| \leq R\}$. Suppose that $f(z)$ has no zeros on the bounding circle $|z| = R$ and that inside the disk it has the zeros $z_1, z_2, \ldots, z_n$ (where a zero of order k is included k times in the list). Suppose, finally, that $f(0) \neq 0$. Then

(1)
$$\log\left| f(0) \cdot \frac{R}{z_1} \cdot \frac{R}{z_2} \cdot \cdots \cdot \frac{R}{z_n} \right| = \frac{1}{2\pi}\int_0^{2\pi} \log|f(Re^{i\theta})|\, d\theta.$$

***Proof*†** If $f(z)$ has no zeros inside the disk, then the equation is merely

(2)
$$\log|f(0)| = \frac{1}{2\pi}\int_0^{2\pi} \log|f(Re^{i\theta})|\, d\theta;$$

that is, the equation is the statement that the value of $\log|f(z)|$ at the center of the disk is equal to its average value on the bounding circle. This can be proved either by observing that $\log|f(z)|$ is the real part of the analytic function $\log f(z)$ and is therefore a harmonic function, or by taking the real part of the Cauchy integral formula

$$\log f(0) = \frac{1}{2\pi i}\int_{|z|=R} \frac{\log f(z)}{z}\, dz = \frac{1}{2\pi}\int_0^{2\pi} \log f(Re^{i\theta})\, d\theta,$$

where $\log f(z)$ is defined in the disk to be

$$\log|f(0)| + \int_0^z \frac{f'(t)}{f(t)}\, dt.$$

Applying this formula (2) to the function

$$F(z) = f(z)\frac{R^2 - \bar{z}_1 z}{R(z - z_1)} \cdot \frac{R^2 - \bar{z}_2 z}{R(z - z_2)} \cdots \frac{R^2 - \bar{z}_n z}{R(z - z_n)}$$

gives

$$\log|F(0)| = \frac{1}{2\pi}\int_0^{2\pi} \log|F(Re^{i\theta})|\, d\theta$$

because $F(z)$ is analytic and has no zeros in the disk. But this is the formula

†See Ahlfors [A3]. This method of proof of Jensen's threorem is to be found in Backlund's 1918 paper on the Lindelöf hypothesis [B3] (see Section 9.4).

of Jensen's theorem (1) because

$$\left|\frac{R^2 - \bar{z}_j \cdot 0}{R(0 - z_j)}\right| = \left|\frac{R}{z_j}\right|$$

and because by a basic formula in the theory of conformal mapping

$$\left|\frac{R^2 - \bar{z}_j z}{R(z - z_j)}\right| = 1 \qquad \text{when} \quad |z| = R.$$

(To prove this formula multiply the numerator by $\bar{z}/R$. This does not change the modulus if $|z| = R$ and it makes the numerator into the complex conjugate of the denominator.) This completes the proof of Jensen's theorem (1).

2.3 A SIMPLE ESTIMATE OF $|\xi(s)|$

Theorem For all sufficiently large values of R the estimate $|\xi(s)| \le R^R$ holds throughout the disk $|s - \frac{1}{2}| \le R$.

Proof It was shown in Section 1.8 that $\xi(s)$ can be expanded as a power series in $(s - \frac{1}{2})$:

$$\xi(s) = a_0 + a_2(s - \tfrac{1}{2})^2 + \cdots + a_{2n}(s - \tfrac{1}{2})^{2n} + \cdots,$$

where

$$a_{2n} = 4 \int_1^\infty \frac{d}{dx}[x^{3/2}\psi'(x)]\, x^{-1/4} \frac{(\frac{1}{2}\log x)^{2n}}{(2n)!}\, dx.$$

The fact that the coefficients a_n are *positive* follows immediately from

$$\begin{aligned}\frac{d}{dx}[x^{3/2}\psi'(x)] &= \frac{d}{dx}\left(-\sum_{n=1}^{\infty} x^{3/2} n^2 \pi e^{-n^2\pi x}\right)\\ &= \sum_{n=1}^{\infty}\left(n^4\pi^2 x - \frac{3}{2}n^2\pi\right) x^{1/2} e^{-n^2\pi x}\end{aligned}$$

because this shows that the integrand in the integral for a_{2n} is positive for $x \ge 1$. Thus the largest value of $\xi(s)$ on the disk $|s - \frac{1}{2}| \le R$ occurs at the point $s = \frac{1}{2} + R$, and to prove the theorem it suffices to show that $\xi(\frac{1}{2} + R) \le R^R$ for all sufficiently large R. Now

$$\xi(s) = \Pi(s/2)\pi^{-s/2}(s - 1)\zeta(s)$$

and $\zeta(s)$ decreases to 1 as $s \to +\infty$, so if R is given and if N is chosen so that $\frac{1}{2} + R \le 2N < \frac{1}{2} + R + 2$, it follows that

$$\begin{aligned}\xi(\tfrac{1}{2} + R) \le \xi(2N) &= (N!)\pi^{-N}(2N - 1)\zeta(2N)\\ &\le N^N \pi^{-0}(2N)\zeta(2)\\ &= \text{const } N^{N+1}\\ &\le \text{const } (\tfrac{1}{2}R + 2)^{(R/2)+3} < R^R\end{aligned}$$

for all sufficiently large R, which completes the proof of the theorem.

2.4 THE RESULTING ESTIMATE OF THE ROOTS ρ

Theorem Let $n(R)$ denote the number of roots ρ of $\xi(\rho) = 0$ which lie inside or on the circle $|s - \frac{1}{2}| = R$ (counted with multiplicities). Then $n(R) \leq 3R \log R$ for all sufficiently large R.

Proof Jensen's theorem applied to $\xi(s)$ on the disk $|s - \frac{1}{2}| \leq 2R$ gives

$$\log \xi\left(\frac{1}{2}\right) + \sum_{|\rho - 1/2| < 2R} \log \frac{2R}{|\rho - \frac{1}{2}|} \leq \log[(2R)^{2R}].$$

The terms of the sum over ρ are all positive and the terms corresponding to roots ρ inside the circle $|\rho - \frac{1}{2}| \leq R$ are all at least $\log 2$; hence,

$$\begin{aligned} n(R) \log 2 &\leq 2R \log 2R - \log \xi(\tfrac{1}{2}) \\ n(R) &\leq \frac{2}{\log 2} R \log R + 2\mathrm{R} - \frac{\log \xi(\frac{1}{2})}{\log 2} \\ &\leq 3R \log R \end{aligned}$$

for all sufficiently large R, as was to be shown. If there are roots ρ on the circle $|s - \frac{1}{2}| = 2R$, so that Jensen's theorem is not applicable, one can apply the above to the circle with radius $R + \epsilon$ and let $\epsilon \longrightarrow 0$.

2.5 CONVERGENCE OF THE PRODUCT

As was noted in Section 1.10, in order to prove the convergence of the product

$$\Pi\left(1 - \frac{s}{\rho}\right) = \prod_{\operatorname{Im} \rho > 0} \left[1 - \frac{s(1 - s)}{\rho(1 - \rho)}\right] \tag{1}$$

for all s, it suffices to prove the convergence of the sum $\sum |\rho(1 - \rho)|^{-1}$. Since all but a finite number of roots ρ satisfy the inequality

$$\frac{1}{|\rho(1 - \rho)|} = \frac{1}{|(\rho - \frac{1}{2})^2 - \frac{1}{4}|} < \frac{1}{|\rho - \frac{1}{2}|^2},$$

it suffices therefore to prove the convergence of the sum $\sum |\rho - \frac{1}{2}|^{-2}$; here the sum can be considered either as a sum over roots ρ in the upper halfplane $\operatorname{Im} \rho > 0$ or as a sum over all roots since the first of these is merely twice the second. The convergence of the product (1) is therefore a consequence of the case $\epsilon = 1$ of the following theorem.

Theorem For any given $\epsilon > 0$ the series

$$\sum \frac{1}{|\rho - \frac{1}{2}|^{1+\epsilon}}$$

converges, where ρ ranges over all roots ρ of $\xi(\rho) = 0$.

[Note that this theorem would follow immediately from Riemann's observation that the vertical density of the roots ρ is a constant times $(\log T)\,dT$ and from the fact that $\int^{\infty} T^{-1-\epsilon}(\log T)\,dT$ converges. This is Riemann's first step in his "proof" of the product formula for ξ.]

Proof Let the roots ρ be numbered $\rho_1, \rho_2, \rho_3, \ldots$ in order of increasing $|\rho - \frac{1}{2}|$. Furthermore, let $R_1, R_2, R_3, \ldots$ be the sequence of positive real numbers defined implicitly by the equation $4R_n \log R_n = n$. Then by the theorem of the preceding section there are at most $3n/4$ roots ρ inside the circle $|s - \frac{1}{2}| = R_n$; hence the nth root is not in this circle, that is, $|\rho_n - \frac{1}{2}| > R_n$. Thus

$$\sum \frac{1}{|\rho_n - \frac{1}{2}|^{1+\epsilon}} \leq \sum \frac{1}{R_n^{1+\epsilon}} = \sum \frac{(4 \log R_n)^{1+\epsilon}}{n^{1+\epsilon}}$$
$$= \sum \frac{1}{n^{1+(\epsilon/2)}} \cdot \frac{(4 \log R_n)^{1+\epsilon}}{n^{\epsilon/2}}.$$

Now $\log n = \log R_n + \log 4 + \log \log R_n > \log R_n$ for n large. Hence $(4 \log R_n)^{1+\epsilon} < 16 (\log n)^2 < n^{\epsilon/2}$ for all sufficiently large n and

$$\sum \frac{1}{|\rho - \frac{1}{2}|^{1+\epsilon}} < \text{const} + \sum \frac{1}{n^{1+(\epsilon/2)}} < \infty$$

as was to be shown.

2.6 RATE OF GROWTH OF THE QUOTIENT

Riemann states that $\log \xi(s) - \sum \log [1 - (s/\rho)]$ grows no faster than $s \log s$, from which he concludes, since it is an even function, that it must be a constant. In this section the weaker result that the growth of its real part is no faster than $|s|^{1+\epsilon}$ will be proved. This still permits one to conclude, as will be shown in the next section, that it is constant.

Theorem Let $\epsilon > 0$ be given. Then

$$\operatorname{Re} \log \frac{\xi(s)}{\prod_{\rho} \left(1 - \frac{s - \frac{1}{2}}{\rho - \frac{1}{2}}\right)} \leq \left| s - \frac{1}{2} \right|^{1+\epsilon}$$

for all sufficiently large $|s - \frac{1}{2}|$.

Proof Let R be given and let the function being estimated be written as a sum of two functions

$$\operatorname{Re} \log \frac{\xi(s)}{\prod \left(1 - \frac{s - \frac{1}{2}}{\rho - \frac{1}{2}}\right)} = u_R(s) + v_R(s),$$

where

$$u_R(s) = \text{Re}\log \frac{\xi(s)}{\prod_{|\rho-1/2|\le 2R}\left(1-\frac{s-\frac{1}{2}}{\rho-\frac{1}{2}}\right)}$$

$$v_R(s) = \text{Re}\log \frac{1}{\prod_{|\rho-1/2|>2R}\left(1-\frac{s-\frac{1}{2}}{\rho-\frac{1}{2}}\right)}.$$

These logarithms are defined only up to multiples of $2\pi i$, but their real parts are well defined except at the points $s = \rho$ for $|\rho - \frac{1}{2}| > 2R$ (at which points u_R is $-\infty$ and v_R is $+\infty$). It will suffice to show that for large R both $u_R(s)$ and $v_R(s)$ are at most $R^{1+\epsilon}$ on $|s - \frac{1}{2}| = R$ since then, when ϵ is decreased slightly to ϵ', it follows that $u_R(s) + v_R(s) \le 2R^{1+\epsilon'} \le |s - \frac{1}{2}|^{1+\epsilon}$ on $|s - \frac{1}{2}| = R$ for all R large enough that $u_R \le R^{1+\epsilon'}$, $v_R \le R^{1+\epsilon'}$, and $2 \le R^{\epsilon-\epsilon'}$.

First consider $u_R(s)$. On the circle $|s - \frac{1}{2}| = 4R$ the factors in the denominator are all at least 1; therefore

$$\begin{aligned} u_R(s) &\le \text{Re}\log \xi(s) = \log|\xi(s)| \\ &\le \log[(4R)^{4R}] = 4R\log 4R \le R^{1+\epsilon} \end{aligned}$$

on the circle $|s - \frac{1}{2}| = 4R$, for large R (large enough that $4\log 4R < R^{\epsilon}$). Now u_R is a harmonic function on the disk $|s - \frac{1}{2}| \le 4R$ except at the points $s = \rho$ in the range $2R < |s - \frac{1}{2}| \le 4R$. But near these singular points $s = \rho$ the value of u_R is near $-\infty$, so the maximum value of the harmonic function u_R on the disk $|s - \frac{1}{2}| \le 4R$ must occur on the outer boundary $|s - \frac{1}{2}| = 4R$. Thus the maximum of u_R on the disk, and in particular on the circle $|s - \frac{1}{2}| = R$, is at most $R^{1+\epsilon}$ as was to be shown.

Now consider $v_R(s)$. For complex x in the disk $|x| \le \frac{1}{2}$ the inequality

$$\begin{aligned} \text{Re}\log\frac{1}{1-x} &= -\text{Re}\log(1-x) = \text{Re}\int_0^x \frac{dt}{1-t} \\ &\le \left|\int_0^x \frac{dt}{1-t}\right| \le |x|\max\frac{1}{|1-t|} = 2|x| \end{aligned}$$

holds. Thus for $|s - \frac{1}{2}| = R$ the inequality

$$\begin{aligned} v_R(s) &= \text{Re}\log \frac{1}{\prod_{|\rho-1/2|>2R}\left(1-\frac{(s-\frac{1}{2})^2}{(\rho-\frac{1}{2})^2}\right)} \\ &\le 2\sum_{|\rho-1/2|>2R} \frac{R^2}{|\rho-\frac{1}{2}|^2} \\ &= 2\sum\left(\frac{R}{|\rho-\frac{1}{2}|}\right)^{1-\epsilon}\left(\frac{R}{|\rho-\frac{1}{2}|}\right)^{1+\epsilon} \\ &\le 2\sum\left(\frac{1}{2}\right)^{1-\epsilon}\frac{R^{1+\epsilon}}{|\rho-\frac{1}{2}|^{1+\epsilon}} \\ &= 2^{\epsilon}R^{1+\epsilon}\sum_{|\rho-1/2|>2R}\frac{1}{|\rho-\frac{1}{2}|^{1+\epsilon}} \end{aligned}$$

holds. Now the sum in this expression converges by the theorem of Section 2.5, and it decreases to zero as R increases. Thus $v_R(s) \leq R^{1+\epsilon}$ on $|s - \frac{1}{2}| = R$ for all sufficiently large R as was to be shown. This completes the proof.

2.7 RATE OF GROWTH OF EVEN ENTIRE FUNCTIONS

Theorem Let $f(s)$ be an analytic function, defined in the entire s-plane, which is even in the sense that $f(-s) \equiv f(s)$ and which grows more slowly than $|s|^2$ in the sense that for every $\epsilon > 0$ there is an R such that $\operatorname{Re} f(s) < \epsilon |s|^2$ at all points s satisfying $|s| \geq R$. Then f must be constant.

Proof The subtle point of the theorem is that only the *upward* growth of the *real part* of f is limited. The main step in the proof is the following lemma, which shows that this implies that the growth of the *modulus* of f is also limited.

Lemma Let $f(s)$ be an analytic function on the disk $\{|s| \leq r\}$, let $f(0) = 0$, and let M be the maximum value of $\operatorname{Re} f(s)$ on the bounding circle $|s| = r$ (and hence on the entire disk). Then for $r_1 < r$ the modulus of f on the smaller disk $\{|s| \leq r_1\}$ is bounded by

$$|f(s)| \leq 2r_1 M/(r - r_1) \qquad (|s| \leq r_1).$$

Proof of the Lemma Consider the function

$$\phi(s) = f(s)/s[2M - f(s)].$$

If $u(s)$ and $v(s)$ denote the real and imaginary parts, respectively, of f, then $|2M - u(s)| \geq M \geq u(s)$ on the circle $|s| = r$; so the modulus of ϕ on this circle is at most

$$|\phi(s)| = \frac{(u^2 + v^2)^{1/2}}{r[(2M - u)^2 + v^2]^{1/2}} \leq \frac{(u^2 + v^2)^{1/2}}{r(u^2 + v^2)^{1/2}} = \frac{1}{r}$$

which implies that $|\phi(s)| \leq r^{-1}$ throughout the disk $\{|s| \leq r\}$. But $f(s)$ can be expressed in terms of $\phi(s)$ as

$$\phi(s)s[2M - f(s)] = f(s), \qquad f(s) = \frac{2Ms\phi(s)}{1 + s\phi(s)}$$

which shows that for $|s| = r_1$ the modulus of $f(s)$ is at most

$$|f(s)| \leq 2Mr_1 r^{-1}/(1 - r_1 r^{-1}) = 2Mr_1/(r - r_1).$$

Hence the same inequality holds throughout the disk $\{|s| \leq r_1\}$ as was to be shown.

Now to prove the theorem let $f(s) = \sum_{n=0}^{\infty} a_n s^n$ be the power series expansion of a function $f(s)$ satisfying the conditions of the theorem. Note first that it can be assumed without loss of generality that $a_0 = 0$ because $f(s)$ satisfies the growth condition of the theorem if and only if $f(s) - f(0)$ does and because $f(s)$ is constant if and only if $f(s) - f(0)$ is. Now Cauchy's integral formula for the coefficients is

$$a_n = \frac{1}{2\pi i} \int_{\partial D} \frac{f(s)\, ds}{s^{n+1}},$$

where D is any domain containing the origin. Let ϵ, R be as in the statement of the theorem and let D be the disk $\{|s| \leq \frac{1}{2}R\}$. Then the above formula gives

$$\begin{aligned} |a_n| &= \left| \frac{1}{2\pi i} \int_0^{2\pi} \frac{f(\frac{1}{2}\mathrm{Re}^{i\theta})}{(\frac{1}{2}\mathrm{Re}^{i\theta})^n} i\, d\theta \right| \\ &\leq \frac{1}{2\pi} \int_0^{2\pi} \frac{2^n\, |f(\frac{1}{2}\mathrm{Re}^{i\theta})|}{R^n}\, d\theta. \end{aligned}$$

The right side is the average value of a function whose value is by the lemma at most

$$\frac{2^n}{R^n} \frac{2(\epsilon R^2)(\frac{1}{2}R)}{R - (\frac{1}{2}R)} = \frac{2^{n+1}\epsilon}{R^{n-2}}.$$

If $n \geq 2$, this is at most $2^{n+1}\epsilon$, and since ϵ is arbitrary, a_n must be zero for $n \geq 2$. Thus $f(s) = a_1 s$. However a_1 must be zero by the evenness condition $f(s) \equiv f(-s)$. Therefore $f(s) \equiv 0$ which is constant, as was to be shown.

2.8 THE PRODUCT FORMULA FOR ξ

The function $F(s) = \xi(s)/\prod_\rho [1 - (s - \frac{1}{2})/(\rho - \frac{1}{2})]$ is analytic in the entire s-plane and is an even function of $s - \frac{1}{2}$. Moreover, it has no zeros, so its logarithm is well defined up to an additive constant $2\pi ni$ (n an integer) by the formula $\log F(s) = \int_0^s F'(z)\, dz/F(z) + \log F(0)$, where $\log F(0)$ is determined to within an additive constant $2\pi ni$. The results of the preceding two sections then combine to give $\log F(s) =$ const, and therefore upon exponentiation

$$\xi(s) = c \prod \left(1 - \frac{s - \frac{1}{2}}{\rho - \frac{1}{2}}\right),$$

where c is a constant. Dividing this by the particular value

$$\xi(0) = c \prod \left(1 - \frac{-\frac{1}{2}}{\rho - \frac{1}{2}}\right)$$

gives

$$\frac{\xi(s)}{\xi(0)} = \prod \left(1 - \frac{s - \frac{1}{2}}{\rho - \frac{1}{2}}\right)\left(1 - \frac{-\frac{1}{2}}{\rho - \frac{1}{2}}\right)^{-1}.$$

The factors on the right are linear functions of s which are 0 when $s = \rho$ and 1 when $s = 0$; hence they are $1 - (s/\rho)$ and the formula is the desired formula

$$\xi(s) = \xi(0) \prod_{\rho} \left(1 - \frac{s}{\rho}\right),$$

where, as always, it is understood that the factors ρ and $1 - \rho$ are paired.†

†The same argument proves the validity of the product formula for the sine

$$\sin \pi s = \pi s \prod_{n=1}^{\infty} \left(1 - \frac{s^2}{n^2}\right)$$

mentioned in Section 1.3. The only other unproved statement in Section 1.3 which is not elementary is the equivalence of the two definitions (2) and (3) of $\Pi(s)$.

Chapter 3

Riemann's Main Formula

3.1 INTRODUCTION

Soon after Hadamard proved the product formula for $\xi(s)$, von Mangoldt [M1] proved Riemann's main formula

$$J(x) = \mathrm{Li}(x) - \sum_\rho \mathrm{Li}(x^\rho) - \log 2 + \int_x^\infty \frac{dt}{t(t^2-1)\log t} \qquad (x > 1). \tag{1}$$

Von Mangoldt also recast this formula in a simpler form which has virtually replaced Riemann's original statement (1) in the subsequent development of the theory. This simpler form of (1) can be derived as follows.

The essence of Riemann's derivation of (1) is the inversion of the relationship

$$\log \zeta(s) = \int_0^\infty x^{-s}\, dJ(x) \tag{2}$$

for $J(x)$ in terms of $\log \zeta(s)$ and then the use of

$$\Pi\left(\frac{s}{2}\right)\pi^{-s/2}(s-1)\zeta(s) = \frac{1}{2}\prod_\rho \left(1 - \frac{s}{\rho}\right) \tag{3}$$

to express $\log \zeta(s)$ in terms of elementary functions and in terms of the roots ρ. Now the function $\log \zeta(s)$ has logarithmic singularities at all the roots ρ, and as a function of a complex variable it is very awkward outside the halfplane $\mathrm{Re}\, s > 1$. On the other hand, its derivative $\zeta'(s)/\zeta(s)$ is analytic in the entire plane except for poles at the roots ρ, the pole 1, and the zeros $-2n$. This might well lead one to begin not with formula (2) but with its derivative

$$\frac{\zeta'(s)}{\zeta(s)} = -\int_0^\infty x^{-s}(\log x)\, dJ(x). \tag{4}$$

The measure $(\log x)\,dJ(x)$ is a point measure which assigns the weight $\log(p^n)\cdot(1/n)$ to prime powers p^n and the weight 0 to all other points. Thus it can be written as a Stieltjes measure $d\psi(x)$, where $\psi(x)$ is the step function which begins at 0 and has a jump of $\log(p^n)\cdot(1/n) = \log p$ at each prime power p^n. In other words

$$\psi(x) = \sum_{p^n<x} \log p$$

except when x is a prime power; at the jumps $x = p^n$ the value of ψ is defined, as usual, to be halfway between the new and old values $\psi(x) = \frac{1}{2}[\psi(x-\epsilon) + \psi(x+\epsilon)]$. This function $\psi(x)$ had already been considered by Chebyshev,† who named it $\psi(x)$ and who proved‡ among other things that the prime number theorem is essentially the same as the statement that $\psi(x) \sim x$ with a relative error which approaches zero as $x \longrightarrow \infty$. In terms of $\psi(x)$ formula (4) becomes

$$(5) \qquad -\frac{\zeta'(s)}{\zeta(s)} = \int_0^\infty x^{-s}\,d\psi(x).$$

In other words, if J is replaced by ψ in the original formula (2), then the awkward function $\log\zeta(s)$ is replaced by the more tractable function $-\zeta'(s)/\zeta(s)$.

Now the argument by which Riemann went from formula (2) to the formula (1) for $J(x)$ can be applied equally well to go from formula (5) to a new formula for $\psi(x)$. The explicit computations of this argument are given in the next section. However, even without the explicit computations, one can guess the formula for $\psi(x)$ as follows. The simplest formulation of Riemann's result is his formula (see Section 1.18)

$$dJ = \left(\frac{1}{\log x} - \sum_\rho \frac{x^{\rho-1}}{\log x} - \frac{1}{x(x^2-1)\log x}\right)dx \qquad (x>1)$$

which gives

$$\begin{aligned} d\psi &= (\log x)\,dJ \\ &= \left(1 - \sum_\rho x^{\rho-1} - \sum_\rho x^{-2n-1}\right)dx \qquad (x>1) \end{aligned}$$

and leads to the guess

$$(6) \qquad \psi(x) = x - \sum_\rho \frac{x^\rho}{\rho} + \sum_n \frac{x^{-2n}}{2n} + \text{const} \qquad (x>1).$$

This is von Mangoldt's reformulation of (1) referred to in the first paragraph. [The value of the constant is given in Section 3.2. It is assumed in von Mangoldt's formula (6), as it is in Riemann's formula (1), that the terms of the sum over ρ are taken in the order of increasing $|\operatorname{Im}\rho|$; these sums converge

†This work of Chebyshev [C3] in 1850 preceded Riemann's paper.

‡Actually Chebyshev does not state this result explicitly, but it follows trivially from the techniques he introduces for deducing estimates of π from estimates of ψ.

only conditionally—even when the terms ρ, $1-\rho$ are paired—so their order is essential.]

The main part of this chapter is devoted to von Mangoldt's proof of the formula (6) for $\psi(x)$. In Section 3.2 the formula is derived from the termwise evaluation of certain definite integrals, and in the following three sections the validity of these termwise evaluations is rigorously proved. Von Mangoldt's proof of Riemann's original formula (1) is outlined in the next two sections, and the last section deals with the numerical evaluation of the constant $\zeta'(0)/\zeta(0)$.

3.2 DERIVATION OF VON MANGOLDT'S FORMULA FOR $\psi(x)$

The technique of Section 1.12 applied to the formula $-\zeta'(s)/\zeta(s) = s\int_0^\infty \psi(x)x^{-s-1}\,dx$ instead of to the formula $\log\zeta(s) = s\int_0^\infty J(x)x^{-s-1}\,dx$ puts $\psi(x)$ in the form of a definite integral

$$\text{(1)}\qquad \frac{1}{2\pi i}\int_{a-i\infty}^{a+i\infty}\left[-\frac{\zeta'(s)}{\zeta(s)}\right]x^s\,\frac{ds}{s}\qquad (a>1).$$

Von Mangoldt's method of proving the formula for $\psi(x)$ is to evaluate this definite integral in two different ways, one of which gives the value $\psi(x)$ and the other $x-\sum(x^\rho/\rho)+\sum(x^{-2n}/2n)+\text{const.}$ (Neither of these evaluations uses Fourier's theorem, so the use of Fourier's theorem in Section 1.12 can be regarded as purely heuristic.)

The first method of evaluating the definite integral (1) is as follows. Beginning with the formula†

$$\text{(2)}\qquad -\frac{\zeta'(s)}{\zeta(s)} = \int_0^\infty x^{-s}\,d\psi(x),$$

let $\Lambda(n)$ denote the weight assigned to the integer n by the measure $d\psi$—that is, $\Lambda(n)$ is zero unless n is a prime power, in which case $\Lambda(n)$ is the log of the prime of which n is a power—so that the integral (2) can be written as a sum

$$\text{(3)}\qquad -\frac{\zeta'(s)}{\zeta(s)} = \sum_{n=2}^{\infty}\Lambda(n)n^{-s}\qquad (\operatorname{Re} s>1).$$

Substitute this formula in (1) and assume termwise integration is valid. This gives the value

$$\sum_{n=2}^{\infty}\Lambda(n)\frac{1}{2\pi i}\int_{a-i\infty}^{a+i\infty}\left(\frac{x}{n}\right)^s\frac{ds}{s}$$

†This formula is essentially the Euler product formula (see the concluding remarks of Section 1.11). In fact, logarithmic differentiation of $\zeta(s)=\prod(1-p^{-s})^{-1}$ gives (3) immediately.

for (1). Now formula (2) of Section 1.14 (with $\beta = 0$) shows that the integral corresponding to n in this sum is 1 if $x/n > 1$ and 0 if $x/n < 1$; hence the sum is just

$$\sum_{n<x} \Lambda(n) = \psi(x)$$

as was to be shown.

To justify this sequence of steps leading to the value $\psi(x)$ for the definite integral (1), note first that the series (3) converges uniformly in any halfplane $\operatorname{Re} s \geq K > 1$ [by comparison with the convergent series $\sum (\log n)n^{-K}$]. This proves both that the termwise differentiation of $-\log \zeta(s) = \sum \log (1 - p^{-s})$ is valid (a series can be differentiated termwise if the result is uniformly convergent) and that the integral over any *finite* path can be computed termwise:

$$\text{(4)} \qquad \frac{1}{2\pi i}\int_{a-ih}^{a+ih}\left[-\frac{\zeta'(s)}{\zeta(s)}\right]x^s\frac{ds}{s} = \sum_{n=1}^{\infty} \Lambda(n)\cdot\frac{1}{2\pi i}\int_{a-ih}^{a+ih}\left(\frac{x}{n}\right)^s\frac{ds}{s}$$

when $a > 1$. Now *if* it can be shown that *the limit as* $h \longrightarrow \infty$ *of the sum on the right is equal to the sum of the limits*, then the fact that the integral (1) has the value $\psi(x)$ will follow immediately from the formula

$$\text{(5)} \qquad \frac{1}{2\pi i}\int_{a-i\infty}^{a+i\infty} y^s \frac{ds}{s} = \begin{cases} 0 & \text{if } 0 < y < 1 \\ \frac{1}{2} & \text{if } y = 1 \\ 1 & \text{if } y > 1 \end{cases} \qquad (a > 0).$$

This formula, which was deduced from Fourier's theorem in Section 1.14, will be proved directly in the next section. To summarize, then, the proof that the definite integral (1) is equal to $\psi(x)$ depends on the proof that the limit as $h \longrightarrow \infty$ of the sum in (4) is the sum of the limits, and on the proof of the integral formula (5).

The second method of evaluating the integral (1) is as follows. Differentiate logarithmically the formula

$$\Pi\left(\frac{s}{2}\right)\pi^{-s/2}(s-1)\zeta(s) = \xi(0)\prod_{\rho}\left(1-\frac{s}{\rho}\right)$$

to find

$$\frac{d}{ds}\log \Pi\left(\frac{s}{2}\right) - \frac{1}{2}\log \pi + \frac{1}{s-1} + \frac{\zeta'(s)}{\zeta(s)} = \sum_{\rho}\frac{1}{1-(s/\rho)}\cdot\left(-\frac{1}{\rho}\right).$$

Using the expression of $\Pi(x)$ as an infinite product [(4) of Section 1.3], and differentiating termwise then gives

$$(6)\qquad -\frac{\zeta'(s)}{\zeta(s)} = \frac{1}{s-1} - \sum_\rho \frac{1}{s-\rho} + \sum_{n=1}^{\infty}\left[-\frac{1}{s+2n} + \frac{1}{2}\log\left(1+\frac{1}{n}\right)\right] - \frac{1}{2}\log\pi.$$

With $s = 0$ this gives

$$-\frac{\zeta'(0)}{\zeta(0)} = -1 - \sum_\rho\left(-\frac{1}{\rho}\right) + \sum_{n=1}^{\infty}\left[-\frac{1}{2n} + \frac{1}{2}\log\left(1+\frac{1}{n}\right)\right] - \frac{1}{2}\log\pi,$$

so subtraction gives

$$-\frac{\zeta'(s)}{\zeta(s)} + \frac{\zeta'(0)}{\zeta(0)} = \left[\frac{1}{s-1} + 1\right] - \sum_\rho\left[\frac{1}{s-\rho} + \frac{1}{\rho}\right] - \sum_{n-1}^{\infty}\left[\frac{1}{s+2n} - \frac{1}{2n}\right]$$

and finally

$$(7)\qquad -\frac{\zeta'(s)}{\zeta(s)} = \frac{s}{s-1} - \sum_\rho \frac{s}{\rho(s-\rho)} + \sum_{n=1}^{\infty}\frac{s}{2n(s+2n)} - \frac{\zeta'(0)}{\zeta(0)}.$$

Substitute this formula in the definite integral (1) and assume that termwise integration is valid. This gives the value

$$\frac{1}{2\pi i}\int_{a-i\infty}^{a+i\infty} x^s\frac{ds}{s-1} - \sum_\rho \frac{1}{2\pi i}\int_{a-i\infty}^{a+i\infty} x^s\frac{ds}{\rho(s-\rho)} + \sum_n \frac{1}{2\pi i}\int_{a-i\infty}^{a+i\infty} x^s\frac{ds}{2n(s+2n)} + \frac{1}{2\pi i}\int_{a-i\infty}^{a+i\infty} x^s\left[-\frac{\zeta'(0)}{\zeta(0)}\right]\frac{ds}{s}$$

for (1). Now the change of variable $t = s - \beta$ in the previous integral formula (5) gives

$$(8)\qquad \frac{1}{2\pi i}\int_{a-i\infty}^{a+i\infty} x^s\frac{ds}{s-\beta} = \frac{1}{2\pi i}\int_{a-\beta-i\infty}^{a-\beta+i\infty} x^\beta x^t\frac{dt}{t} = x^\beta\frac{1}{2\pi i}\int_{\mathrm{Re}(a-\beta)-i\infty}^{\mathrm{Re}(a-\beta)+i\infty} x^t\frac{dt}{t} = x^\beta$$

provided $x > 1$ and $a > \mathrm{Re}\,\beta$. (The middle equation here, in which the limits of integration are switched from $a - \beta \pm i\infty$ to $\mathrm{Re}(a - \beta) \pm i\infty$, is not trivial because these two integrals when written as limits as $h \to \infty$ are different. However, the difference between them is two integrals of $x^t\,dt/t$ over intervals of the form $[a \pm ib, a \pm i(b + c)]$, where c is fixed and $b \to \infty$; thus the difference is less than a constant times $c/b \to 0$.) Thus the value of

(1) reduces, when $x > 1$, to

$$x - \sum_\rho \frac{x^\rho}{\rho} + \sum_n \frac{x^{-2n}}{2n} - \frac{\zeta'(0)}{\zeta(0)} \qquad (x > 1)$$

as desired.

To justify this sequence of steps leading to the value $x - \Sigma(x^\rho/\rho) + \Sigma(x^{-2n}/2n) - \zeta'(0)/\zeta(0)$ of the definite integral (1), note first that both of the infinite series in (6) converge uniformly in any disk $|s| \le K$. (The series in n converges uniformly because

$$\begin{aligned}
&|(s + 2n)^{-1} - \tfrac{1}{2}\log(1 + n^{-1})| \\
&\quad = |(s + 2n)^{-1} - (2n)^{-1} + (2n)^{-1} \\
&\qquad - \tfrac{1}{2}(n^{-1} - \tfrac{1}{2}n^{-2} + \tfrac{1}{3}n^{-3} - \cdots)| \\
&\quad \le |s(s + 2n)^{-1}(2n)^{-1}| + |\tfrac{1}{4}n^{-2} - \tfrac{1}{6}n^{-3} + \cdots| \\
&\quad \le K(2n)^{-2} + n^{-2} \le \text{const}/n^2
\end{aligned}$$

for all sufficiently large n, and the series in ρ converges uniformly because when the terms ρ and $1 - \rho$ are paired

$$\begin{aligned}
&|(s - \rho)^{-1} + [s - (1 - \rho)]^{-1}| \\
&\quad = \left|\left[\left(s - \frac{1}{2}\right) - \left(\rho - \frac{1}{2}\right)\right]^{-1} + \left[\left(s - \frac{1}{2}\right) + \left(\rho - \frac{1}{2}\right)\right]^{-1}\right| \\
&\quad = \left|\frac{2(s - \frac{1}{2})}{(s - \frac{1}{2})^2 - (\rho - \frac{1}{2})^2}\right| \le \text{const}\left|\rho - \frac{1}{2}\right|^{-2}
\end{aligned}$$

for all sufficiently large ρ once K is fixed and because $\sum |\rho - \frac{1}{2}|^{-2}$ converges by the theorem of Section 2.5.) This proves that the termwise differentiation by which (6) was obtained is valid. Then it follows by an elementary theorem $[\sum (a_n + b_n) = \sum a_n + \sum b_n$ when $\sum a_n, \sum b_n$ both converge$]$ that (7) is valid—except at the zeros and poles 1, ρ, $-2n$ of ζ—and that the series it contains both converge uniformly in $|s| \le K$. Thus it can be integrated termwise over *finite* intervals and the integral (1) is therefore equal to

$$\text{(9)} \qquad \begin{aligned}
&\lim_{h\to\infty} \frac{1}{2\pi i}\int_{a-ih}^{a+ih}\left[\frac{s}{s-1} - \sum_\rho \frac{s}{\rho(s-\rho)} + \sum_n \frac{s}{2n(s+2n)} - \frac{\zeta'(0)}{\zeta(0)}\right] x^s \frac{ds}{s} \\
&\quad = x - \lim_{h\to\infty} \sum_\rho \frac{1}{2\pi i}\int_{a-ih}^{a+ih} \frac{x^s\,ds}{\rho(s-\rho)} \\
&\qquad + \lim_{h\to\infty} \sum_n \frac{1}{2\pi i}\int_{a-ih}^{a+ih} \frac{x^s\,ds}{2n(s+2n)} - \frac{\zeta'(0)}{\zeta(0)}
\end{aligned}$$

[where use is made of the fact that the limit of a finite sum is the sum of the limits and the first and last terms are evaluated using (8) and (5)]. Now *if* the limits of these two sums are equal to the sums of the limits, then the rest of the argument is elementary and the value $x - \sum (x^\rho/\rho) + \sum (x^{-2n}/2n) - \zeta'(0)/\zeta(0)$ for (1) will be proved. Thus, in addition to the basic formula (5), the rigorous proof that the definite integral (1) has this value depends on the validity of the termwise evaluation of the two limits in (9).

Thus von Mangoldt's formula

$$\psi(x) = x - \sum_\rho \frac{x^\rho}{\rho} + \sum_n \frac{x^{-2n}}{2n} - \frac{\zeta'(0)}{\zeta(0)} \qquad (x > 1)$$

depends on the validity of the integral formula (5) and of the three interchanges of $\lim_{h\to\infty}$ with infinite sums in (4) and (9). The following three sections are devoted to proving that all are indeed valid. The numerical value of the constant $\zeta'(0)/\zeta(0) = \log 2\pi$ is found in Section 3.8.

3.3 THE BASIC INTEGRAL FORMULA

This section is devoted to the evaluation of

(1) $$\lim_{h\to\infty} \frac{1}{2\pi i}\int_{a-ih}^{a+ih} \frac{x^s\,ds}{s} \qquad (x > 0, \quad a > 0).$$

Since the arguments of Section 3.2 deal with infinite sums of such limits, it will be necessary to find, in addition to the limit (1), the *rate* at which this limit is approached. For the case $0 < x < 1$ this is accomplished by the estimate

(2) $$\left|\frac{1}{2\pi i}\int_{a-ih}^{a+ih} \frac{x^s\,ds}{s}\right| \le \frac{x^a}{\pi h\,|\log x|} \qquad (a > 0, \quad 0 < x < 1),$$

which can be proved as follows. Because $a > 0$, the function x^s/s has no singularity in the rectangle $\{a \le \operatorname{Re} s \le K, -h \le \operatorname{Im} s \le h\}$, where K is a large constant. Hence by Cauchy's theorem the integral of $x^s\,ds/s$ around the boundary of this rectangle is zero, which gives

$$\frac{1}{2\pi i}\int_{a-ih}^{a+ih} \frac{x^s\,ds}{s} = -\frac{1}{2\pi i}\int_{a+ih}^{K+ih} \frac{x^s\,ds}{s} + \frac{1}{2\pi i}\int_{a-ih}^{K-ih} \frac{x^s\,ds}{s} + \frac{1}{2\pi i}\int_{K-ih}^{K+ih} \frac{x^s\,ds}{s}.$$

The last integral has modulus at most $(2\pi)^{-1}(x^K/K)(2h)$. Each of the other two integrals on the right has modulus at most

$$\frac{1}{2\pi}\int_a^K \frac{x^\sigma\,d\sigma}{h} = \frac{1}{2\pi}\left[\frac{x^\sigma}{h\log x}\right]_{\sigma=a}^{\sigma=K}$$

which then gives

$$\left|\frac{1}{2\pi i}\int_{a-ih}^{a+ih} \frac{x^s\,ds}{s}\right| \le \frac{1}{\pi}\,\frac{|x^K - x^a|}{h\,|\log x|} + \frac{1}{2\pi}\,\frac{x^K}{K}2h.$$

When $0 < x < 1$, the limit of x^K as $K \to \infty$ is zero and (2) follows. Thus, in particular, the limit (1) is zero when $0 < x < 1$.

In the case $x = 1$ no estimate of the rate of approach to the limit will be needed and it suffices to note that (1) is

$$\lim_{h\to\infty} \frac{1}{2\pi}\int_{-h}^{h} \frac{dt}{a+it} = \lim_{h\to\infty}\left(\frac{1}{2\pi}\int_{-h}^{h}\frac{a\,dt}{a^2+t^2} - i\frac{1}{2\pi}\int_{-h}^{h}\frac{t\,dt}{a^2+t^2}\right)$$
$$= \lim_{h\to\infty}\frac{1}{2\pi}\int_{-h/a}^{h/a}\frac{du}{1+u^2} = \frac{1}{2}$$

using the well-known† formula $\int_{-\infty}^{\infty}(1+u^2)^{-1}\,du = \pi$. Thus for $x = 1$ the limit (1) is $\frac{1}{2}$.

For $x > 1$ the estimate analogous to (2) is obtained by considering the integral of $x^s\,ds/(2\pi i s)$ around the boundary of a rectangle of the form $\{-K \le \operatorname{Re} s \le a,\ -h \le \operatorname{Im} s \le h\}$. Since x^s is analytic in this rectangle and $s = 0$ lies inside the rectangle, the Cauchy integral formula states that this integral is $x^0 = 1$, hence

$$\frac{1}{2\pi i}\int_{a-ih}^{a+ih}\frac{x^s\,ds}{s} + \frac{1}{2\pi i}\int_{a+ih}^{-K+ih}\frac{x^s\,ds}{s}$$
$$+ \frac{1}{2\pi i}\int_{-K+ih}^{-K-ih}\frac{x^s\,ds}{s} + \frac{1}{2\pi i}\int_{-K-ih}^{a-ih}\frac{x^s\,ds}{s} = 1,$$

$$\left|\frac{1}{2\pi i}\int_{a-ih}^{a+ih}\frac{x^s\,ds}{s} - 1\right| \le \frac{1}{2\pi}\int_{-K}^{a}\frac{x^\sigma\,d\sigma}{h} + \frac{1}{2\pi}\frac{x^{-K}}{K}\cdot 2h + \frac{1}{2\pi}\int_{-K}^{a}\frac{x^\sigma\,d\sigma}{h}$$
$$= \frac{1}{\pi}\frac{x^a - x^{-K}}{h\log x} + \frac{1}{\pi}\frac{x^{-K}h}{K}.$$

Letting $K \longrightarrow \infty$ then gives

(3) $$\left|\frac{1}{2\pi i}\int_{a-ih}^{a+ih}\frac{x^s\,ds}{s} - 1\right| \le \frac{x^a}{\pi h \log x} \qquad (a > 0,\quad x > 1)$$

which is the desired estimate. In particular, the limit (1) is 1 when $x > 1$.

One other estimate of integrals is used in the proof of von Mangoldt's formula, namely, von Mangoldt's estimate

(4) $$\left|\frac{1}{2\pi i}\int_{a+ic}^{a+id}\frac{x^s\,ds}{s}\right| \le K\frac{x^a}{(a+c)\log x} \qquad (x > 1,\quad a > 0,\quad d > c \ge 0),$$

where K is a constant which may be taken to be $(4+\pi)/(2\pi\sqrt{2})$. To prove this formula, note that integration by parts

$$\int\frac{x^s\,ds}{s} = \frac{x^s}{s\log x} + \int\frac{x^s\,ds}{s^2\log x}$$

gives

$$\left|\int_{a+ic}^{a+id}\frac{x^s\,ds}{s}\right| \le \left|\frac{x^{a+id}}{(a+id)\log x}\right| + \left|\frac{x^{a+ic}}{(a+ic)\log x}\right|$$
$$+ \frac{x^a}{\log x}\left|\int_c^d\frac{x^{it}\,dt}{(a+it)^2}\right|.$$

†See Edwards [E1, p. 65].

Since $|x^{it}| = 1$ and $|a + id| \geq |a + ic| \geq (a + c)/\sqrt{2}$, the first two terms are each less then $\sqrt{2}\, x^a/(a + c) \log x$ and the integral in the third term is less than

$$\int_c^\infty \frac{dt}{a^2 + t^2} = \int_0^\infty \frac{dt}{a^2 + (c + t)^2} \leq \int_0^\infty \frac{dt}{a^2 + c^2 + t^2}$$
$$= \int_0^\infty \frac{(a^2 + c^2)^{1/2}\, du}{a^2 + c^2 + [u(a^2 + c^2)^{1/2}]^2} = \frac{1}{(a^2 + c^2)^{1/2}} \int_0^\infty \frac{du}{1 + u^2}$$
$$\leq \frac{\sqrt{2}}{a + c} \cdot \frac{\pi}{2}.$$

Thus

$$\left| \frac{1}{2\pi i} \int_{a+ic}^{a+id} \frac{x^s\, ds}{s} \right| \leq \frac{1}{2\pi} \left(2\sqrt{2} + \frac{\pi}{\sqrt{2}} \right) \frac{x^a}{(a + c) \log x}$$

which is the desired result.

3.4 THE DENSITY OF THE ROOTS

This section is devoted to the proof of the following theorem:

Theorem The vertical density of the roots ρ of $\xi(\rho) = 0$ is less than $2 \log T$ for large T. More specifically, there is an H such that for $T \geq H$ the number of roots ρ with imaginary parts in the range $T \leq \operatorname{Im} \rho \leq T + 1$ is less than $2 \log T$.

Proof Von Mangoldt's proof of this fact is based on Hadamard's proof of the product formula and on a strong version of Stirling's formula which was published by Stieltjes in 1889.

Hadamard's theorem that the series $\sum |\rho - \frac{1}{2}|^{-2}$ converges implies (see Sect. 3.2) that the termwise integration of the series $\xi'(s)/\xi(s) = \sum (s - \rho)^{-1}$ is valid over any finite segment. Hence, in particular,

$$\int_{2+iT}^{2+i(T+1)} \frac{\xi'(s)}{\xi(s)}\, ds = \sum_\rho \int_{2+iT}^{2+i(T+1)} \frac{ds}{s - \rho}$$

for any T. Now for any fixed ρ the imaginary part of the integral on the right

$$\operatorname{Im}\left(\int_{2+iT}^{2+i(T+1)} \frac{ds}{s - \rho} \right)$$

is equal (because $dz/z = d \log r + i\, d\theta$) to the angle subtended by the segment $[2 + iT, 2 + i(T + 1)]$ at the point ρ. Thus it is always positive—because the roots ρ are to the left of $\operatorname{Re} s = 2$—and if ρ lies in the range $T \leq \operatorname{Im} \rho \leq T + 1$, then it is at least the angle subtended by the segment $[2 + iT, 2 + i(T + 1)]$ at the point iT, which is $\operatorname{Arctan} \frac{1}{2}$. Thus if n denotes the number of roots ρ in $T \leq \operatorname{Im} \rho \leq T + 1$, it follows that

$$n \cdot \operatorname{Arctan} \frac{1}{2} \leq \operatorname{Im} \int_{2+iT}^{2+i(T+1)} \frac{\xi'(s)}{\xi(s)}\, ds. \tag{1}$$

Thus an upper bound on n will result from an upper bound on the integral on the right.

On the other hand, the integral of $\xi'(s)\, ds/\xi(s)$ over the interval from $2 + iT$ to $2 + i(T + 1)$ is, by the fundamental theorem of calculus, equal to the amount by which the function

$$\log \xi(s) = \log \Pi(s/2) - (s/2) \log \pi + \log (s - 1) + \log \zeta(s)$$

changes between these two points. It is in the estimation of $\log \Pi(s/2)$ that von Mangoldt uses Stieltjes' version of Stirling's formula. Specifically, he uses the fact that the modulus of the error in the approximation

$$\log \Pi(z) \sim (z + \tfrac{1}{2}) \log z - z + \tfrac{1}{2} \log 2\pi$$

is at most $(6\,|z|)^{-1}$ for z in the halfplane $\operatorname{Re} z \geq 0$. (For a complete statement and proof of Stieltjes' result see Section 6.3.) Thus

$$\begin{aligned}
\log \xi(s) &\sim \left(\frac{s}{2} + \frac{1}{2}\right) \log s - \left(\frac{s}{2} + \frac{1}{2}\right) \log 2 - \frac{s}{2} + \frac{1}{2} \log 2\pi \\
&\quad - \frac{s}{2} \log \pi + \log(s - 1) + \log \zeta(s) \\
&= \frac{s + 1}{2} \log s - \frac{s}{2} \log 2\pi - \frac{s}{2} \\
&\quad + \log (s - 1) + \log \zeta(s) + \text{const}
\end{aligned}$$

with an error of at most $(6\,|s|)^{-1}$. As was noted in Section 1.13, formula (3), the modulus of $\log \zeta(s)$ is at most $\log \zeta(2)$ on the line $\operatorname{Re} s = 2$, so neglecting this term introduces an error of at most $\log \zeta(2) = \log (\pi^2/6) < 1$. Thus the change in $\log \xi(s)$ between $2 + i(T + 1)$ and $2 + iT$ is approximately

$$\begin{aligned}
&\frac{3 + i(T + 1)}{2} \log[2 + i(T + 1)] - \frac{3 + iT}{2} \log(2 + iT) \\
&\quad - \frac{i}{2} \log 2\pi - \frac{i}{2} + \log \frac{1 + i(T + 1)}{1 + iT} \\
&= \frac{i}{2} \log[2 + i(T + 1)] + \frac{3 + iT}{2} \log\left(1 + \frac{i}{2 + iT}\right) \\
&\quad + \text{const} + \log\left(1 + \frac{1}{1 + iT}\right)
\end{aligned}$$

and the error has modulus at most $2(6T)^{-1} + 2$. Neglecting terms which are bounded for large T puts this estimate in the form

$$\frac{i}{2} \log(iT) + \frac{3 + iT}{2} \cdot \frac{i}{2 + iT} \sim \frac{i}{2} \log T.$$

Thus the error in the approximation

$$\int_{2+iT}^{2+i(T+1)} \frac{\xi'(s)}{\xi(s)}\, ds \sim \frac{i}{2} \log T$$

remains bounded in modulus as $T \to \infty$—say by K—hence (1) gives

$$n \cdot \operatorname{Arctan} \tfrac{1}{2} \leq \tfrac{1}{2} \log T + K,$$

$$n \leq \frac{\frac{1}{2} \log T + K}{\pi/8} < 2 \log T$$

for all sufficiently large T, as was to be shown.

3.5 PROOF OF VON MANGOLDT'S FORMULA FOR $\psi(x)$

The derivation of von Mangoldt's formula

(1) $$\psi(x) = x - \sum_\rho \frac{x^\rho}{\rho} + \sum_n \frac{x^{-2n}}{2n} - \frac{\zeta'(0)}{\zeta(0)} \qquad (x > 1)$$

which was given in Section 3.2 depended on the integral formula which was proved in Section 3.3 and on three termwise integrations. This section is devoted to proving that these termwise integrations are valid. In all three cases, the series converge uniformly on any finite segment $[a - ih, a + ih]$ $(a > 1)$, so the integral can be computed termwise on finite segments and the problem is to show that the limit of their sum as $h \to \infty$ is equal to the sum of their limits.

Consider first the limit

(2) $$\lim_{h\to\infty} \sum_{n=1}^{\infty} \Lambda(n) \frac{1}{2\pi i} \int_{a-ih}^{a+ih} \left(\frac{x}{n}\right)^s \frac{ds}{s}.$$

The limit of a finite sum is the sum of the limits; hence one can disregard the finite number of terms in which $n \leq x$ and consider only the terms $n > x$. For these, the estimate (2) of Section 3.3 gives

$$\left| \Lambda(n) \frac{1}{2\pi i} \int_{a-ih}^{a+ih} \left(\frac{x}{n}\right)^s \frac{ds}{s} \right| \leq \log n \frac{x^a}{n^a \pi h (\log n - \log x)}$$

$$\leq \text{const} \frac{1}{n^a h}.$$

Hence their sum over $n > x$ is at most a constant times h^{-1} and therefore approaches zero as $h \to \infty$. This shows that the limit (2) can be evaluated termwise.

Consider next the limit

(3) $$\lim_{h\to\infty} \sum_{n=1}^{\infty} \frac{x^{-2n}}{2n} \frac{1}{2\pi i} \int_{a-ih}^{a+ih} \frac{x^{s+2n}\, ds}{s + 2n} \qquad (x > 1).$$

The limit of the nth term of this series is $x^{-2n}/2n$ by (3) of Section 3.3; hence the sum of the limits converges. Now by (4) of Section 3.3 the nth term is at most

$$\frac{x^{-2n}}{2n} \cdot 2\left|\frac{1}{2\pi i}\int_{a+2n}^{a+2n+ih} \frac{x^t\,dt}{t}\right| \leq \frac{x^{-2n}}{n} \cdot K\frac{x^{a+2n}}{(a+2n)\log x}$$
$$\leq \frac{\text{const}}{n^2}$$

for all h. Thus the series (3) converges uniformly in h and one can pass to the limit $h \to \infty$ termwise, as was to be shown. [Given $\epsilon > 0$, choose N large enough that the sum (3) differs by at most ϵ from the sum of the first N terms for all h. By enlarging N if necessary, one can also assume that the sum of the limits $x^{-2n}/2n$ differs by at most ϵ from the sum of the first N of them. Now choose H large enough that each of the first N terms of (3) differs by at most ϵ/N from its limit when $h \geq H$. Then the sum (3) differs by at most 3ϵ from the sum of the limits provided only that $h \geq H$. Since ϵ is arbitrary this proves the desired result.]

Consider finally the limit

$$\lim_{h\to\infty} \sum_{\rho} \frac{x^\rho}{\rho} \frac{1}{2\pi i}\int_{a-ih}^{a+ih} \frac{x^{s-\rho}}{s-\rho}\,ds. \tag{4}$$

This limit has now been shown to *exist* because it has been shown that the limit (9) of Section 3.2 exists [and is equal to $\psi(x)$] and it has been shown that the limit of the sum over n in (9) also exists. It has also been shown that the individual terms of (4) approach limits x^ρ/ρ as $h \to \infty$, but it has *not* been shown that the sum of these limits converges; in fact, the proof that $\sum x^\rho/\rho$ converges when summed in the order of increasing $|\text{Im}\,\rho|$ is the major difficulty in the proof of von Mangoldt's formula. Broadly speaking, von Mangoldt overcomes this difficulty by approaching the limit (4) "diagonally," that is, by considering the limit

$$\lim_{h\to\infty} \sum_{|\text{Im}\,\rho|\leq h} \frac{x^\rho}{\rho} \frac{1}{2\pi i}\int_{a-ih}^{a+ih} \frac{x^{s-\rho}}{s-\rho}\,ds \tag{5}$$

as an intermediate step between the limit (4) which is known to exist and the sum of the limits

$$\lim_{h\to\infty} \sum_{|\text{Im}\,\rho|\leq h} \frac{x^\rho}{\rho} \tag{6}$$

which is to be shown to exist and be equal to (4).

Specifically, consider for each h the differences

$$\sum_{\rho} \frac{x^\rho}{\rho} \frac{1}{2\pi i}\int_{a-ih}^{a+ih} \frac{x^{s-\rho}}{s-\rho}\,ds - \sum_{|\text{Im}\,\rho|\leq h} \frac{x^\rho}{\rho} \frac{1}{2\pi i}\int_{a-ih}^{a+ih} \frac{x^{s-\rho}}{s-\rho}\,ds \tag{7}$$

and

$$\text{(8)}\qquad \sum_{|\text{Im}\,\rho|\le h} \frac{x^\rho}{\rho}\frac{1}{2\pi i}\int_{a-ih}^{a+ih}\frac{x^{s-\rho}}{s-\rho}\,ds - \sum_{|\text{Im}\,\rho|\le h}\frac{x^\rho}{\rho}.$$

It will be shown that both these differences approach zero as $h \to \infty$. Then, since the limit (4) exists, it follows first that the "diagonal" limit (5) exists and is equal to it [because (7) goes to zero] and hence that the limit (6) exists and is equal to it [because (8) goes to zero hence (6) equals (5)] as desired.

Consider first the estimate of (7). Let $\rho = \beta + i\gamma$ denote a typical root. Then by (4) of Section 3.3 the modulus of (7) is at most

$$\begin{aligned}
&\sum_{|\gamma|>h}\left|\frac{x^\rho}{\rho}\right|\left|\frac{1}{2\pi i}\int_{a-ih}^{a+ih}\frac{x^{s-\rho}}{s-\rho}\,ds\right| \\
&\quad\le 2\sum_{\gamma>h}\frac{x^\beta}{\gamma}\left|\frac{1}{2\pi i}\int_{a-\beta+i(\gamma-h)}^{a-\beta+i(\gamma+h)}\frac{x^t\,dt}{t}\right| \\
&\quad\le 2\sum_{\gamma>h}\frac{x^\beta}{\gamma}\cdot K\frac{x^{a-\beta}}{(a-\beta+\gamma-h)\log x} \\
&\quad\le 2K\frac{x^a}{\log x}\sum_{\gamma>h}\frac{1}{\gamma(\gamma-h+c)}
\end{aligned}$$

where $c = a - 1 > 0$ so that $c \le a - \beta$ for all roots ρ. Now if the roots γ beyond h are grouped in intervals $h < \gamma \le h + 1, h + 1 < \gamma \le h + 2, h + 2 < \gamma \le h + 3, \ldots$, then (assuming h is large enough that the estimate of Section 3.4 applies beyond h) the interval $h + j < \gamma \le h + j + 1$ contains at most $2 \log (h + j)$ of the γ's and the modulus of (7) is at most a constant times

$$\sum_{j=0}^{\infty}\frac{\log(h+j)}{(h+j)(j+c)},$$

and it remains only to show that this sum approaches zero as $h \to \infty$. One can do this by choosing h large enough that $\log (h + j) < (h + j)^{1/2}$ for all $j \ge 0$; then the summand is at most one over $(h + j)^{1/4}(c + j)^{1/4}\cdot(c + j)$, so the sum is at most a constant times $h^{-1/4} \to 0$ as was to be shown.

Consider now the estimate of (8). The fact that the terms corresponding to ρ and $\bar{\rho}$ give equal contributions and the estimates (3) and (4) of Section 3.3 show that the modulus of (8) is at most

$$\begin{aligned}
&2\sum_{0<\gamma\le h}\left|\frac{x^\rho}{\rho}\right|\left|\frac{1}{2\pi i}\int_{a-\beta-i\gamma-ih}^{a-\beta-i\gamma+ih}\frac{x^t\,dt}{t} - 1\right| \\
&\quad\le 2\sum_{0<\gamma\le h}\frac{x^\beta}{\gamma}\left|\frac{1}{2\pi i}\int_{a-\beta-i(h+\gamma)}^{a-\beta+i(h+\gamma)}\frac{x^t\,dt}{t} - 1\right| \\
&\quad\quad + 2\sum_{0<\gamma\le h}\frac{x^\beta}{\gamma}\left|\frac{1}{2\pi i}\int_{a-\beta+i(h-\gamma)}^{a-\beta+i(h+\gamma)}\frac{x^t\,dt}{t}\right|
\end{aligned}$$

$$\leq 2 \sum_{0<\gamma\leq h} \frac{x^\beta}{\gamma} \frac{x^{a-\beta}}{\pi(h+\gamma)\log x}$$
$$+ 2 \sum_{0<\gamma\leq h} \frac{x^\beta}{\gamma} K \frac{x^{a-\beta}}{(a-\beta+h-\gamma)\log x}$$
$$\leq \frac{2x^a}{\pi \log x} \sum_{0<\gamma\leq h} \frac{1}{\gamma(h+\gamma)}$$
$$+ \frac{2Kx^a}{\log x} \sum_{0<\gamma\leq h} \frac{1}{\gamma(c+h-\gamma)},$$

where $c = a - 1 > 0$ as before. Thus it suffices to prove that these two sums over γ approach zero as $h \longrightarrow \infty$. Let H be a large integer such that the estimate of Section 3.4 applies beyond H and let the roots be grouped in intervals $H \leq \gamma < H+1$, $H+1 \leq \gamma < H+2, \ldots$. Then the interval $H+j \leq \gamma \leq H+j+1$ contains at most $2\log(H+j)$ of the γ's and

$$\sum_{0<\gamma\leq h} \frac{1}{\gamma(h+\gamma)} \leq \sum_{0<\gamma\leq H} \frac{1}{\gamma(h+\gamma)} + \sum_{0\leq j\leq h-H} \frac{2\log(H+j)}{(H+j)(h+H+j)}.$$

The first sum has a fixed finite number of terms and therefore clearly has the limit zero as $h \longrightarrow \infty$. The second sum is at most

$$2 \sum_{0\leq j\leq h-H} (\log h)\left[\frac{1}{h}\left(\frac{1}{H+j} - \frac{1}{h+H+j}\right)\right]$$
$$\leq \frac{\log h}{h} 2 \sum_{0\leq j\leq h-H} \frac{1}{H+j}$$
$$\leq 2\frac{\log h}{h} \int_{H-1}^{h} \frac{dt}{t} \leq 2\frac{(\log h)^2}{h}$$

which approaches zero as $h \longrightarrow \infty$. A similar estimate shows that

$$\sum_{H\leq\gamma\leq h} \frac{1}{\gamma(c+h-\gamma)}$$

approaches zero as $h \longrightarrow \infty$ and completes the proof that (6) equals (4).

Thus the limits (2), (3), and (4) can be evaluated termwise—provided the sum of the limits of (4) is defined in the sense of (6)—and the derivation of Section 3.2 proves von Mangoldt's formula (1).

3.6 RIEMANN'S MAIN FORMULA

There are at least two reasons why Riemann's formula for $J(x)$ [Section 3.1, formula (1)] is generally neglected today. First, it contains essentially the same information as von Mangoldt's formula for $\psi(x)$ but is less "natural" than this formula in the sense that it is harder to prove and harder to generalize. Secondly, Riemann's reason for establishing the formula in the first place—to show an explicit analytic connection between the arithmetic function $\pi(x)$ and the empirically derived approximation $\mathrm{Li}(x)$—was rendered superfluous by Chebyshev's observation that the prime number theorem $\pi(x) \sim \mathrm{Li}(x)$ can be deduced from the more natural theorem $\psi(x) \sim x$. Thus, in all respects, the formula for $\psi(x)$ is preferable to that for $J(x)$.

Nonetheless, it is the formula for $J(x)$ that was stated† by Riemann and for this reason, if for no other, it is of great interest to know whether or not the formula is valid. Von Mangoldt proved that it is. Von Mangoldt did not, however, follow Riemann's method of proving the formula for $J(x)$; once the product formula for $\xi(s)$ was established by Hadamard, the only real difficulty which remained in Riemann's derivation of the formula for $J(x)$ was the proof that the termwise integration of the sum over ρ (Section 1.15) is valid, but von Mangoldt does not justify this termwise integration directly and proves the formula for $J(x)$ by a quite different method. It may be that he did this simply as a matter of convenience or it may be that he was in fact unable to derive estimates which would justify the termwise integration of Section 1.15 directly. In any event, Landau [L2] in 1908 proved in a more or less direct manner that termwise integration is valid.

3.7 VON MANGOLDT'S PROOF OF RIEMANN'S MAIN FORMULA

For $r > 0$ consider the definite integral

(1) $$\frac{1}{2\pi i}\int_{a-i\infty}^{a+i\infty}\left[-\frac{\zeta'(s+r)}{\zeta(s+r)}\right]x^s\,ds \qquad (a > 1, \quad x > 1).$$

When the formula $-\zeta'(s+r)/\zeta(s+r) = \sum \Lambda(n)n^{-s-r}$ is substituted in this equation and integration is carried out termwise, one obtains the value $\sum_{n<x} \Lambda(n)n^{-r}$ where, as usual, at any point $x = p^n$ where the value jumps, it is defined to split the difference. Von Mangoldt denotes this function by $\psi(x, r)$. An equivalent definition of $\psi(x, r)$ is

(2) $$\psi(x, r) = \int_0^x x^{-r}\,d\psi(x).$$

On the other hand, the derivation of the formula

$$-\frac{\zeta'(s)}{\zeta(s)} = \frac{s}{s-1} - \sum_\rho \frac{s}{\rho(s-\rho)} + \sum_n \frac{s}{2n(s+2n)} - \frac{\zeta'(0)}{\zeta(0)}$$

†Landau [L3] began a tradition of referring to this and other statements of Riemann as "conjectures," which gives the very mistaken impression that Riemann had some doubts about them. The only conjecture which Riemann makes in the paper is that $\mathrm{Re}\,\rho = \frac{1}{2}$.

of Section 3.2 is easily modified to give the more general formula

$$-\frac{\zeta'(s+r)}{\zeta(s+r)} = -\frac{s}{(r-1)(s+r-1)} + \sum_\rho \frac{s}{(r-\rho)(s+r-\rho)} + \sum_n \frac{s}{(r+2n)(s+r+2n)} - \frac{\zeta'(r)}{\zeta(r)}$$

which is valid for $r > 0$ except at $r = 1$. When this expression is substituted in (1) and the integration is carried out termwise, the result is

$$\frac{x^{1-r}}{1-r} - \sum_\rho \frac{x^{\rho-r}}{\rho - r} + \sum_n \frac{x^{-2n-r}}{2n+r} - \frac{\zeta'(r)}{\zeta(r)}. \tag{3}$$

The estimates of Section 3.5 can be modified to show that for fixed $r > 0$, $x > 1$, $r \neq 1$ the termwise integrations are valid and hence that

$$\int_0^x x^{-r}\, d\psi(x) = \frac{x^{1-r}}{1-r} - \sum_\rho \frac{x^{\rho-r}}{\rho-r} + \sum_n \frac{x^{-2n-r}}{2n+r} - \frac{\zeta'(r)}{\zeta(r)}.$$

The first and last terms on the right have poles at $r = 1$ which cancel each other, so the entire right side defines a continuous function of r as can be seen by rewriting it in the form

$$\int_0^x x^{-r}\, d\psi(x) = \left[\frac{x^{1-r}}{1-r} - \frac{\zeta'(r)}{\zeta(r)}\right] - x^{-r} \sum_\rho \frac{x^\rho}{\rho} - x^{-r} \sum_\rho \frac{r x^\rho}{\rho(\rho - r)} + \sum_n \frac{x^{-2n-r}}{2n+r} \qquad (x > 1) \tag{4}$$

and noting that near any value of r these series converge uniformly in r and hence define continuous functions of r. Thus the formula is valid for $r = 1$ as well.

Now integrate both sides dr from $r = 0$ to $r = \infty$. On the left one obtains

$$\int_0^\infty \int_0^x x^{-r}\, d\psi(x)\, dr = \int_0^x \left(\int_0^\infty x^{-r}\, dr\right) d\psi(x) = \int_0^x \frac{1}{\log x}\, d\psi(x) = \int_0^x dJ(x) = J(x).$$

(Or, less elegantly,

$$\int_0^\infty \sum_{n<x} \Lambda(n) n^{-r}\, dr = \sum_{n<x} \frac{\Lambda(n)}{\log n} = J(x)$$

with the usual adjustment if x is a prime power.) The sum over n on the right can be integrated termwise by the Lebesgue dominated convergence theorem because it is dominated by $x^{-r} \sum_n x^{-2n}$. The result is

$$\sum_n \int_0^\infty \frac{x^{-2n-r}\, dr}{2n+r} = \sum_n \int_{2n}^\infty \frac{x^{-u}\, du}{u} = \sum_n \int_{2n}^\infty \int_x^\infty t^{-u-1}\, dt\, du = \sum_n \int_x^\infty \frac{t^{-2n-1}}{\log t}\, dt = \int_x^\infty \frac{dt}{t(t^2-1)\log t}.$$

The first sum over ρ does not involve r and can therefore be integrated termwise. The second sum over ρ can be integrated termwise by the Lebesgue dominated convergence theorem because it is dominated by $rx^{1-r}\sum|\rho|^{-2}$. The result of these two termwise integrations is

$$-\sum_{\rho}\left[\int_0^\infty x^{-r}\frac{x^\rho}{\rho}\,dr + \int_0^\infty x^{-r}\frac{rx^\rho}{\rho(\rho-r)}\,dr\right]$$
$$= -\sum_{\rho}\int_0^\infty \frac{x^{\rho-r}}{\rho-r}\,dr = -\sum_{\operatorname{Im}\rho>0}[\operatorname{Li}(x^\rho)+\operatorname{Li}(x^{1-\rho})]$$

as will now be shown. [Here, as in (4), the sum over ρ is to be taken in the order of increasing $|\operatorname{Im}\rho|$.]

The formula needed above is

(5) $$\int_0^\infty \frac{x^{\rho-r}}{\rho-r}\,dr = \operatorname{Li}(x^\rho) \mp i\pi,$$

where $x > 1$, $\operatorname{Re}\rho > 0$, and† where the sign of $i\pi$ is opposite to that of $\operatorname{Im}\rho$. This can be proved by setting $t = (\rho - r)\log x$, $dt = -(\log x)\,dr$ to put the integral in the form

$$-\int_\infty^0 \frac{x^{\rho-r}(\log x)\,dr}{(\rho-r)\log x} = \int_{\rho\log x-\infty\log x}^{\rho\log x}\frac{e^t\,dt}{t}.$$

Since the integrand vanishes very rapidly near $-\infty$, the lower limit of integration can be taken as $-\infty$; hence

$$\int_0^\infty \frac{x^{\rho-r}\,dr}{\rho-r} = \int_{-\infty}^{\rho\log x}\frac{e^t\,dt}{t},$$

where the path of integration passes above the singularity at $t = 0$ if $\operatorname{Im}\rho > 0$, below if $\operatorname{Im}\rho < 0$. Now, if it is stipulated that the path of integration in the integral

$$\int_{-\infty}^{\beta\log x}\frac{e^t\,dt}{t}$$

must enter the halfplane $\operatorname{Re} t > 0$ by crossing the *positive* imaginary axis, then this integral defines an analytic function of β in $\operatorname{Re}\beta > 0$ which for real β is equal to

$$\int_0^{x^\beta}\frac{du}{\log u} = \operatorname{Li}(x^\beta) - i\pi$$

(see Section 1.15). Hence the same is true by analytic continuation for all

†Thus, as with Riemann, the possibility $\operatorname{Re}\rho = 0$ is not included. The extension of the formula to cover this case is trivial, but, as Hadamard showed (see Section 4.2), none of the ρ's lies on the imaginary axis.

β in Re $\beta > 0$, and

$$\int_0^\infty \frac{x^{\rho-r}}{\rho - r}\,dr = \mathrm{Li}(x^\rho) - i\pi$$

for Im $\rho > 0$ follows. The case Im $\rho < 0$ is analogous and (5) is proved.

It remains only to show that the remaining pair of terms in (4) when integrated dr from 0 to ∞ give the remaining terms in Riemann's formula, that is, to show that

$$\int_0^\infty \left[\frac{x^{1-r}}{1-r} - \frac{\zeta'(r)}{\zeta(r)}\right] dr = \mathrm{Li}(x) + \log \xi(0).$$

This can be proved as follows.

The two terms can be integrated separately

$$\int_0^\infty \left[\frac{x^{1-r}}{1-r} - \frac{\zeta'(r)}{\zeta(r)}\right] dr - \int_0^\infty \frac{x^{1-r}}{1-r}\,dr - \int_0^\infty \frac{\zeta'(r)}{\zeta(r)}\,dr$$

provided the path from 0 to ∞ is perturbed slightly to avoid the singularity at $r = 1$, say by passing slightly above it. The first integral is then the limiting case $\rho = 1 - i\epsilon$ of formula (5) and is therefore $\mathrm{Li}(x) + i\pi$. The second integral can be evaluated by integrating

$$\frac{d}{dr}\log \xi(r) = \frac{d}{dr}\log\left[\pi^{-r/2}\,\Pi\left(\frac{r}{2}\right)\right] + \frac{1}{r-1} + \frac{\zeta'(r)}{\zeta(r)}$$

from 0 to K passing above $r = 1$ to obtain

$$\log \xi(K) - \log \xi(0) = \log \pi^{-K/2}\,\Pi\left(\frac{K}{2}\right) + \log(K-1) - i\pi$$
$$+ \int_0^K \frac{\zeta'(r)}{\zeta(r)}\,dr,$$

$$-\int_0^K \frac{\zeta'(r)}{\zeta(r)}\,dr = \log \xi(0) - i\pi + \log \zeta(K).$$

The limit as $K \to \infty$ is thus $\log \xi(0) - i\pi$ and the desired formula follows.

In summary, then, when (4) is integrated dr from 0 to ∞, the result is Riemann's main formula

$$J(x) = \mathrm{Li}(x) - \sum_\rho \mathrm{Li}(x^\rho)$$
$$+ \int_x^\infty \frac{dt}{t(t^2-1)\log t} + \log \xi(0) \qquad (x > 1)$$

which is thereby proved.

3.8 NUMERICAL EVALUATION OF THE CONSTANT

Von Mangoldt found that the numerical value of the constant in the formula for $\psi(x)$ is

$$\zeta'(0)/\zeta(0) = \log 2\pi. \tag{1}$$

The series $\Sigma(x^{-2n}/2n)$ can also be summed using the series expansion

$$\log[1/(1-x)] = x + \tfrac{1}{2}x^2 + \tfrac{1}{3}x^3 + \cdots$$

to put the formula for $\psi(x)$ in the form

$$\psi(x) = x - \sum_\rho \frac{x^\rho}{\rho} + \frac{1}{2}\log\left(\frac{x^2}{x^2-1}\right) - \log 2\pi \qquad (x > 1),$$

where the sum over ρ is in the order of increasing $|\operatorname{Im}\rho|$.

The value (1) of the constant can be obtained as follows.† Consider the function

$$\frac{\zeta(s)}{\Pi(-s)} = \frac{1}{2\pi i}\int_{+\infty}^{+\infty} \frac{(-x)^s}{e^x - 1}\cdot\frac{dx}{x}, \tag{2}$$

where the path of integration is as in Section 1.4. It will first be shown that the derivative of this function is zero at $s = 1$, that is, it will be shown that

$$\frac{1}{2\pi i}\int_{+\infty}^{+\infty} \frac{(-x)\log(-x)}{e^x - 1}\frac{dx}{x} = 0. \tag{3}$$

Let the path of integration be written as a sum of three parts as in Section 1.4 so that this integral becomes

$$\begin{aligned} &\frac{1}{2\pi i}\int_\infty^\epsilon \frac{(-x)(\log x - i\pi)}{e^x - 1}\frac{dx}{x} \\ &\qquad + \frac{1}{2\pi i}\int_{|x|=\epsilon} \frac{(-x)(\log\epsilon + i\theta - i\pi)}{e^x - 1}\frac{dx}{x} \\ &\qquad + \frac{1}{2\pi i}\int_\epsilon^\infty \frac{(-x)(\log x + i\pi)}{e^x - 1}\frac{dx}{x} \\ &\qquad = -\int_\epsilon^\infty \frac{dx}{e^x - 1} - \frac{\log\epsilon}{2\pi i}\int_{|x|=\epsilon} \frac{x}{e^x - 1}\cdot\frac{dx}{x} \\ &\qquad - \frac{1}{2\pi i}\int_{-\pi}^{\pi} \frac{x}{e^x - 1}\,\phi\, d\phi, \end{aligned}$$

where $x = \epsilon e^{i(\phi+\pi)}$ in the last integral. Since $x(e^x - 1)^{-1}$ is 1 at $x = 0$, the middle integral is $(-\log\epsilon)$ by the Cauchy integral formula and the last integral approaches zero as $\epsilon \downarrow 0$.

†For an alternative proof see Section 6.8.

The first integral can be evaluated directly

$$-\int_\epsilon^\infty \frac{dx}{e^x - 1} = -\int_\epsilon^\infty \left(\sum_{n=1}^\infty e^{-nx}\right) dx = -\left[\sum_{n=1}^\infty \frac{e^{-nx}}{-n}\right]\Bigg|_\epsilon^\infty = -\sum \frac{(e^{-\epsilon})^n}{n}$$

$$= \log(1 - e^{-\epsilon}) = \log\left(\epsilon - \frac{\epsilon^2}{2} + \frac{\epsilon^3}{6} - \cdots\right)$$

$$= \log \epsilon + \log\left(1 - \frac{\epsilon}{2} + \cdots\right).$$

Thus the log ϵ's cancel and the limit as $\epsilon \downarrow 0$ of the remaining terms is zero, which proves (3). Now by the functional equation the function (2) can also be written in the form

$$\zeta(s)/\Pi(-s) = (2\pi)^{s-1}\zeta(1 - s)2 \sin(\pi s/2).$$

Since its derivative is zero at $s = 1$, its logarithmic derivative

$$\log(2\pi) - \frac{\zeta'(1 - s)}{\zeta(1 - s)} + \frac{\pi}{2} \frac{\cos(\pi s/2)}{\sin(\pi s/2)}$$

must also be zero at $s = 1$, which gives (1).

Since the logarithmic derivative of $\xi(s)$ is on the one hand

$$\sum_\rho \frac{d}{ds} \log\left(1 - \frac{s}{\rho}\right) = \sum_\rho \frac{1}{s - \rho}$$

and on the other hand

$$\frac{d}{ds} \log \Pi\left(\frac{s}{2}\right) - \frac{1}{2} \log \pi + \frac{1}{s - 1} + \frac{\zeta'(s)}{\zeta(s)},$$

the sum of the series $\sum (1/\rho)$ is the value of

$$-\frac{1}{2} \frac{\Pi'(s/2)}{\Pi(s/2)} + \frac{1}{2} \log \pi + \frac{1}{1 - s} - \frac{\zeta'(s)}{\zeta(s)}$$

at $s = 0$. Now logarithmic differentiation of the product formula for $\Pi(x)$ [(4) of Section 1.3] gives

$$\frac{\Pi'(s)}{\Pi(s)} = \sum_{n=1}^\infty \left[-\frac{1}{s + n} - \log n + \log(n + 1)\right]$$

$$-\frac{\Pi'(0)}{\Pi(0)} = \lim_{n\to\infty} \left[1 + \frac{1}{2} + \frac{1}{3} + \cdots + \frac{1}{n} - \log(n + 1)\right].$$

The number on the right side of this equation is by definition *Euler's constant*, and is traditionally denoted γ. Thus

$$\sum_\rho \frac{1}{\rho} = \frac{1}{2}\gamma + \frac{1}{2} \log \pi + 1 - \log 2\pi. \tag{4}$$

This formula was known to Riemann, who used it in his computations of the roots ρ (see Section 7.6 below). From this it is clear that the formula (1) was also known to Riemann.

Chapter 4

The Prime Number Theorem

4.1 INTRODUCTION

The prime number theorem is the statement that the relative error in the approximation $\pi(x) \sim \mathrm{Li}(x)$ approaches zero as $x \longrightarrow \infty$. Following the work of Chebyshev it was well known that the prime number theorem could be deduced from the theorem that the relative error in the approximation $\psi(x) \sim x$ approaches zero as $x \longrightarrow \infty$. But von Mangoldt's formula for $\psi(x)$ shows that $\psi(x) \sim x$ has this property if and only if

$$\lim_{x\to\infty} \frac{-\sum_{\rho} (x^{\rho}/\rho) + \sum_{n} (x^{-2n}/2n) + \text{const}}{x} = 0$$

or, what is the same, if and only if

$$\lim_{x\to\infty} \sum_{\rho} \frac{x^{\rho-1}}{\rho} = 0. \tag{1}$$

If the limit of this sum could be taken termwise, then it would suffice to prove that $x^{\rho-1} \longrightarrow 0$ for all ρ or, what is the same, that $\mathrm{Re}\,\rho < 1$ for all ρ. Since $\mathrm{Re}\,\rho \leq 1$ for all ρ (by the Euler product formula—see Section 1.9), this amounts to proving that there are no roots ρ on the line $\mathrm{Re}\, s = 1$. Thus, given von Mangoldt's 1894 formula for $\psi(x)$, the proof of the prime number theorem can be reduced to proving that there are no roots ρ on the line $\mathrm{Re}\; s = 1$ and to proving that the above limit can be evaluated termwise.

Both Hadamard [H2] and de la Vallée Poussin [V1] succeeded in 1896 in filling in these remaining steps in the proof of the prime number theorem. They both circumvented proving that the limit (1) can be evaluated termwise and instead they each derived a variation of von Mangoldt's formula, namely,

$$(2) \qquad \int_0^x t^{-2}\psi(t)\,dt = \log x - \sum_\rho \frac{x^{\rho-1}}{\rho(\rho-1)} - \sum_n \frac{x^{-2n-1}}{2n(2n+1)} + \frac{1}{x}\frac{\zeta'(0)}{\zeta(0)} + \text{const} \qquad (x > 1)$$

in de la Vallée Poussin's case† and

$$(3) \qquad \int_0^x t^{-1}\psi(t)\,dt = x - \sum_\rho \frac{x^\rho}{\rho^2} - \sum_n \frac{x^{-2n}}{(2n)^2} - \frac{\zeta'(0)}{\zeta(0)}\log x + \text{const} \qquad (x > 1)$$

in Hadamard's case. Either of these formulas is quite easy to prove using von Mangoldt's methods—easier‡ in fact than von Mangoldt's formula for $\psi(x)$—and if it is known that $\operatorname{Re}\rho < 1$ for all roots ρ, then either of them can be used to conclude by straightforward estimates that $\psi(x) \sim x$. Thus, although it certainly required insight to see that a formula such as (2) or (3) could be used, the substantial step beyond von Mangoldt's work which was required for the proof of the prime number theorem was the proof that there are no roots ρ on the line $\operatorname{Re} s = 1$.

Hadamard's proof that there are no roots ρ on $\operatorname{Re} s = 1$ is given in Section 4.2. De la Vallée Poussin admitted that Hadamard's proof was the simpler of the two, and although simpler proofs have since been found (see Section 5.2), Hadamard's is perhaps still the most straightforward and natural proof of this fact. Section 4.3 is devoted to a proof that $\psi(x) \sim x$. This proof follows the same general line of argument as was followed by both Hadamard and de la Vallée Poussin, but it is somewhat simpler in that it is based on the formula

$$(4) \qquad \int_0^x \psi(t)\,dt = \frac{x^2}{2} - \sum_\rho \frac{x^{\rho+1}}{\rho(\rho+1)} - \sum_n \frac{x^{-2n+1}}{2n(2n-1)} - x\frac{\zeta'(0)}{\zeta(0)} + \text{const} \qquad (x > 1)$$

rather than on the analogous, but somewhat more complicated, formulas (2) or (3). Finally, Section 4.4 gives the very simple deduction of $\pi(x) \sim \operatorname{Li}(x)$ from $\psi(x) \sim x$.

†For the value of the constant, which is $-1 - \gamma$, see footnote †, p. 94.

‡At the time, de la Vallee Poussin was under the mistaken impression that von Mangoldt's proof was fallacious, so of course he gave his own proof. Hadamard too preferred to avoid making appeal to von Mangoldt's more difficult estimates and produced his own proof.

4.2 HADAMARD'S PROOF THAT Re $\rho < 1$ FOR ALL ρ

The representation

$$\text{(1)} \qquad \log \zeta(s) = \int_0^\infty x^{-s}\, dJ(x) = \sum_p \frac{1}{p^s} + \frac{1}{2}\sum_p \frac{1}{p^{2s}} + \frac{1}{3}\sum_p \frac{1}{p^{3s}} + \cdots$$

is valid throughout the halfplane Re $s > 1$. The presence of zeros ρ of $\zeta(s)$ on the line Re $s = 1$ would imply the presence of points $s = \sigma + it$ slightly to the right of Re $s = 1$ where Re $\log \zeta(s)$ was near $-\infty$. The series in (1) has the property that the sum of the terms after the first is bounded by the number

$$B = \frac{1}{2}\sum_p \frac{1}{p^2} + \frac{1}{3}\sum_p \frac{1}{p^3} + \frac{1}{4}\sum_p \frac{1}{p^4} + \cdots$$

for the entire halfplane Re $s \geq 1$ including Re $s = 1$; hence

$$\operatorname{Re} \log \zeta(\sigma + it) \geq \sum_p \frac{\cos(t \log p)}{p^\sigma} - B$$

can approach $-\infty$ as $\sigma \downarrow 1$ only if the first term approaches $-\infty$. In short, if $1 + it$ were a zero ρ of $\zeta(s)$, then it would follow that

$$\text{(2)} \qquad \lim_{\sigma \downarrow 1} \sum_p \frac{\cos(t \log p)}{p^\sigma} = -\infty$$

for this value of t. The objective is to show that this is impossible.

Now the fact that

$$\lim_{s \to 1} (s - 1)\zeta(s) = \frac{\xi(1)}{\Pi(\frac{1}{2})\pi^{-1/2}} = \frac{(\frac{1}{2})}{\frac{1}{2}\pi^{1/2}\pi^{-1/2}} = 1$$

implies that

$$\lim_{\sigma \downarrow 1} [\operatorname{Re} \log \zeta(\sigma) + \log(\sigma - 1)] = 0,$$

$$\lim_{\sigma \downarrow 1} \left[\sum_p \frac{1}{p^\sigma} + \text{bounded} + \log(\sigma - 1)\right] = 0,$$

which means that for σ slightly larger than one

$$\text{(3)} \qquad \sum_p \frac{1}{p^\sigma} \sim -\log(\sigma - 1) \sim +\infty.$$

On the other hand, the derivation of (2) is easily strengthened to give

$$\lim_{\sigma \downarrow 1} \left[\sum_p \frac{\cos(t \log p)}{p^\sigma} + \text{bounded} - \operatorname{Re} \log(\sigma - 1)\right] = 0,$$

$$\text{(4)} \qquad \sum_p \frac{\cos(t \log p)}{p^\sigma} \sim \log(\sigma - 1) \sim -\infty.$$

If this were true then, because of (3), one would expect that the number $\cos(t \log p)$ would have to be nearly -1 for the overwhelming majority of primes p. This would imply a surprising regularity in the distribution of the numbers $\log p$, namely, that most of them lie near the points of the arithmetic progression $(2n + 1)t^{-1}\pi$ for this particular t. However, such a regularity cannot exist because it would imply that $\cos(2t \log p)$ was nearly $+1$ for the overwhelming majority of primes p, hence

$$\sum_p \frac{\cos(2t \log p)}{p^\sigma} \sim +\infty,$$

$$\operatorname{Re} \log \zeta(\sigma + 2it) \sim +\infty$$

which would imply the existence of a pole of $\zeta(s)$ at $s = 1 + 2it$. Since $\zeta(s)$ has no poles other than $s = 1$, this line of argument might be expected to yield a proof of the impossibility of (4) as desired. The actual proof requires little more than a quantitative description of the "overwhelming majority" of the primes p for which $t \log p \sim (2n + 1)\pi$ would have to be true.

Assume that $1 + it$ is a zero of $\zeta(s)$. Then $\zeta(s)/(s - 1 - it)$ would be analytic near $s = 1 + it$ so the real part of its log would be bounded above, say by K; hence with $s = \sigma + it$

$$\sum_p \frac{\cos(t \log p)}{p^\sigma} + \frac{1}{2} \sum_p \frac{\cos(2t \log p)}{p^{2\sigma}} + \frac{1}{3} \sum_p \frac{\cos(3t \log p)}{p^{3\sigma}}$$
$$+ \cdots - \operatorname{Re} \log(\sigma - 1) < K,$$

$$\sum_p \frac{\cos(t \log p)}{p^\sigma} < \log(\sigma - 1) + K + B \tag{5}$$

for all $\sigma > 1$ near 1. This is a quantitative version of (4); the objective is to show that no t has this property.

Let δ be a small positive number (for the sake of definiteness, $\delta = \pi/8$ will work in the following proof) and let the terms of the sum on the left side of (5) be divided into those terms corresponding to primes p for which there is an n such that

$$|(2n + 1)\pi - t \log p| < \delta \tag{6}$$

and those terms for which p does not satisfy this condition. In the terms of the second group $\cos(t \log p) = \cos(\pi - \alpha) = -\cos \alpha$ where $\delta \le |\alpha| \le \pi$, hence $\cos(t \log p) > -\cos \delta$ for these terms and (5) implies

$$-S' - (\cos \delta)S'' < \log(\sigma - 1) + K + B$$

where S' denotes the sum of $p^{-\sigma}$ over all primes p which satisfy (6) and S'' denotes the sum of $p^{-\sigma}$ over all primes p which do not. Now (3) says that $S' + S'' \sim -\log(\sigma - 1)$, and the derivation of (3) shows easily that there is a K' such that $S' + S'' < -\log(\sigma - 1) + K'$ for all $\sigma > 1$ near 1. Thus (5) implies

$$-S' - (\cos\delta)S'' < -S' - S'' + K' + K + B,$$
$$(1 - \cos\delta)S'' < \text{const}$$

for all $\sigma > 1$ near 1. Since $S' + S''$ becomes infinite as $\sigma \downarrow 1$, this shows that (5) implies

$$\lim_{\sigma\downarrow 1} \frac{S''}{S' + S''} = 0, \qquad \lim_{\sigma\downarrow 1} \frac{S'}{S' + S''} = 1 \tag{7}$$

which is a specific sense in which the "overwhelming majority" of primes must satisfy (6) if (5) is true.

However, since $1 + 2it$ is not a pole of $\zeta(s)$, the real part of $\log\zeta(s)$ is bounded above near $s = 1 + 2it$, say by K'', hence

$$\sum_p \frac{\cos(2t\log p)}{p^\sigma} < K'' + B. \tag{8}$$

When the terms in this sum are split into those terms for which p satisfies (6) and those terms for which p does not, the terms in the first group have cos $(2t\log p) = \cos(2\alpha)$, where $|\alpha| < 2\delta$; hence $\cos(2t\log p) > \cos 2\delta > 0$, so (8) implies

$$S'\cos 2\delta - S'' < K'' + B,$$
$$\frac{S''}{S' + S''} > \cos 2\delta \frac{S'}{S' + S''} - \frac{\text{const}}{S' + S''}$$

which contradicts (7). Thus (5) is impossible and the proof is complete.

4.3 PROOF THAT $\psi(x) \sim x$

Since von Mangoldt's formula

$$\psi(x) = x - \sum \frac{x^\rho}{\rho} + \sum \frac{x^{-2n}}{2n} - \frac{\zeta'(0)}{\zeta(0)} \qquad (x > 1)$$

is obtained by evaluating the definite integral

$$\frac{1}{2\pi i}\int_{a-i\infty}^{a+i\infty}\left[-\frac{\zeta'(s)}{\zeta(s)}\right]\frac{x^s\,ds}{s} \qquad (x > 1,\quad a > 1)$$

in two different ways, it is natural to expect that the antiderivative of von Mangoldt's formula, namely,

$$\int_0^x \psi(t)\,dt = \frac{x^2}{2} - \sum \frac{x^{\rho+1}}{\rho(\rho+1)} - \sum \frac{x^{-2n+1}}{2n(2n-1)}$$
$$- \frac{\zeta'(0)}{\zeta(0)}x + \text{const} \qquad (x > 1)$$

could be obtained by evaluating the antiderivative of this definite integral, namely,

$$\frac{1}{2\pi i}\int_{a-i\infty}^{a+i\infty}\left[-\frac{\zeta'(s)}{\zeta(s)}\right]\frac{x^{s+1}\,ds}{s(s+1)} \qquad (x>1, \quad a>1) \tag{1}$$

in two different ways.

The first way of evaluating (1) is to use the expansion $-\zeta'(s)/\zeta(s) = \sum \Lambda(n)n^{-s}$. If termwise integration of this expansion is valid, then (1) is

$$\sum_{n=1}^{\infty}\Lambda(n)\frac{1}{2\pi i}\int_{a-i\infty}^{a+i\infty}\frac{x^{s+1}\,ds}{n^s s(s+1)}. \tag{2}$$

The partial fractions expansion

$$\frac{1}{s(s+1)} = \frac{1}{s} - \frac{1}{s+1} \tag{3}$$

gives

$$\frac{x^{s+1}}{n^s s(s+1)} = \frac{x}{s}\left(\frac{x}{n}\right)^s - \frac{n}{s+1}\left(\frac{x}{n}\right)^{s+1}$$

$$\frac{1}{2\pi i}\int_{a-i\infty}^{a+i\infty}\frac{x^{s+1}\,ds}{n^s s(s+1)} = \frac{x}{2\pi i}\int_{a-i\infty}^{a+i\infty}\left(\frac{x}{n}\right)^s\frac{ds}{s} - \frac{n}{2\pi i}\int_{a+1-i\infty}^{a+1+i\infty}\left(\frac{x}{n}\right)^u\frac{du}{u}$$

$$= \begin{cases} x-n & \text{if } n \le x, \\ 0 & \text{if } n \ge x, \end{cases}$$

so that if termwise integration is valid, then (1) is equal to†

$$\sum_{n\le x}\Lambda(n)(x-n) = \int_0^x (x-t)\,d\psi(t) = \int_0^x \psi(t)\,dt.$$

The proof that termwise integration is valid, and hence that (1) is equal to $\int_0^x \psi(t)\,dt$, is easily accomplished by writing the integrand of (1) in the form

$$-\frac{\zeta'(s)}{\zeta(s)}\frac{x^{s+1}}{(s+1)s} = \left[x\sum_{n=1}^{\infty}\Lambda(n)\left(\frac{x}{n}\right)^s\frac{1}{s}\right] - \left[\sum_{n=1}^{\infty}\Lambda(n)\left(\frac{x}{n}\right)^{s+1}\frac{n}{s+1}\right]$$

to express (1) as the difference of two integrals of infinite sums over n, and by then observing that von Mangoldt's method shows immediately that each of these two integrals can be evaluated termwise.

The second way of evaluating (1) is to use the expansion

$$-\frac{\zeta'(s)}{\zeta(s)} = \frac{s}{s-1} - \sum_{\rho}\frac{s}{\rho(s-\rho)} + \sum_{n=1}^{\infty}\frac{s}{2n(s+2n)} - \frac{\zeta'(0)}{\zeta(0)} \tag{4}$$

of Section 3.2, (7). When this is used in conjunction with (3) in expressing the integrand of (1), the term $1/s$ divides evenly, but the term $1/(s+1)$ does not.

†Note that the case $x = n$ presents no difficulties because then $x - n = 0$. This occurs because the antiderivative of a function with jump discontinuities has no jump discontinuities.

To simplify the resulting expression note that

$$-\frac{\zeta'(s)}{\zeta(s)}+\frac{\zeta'(-1)}{\zeta(-1)}=\frac{s}{s-1}-\frac{-1}{-1-1}-\sum_\rho\left[\frac{s}{\rho(s-\rho)}-\frac{-1}{\rho(-1-\rho)}\right]$$
$$+\sum_n\left[\frac{s}{2n(s+2n)}-\frac{-1}{2n(-1+2n)}\right]$$
$$-\frac{\zeta'(s)}{\zeta(s)}=\frac{s+1}{2(s-1)}-\sum_\rho\frac{s+1}{(\rho+1)(s-\rho)}$$
$$+\sum_n\frac{s+1}{(2n-1)(s+2n)}-\frac{\zeta'(-1)}{\zeta(-1)}$$

so that the integrand in (1) is equal to

$$-\frac{\zeta'(s)}{\zeta(s)}x^{s+1}\left(\frac{1}{s}-\frac{1}{s+1}\right)=\frac{x^{s+1}}{s-1}-\sum_\rho\frac{x^{s+1}}{\rho(s-\rho)}+\sum_n\frac{x^{s+1}}{2n(s+2n)}$$
$$-\frac{\zeta'(0)}{\zeta(0)}\frac{x^{s+1}}{s}-\frac{x^{s+1}}{2(s-1)}+\sum_\rho\frac{x^{s+1}}{(\rho+1)(s-\rho)}$$
$$-\sum_n\frac{x^{s+1}}{(2n-1)(s+2n)}+\frac{\zeta'(-1)}{\zeta(-1)}\frac{x^{s+1}}{s+1}.$$

Von Mangoldt's estimates (Section 3.5) with only slight modifications show that both of the sums over ρ and both of the sums over n can be integrated termwise. This gives for (1) the value

$$x^2-\sum_\rho\frac{x^{\rho+1}}{\rho}+\sum_n\frac{x^{1-2n}}{2n}-\frac{\zeta'(0)}{\zeta(0)}x-\frac{x^2}{2}+\sum_\rho\frac{x^{\rho+1}}{\rho+1}-\sum_n\frac{x^{1-2n}}{2n-1}+\frac{\zeta'(-1)}{\zeta(-1)}$$
$$=\frac{x^2}{2}-\sum_\rho\frac{x^{\rho+1}}{\rho(\rho+1)}-\sum_n\frac{x^{1-2n}}{2n(2n-1)}-\frac{\zeta'(0)}{\zeta(0)}x+\frac{\zeta'(-1)}{\zeta(-1)}$$

which by the above is also equal to $\int_0^x\psi(t)\,dt$. This completes the proof of the formula†

$$\int_0^x\psi(t)\,dt=\frac{x^2}{2}-\sum_\rho\frac{x^{\rho+1}}{\rho(\rho+1)}-\sum_n\frac{x^{-2n+1}}{2n(2n-1)}-\frac{\zeta'(0)}{\zeta(0)}x+\frac{\zeta'(-1)}{\zeta(-1)}$$

which holds for all $x>1$.

†This is a special case ($u=0$, $v=-1$) of the formula

$$y^u\int_0^y t^{-u}\,d\psi(t)-y^v\int_0^y t^{-v}\,d\psi(t)=-\left[y^u\frac{\zeta'(u)}{\zeta(u)}-y^v\frac{\zeta'(v)}{\zeta(v)}\right]+\frac{u-v}{(u-1)(v-1)}y$$
$$-(u-v)\sum_\rho\frac{y^\rho}{(u-\rho)(v-\rho)}$$
$$-(u-v)\sum_n\frac{y^{-2n}}{(u+2n)(v+2n)}\qquad(y>1)$$

which de la Vallée Poussin proved in 1896 by using the above method in conjunction with elementary arguments—not those of von Mangoldt—to justify the termwise integrations. However, he used the case $u=1$, $v=0$, which gives (2) of Section 4.1, in his proof of the prime number theorem. Note that the constant in formula (4) of Section 4.1 has been shown to be $\zeta'(-1)/\zeta(-1)$.

Using this formula for $\int_0^x \psi(t)\,dt$ it is quite easy to show that

$$\int_0^x \psi(t)\,dt \sim \frac{x^2}{2}, \tag{5}$$

where, as usual, the symbol $\sim$ means that the relative error in the approximation approaches zero as $x \to \infty$. One need only note that $\int_0^x \psi(t)\,dt - (x^2/2)$ divided by $x^2/2$ is, by the formula, equal to $2\Sigma x^{\rho-1}/\rho(\rho+1)$ plus terms which go to zero as $x \to \infty$. Since $\Sigma \rho^{-1}(\rho+1)^{-1}$ converges absolutely and since $|x^{\rho-1}| \leq 1$ (because $\operatorname{Re}\rho \leq 1$), it follows that the series $\Sigma x^{\rho-1}/\rho(\rho+1)$ converges uniformly in x and hence that the limit as $x \to \infty$ can be evaluated termwise; since each term goes to zero (because $\operatorname{Re}\rho < 1$), it follows that this limit is zero and hence that the relative error in (5) approaches zero as $x \to \infty$.

To deduce $\psi(x) \sim x$, let $\epsilon > 0$ be given and let X be such that

$$(1 - \epsilon)\frac{x^2}{2} < \int_0^x \psi(t)\,dt < (1+\epsilon)\frac{x^2}{2}$$

for all $x \geq X$. Then for $y > x \geq X$ it follows that

$$\int_0^y \psi(t)\,dt - \int_0^x \psi(t)\,dt \tag{6}$$

is at least

$$(1-\epsilon)\frac{y^2}{2} - (1+\epsilon)\frac{x^2}{2} = (1-\epsilon)\frac{y^2 - x^2}{2} - 2\epsilon\frac{x^2}{2}$$

and at most

$$(1+\epsilon)\frac{y^2}{2} - (1-\epsilon)\frac{x^2}{2} = (1+\epsilon)\frac{y^2 - x^2}{2} + 2\epsilon\frac{x^2}{2}.$$

On the other hand, ψ is an increasing function, so (6) is at least

$$\int_x^y \psi(t)\,dt \geq (y - x)\psi(x)$$

and at most

$$\int_x^y \psi(t)\,dt \leq (y - x)\psi(y).$$

Combining these inequalities gives

$$(y - x)\psi(x) \leq (6) \leq (1+\epsilon)\frac{y^2 - x^2}{2} + 2\epsilon\frac{x^2}{2},$$

$$(y - x)\psi(y) \geq (6) \geq (1-\epsilon)\frac{y^2 - x^2}{2} - 2\epsilon\frac{x^2}{2},$$

for all $y > x \geq X$. With $y = \beta x$ these inequalities give

$$\frac{\psi(x)}{x} \leq (1+\epsilon)\frac{\beta+1}{2} + \frac{\epsilon}{\beta - 1},$$

$$\frac{\psi(y)}{y} \geq (1-\epsilon)\frac{\beta+1}{2\beta} - \frac{\epsilon}{\beta(\beta-1)}.$$

Given any $\beta > 1$, $\epsilon > 0$, these inequalities are satisfied for all sufficiently large x, y. Now the quantity on the right side of the first inequality can be made less than any number greater than 1 by first choosing $\beta > 1$ near enough to 1 that the first term is very near 1 and by then choosing ϵ so small that the second term is still small; this shows that $\psi(x)/x$ is less than any number greater than 1 for x sufficiently large. Similarly the second inequality can be used to show that $\psi(y)/y$ is greater than any number less than 1 for y sufficiently large. This completes the proof that $\psi(x) \sim x$.

4.4 PROOF OF THE PRIME NUMBER THEOREM

Once the estimate $\psi(x) \sim x$ has been proved, the prime number theorem $\pi(x) \sim \mathrm{Li}(x)$ is easily deduced. The technique used below is essentially the technique used by Chebyshev in 1850 [C3] to deduce his estimate of $\pi(x)$ from his estimate of $\psi(x)$.

Let $\theta(x)$ denote† the sum of the logarithms of all the primes p less than x, with the usual understanding that if x itself is a prime, then $\theta(x) = \frac{1}{2}[\theta(x+\epsilon) + \theta(x-\epsilon)]$. Then the relationship of θ and ψ is analogous to the relationship of π and J, and in analogy to the formula (1) of Section 1.17 relating π and J there is the formula

$$\psi(x) = \theta(x) + \theta(x^{1/2}) + \theta(x^{1/3}) + \theta(x^{1/4}) + \cdots$$

relating θ and ψ. The series on the right has the property that each term is larger than the following term and that all terms $\theta(x^{1/n})$ in which $x^{1/n} < 2$ are zero. Thus there are at most $\log x/\log 2$ nonzero terms and

$$\theta(x) < \psi(x) < \theta(x) + \theta(x^{1/2}) \log x/\log 2$$

which gives

$$\psi(x) - \theta(x^{1/2}) \log x/\log 2 < \theta(x) < \psi(x). \tag{1}$$

The inequality on the right together with $\psi(x) \sim x$ shows that $\theta(x)/x^{1+\epsilon} \to 0$ as $x \to \infty$. Hence $\theta(x^{1/2}) \log x/x = [\theta(x^{1/2})/(x^{1/2})^{1+\epsilon}][\log x/(x^{1/2})^{1-\epsilon}]$ goes to zero as $x \to \infty$ and $\theta(x) \sim x$ follows from (1) and $\psi(x) \sim x$.

Now let $\epsilon > 0$ be given and let X be such that $(1-\epsilon)x \leq \theta(x) \leq (1+\epsilon)x$ whenever $x \geq X$. Then for $y > x \geq X$ it follows that

$$\begin{aligned}\pi(y) - \pi(x) &= \int_x^y \frac{d\theta(t)}{\log t}\\ &= \left[\frac{\theta(t)}{\log t}\right]_x^y + \int_x^y \frac{\theta(t)\,dt}{(\log t)^2 t}\end{aligned}$$

†This notation, introduced by Chebyshev, is now standard.

is at most

$$\frac{(1+\epsilon)y}{\log y} - \frac{(1-\epsilon)x}{\log x} + \int_x^y \frac{(1+\epsilon)t\,dt}{(\log t)^2 t}$$

$$= 2\epsilon\frac{x}{\log x} + (1+\epsilon)\left\{\left[\frac{t}{\log t}\right]_x^y + \int_x^y \frac{t\,dt}{(\log t)^2 t}\right\}$$

$$= 2\epsilon\frac{x}{\log x} + (1+\epsilon)\left\{\int_x^y \frac{dt}{\log t}\right\}$$

$$= 2\epsilon\frac{x}{\log x} + (1+\epsilon)[\mathrm{Li}(y) - \mathrm{Li}(x)]$$

and is at least

$$-2\epsilon\frac{x}{\log x} + (1-\epsilon)[\mathrm{Li}(y) - \mathrm{Li}(x)].$$

Thus for fixed x it follows that $\pi(y)/\mathrm{Li}(y)$ is at most

$$1 + \epsilon + \frac{\pi(x) + 2\epsilon x(\log x)^{-1} - (1+\epsilon)\,\mathrm{Li}(x)}{\mathrm{Li}(y)} \leq 1 + 2\epsilon$$

for all sufficiently large y and similarly that it is at least $1 - 2\epsilon$ for all sufficiently large y. Since ϵ was arbitrary, this proves the prime number theorem $\mathrm{Li}(y) \sim \pi(y)$.

Chapter 5

De la Vallée Poussin's Theorem

5.1 INTRODUCTION

After it was proved that the relative error in the approximation $\pi(x) \sim \mathrm{Li}(x)$ approaches zero as x approaches infinity, the next step was an estimate of the *rate* at which it approaches zero. De la Vallée Poussin [V2] proved in 1899 that there is a $c > 0$ such that the relative error approaches zero at least as fast as $\exp[-(c \log x)^{1/2}]$ does; that is,

$$\left|\frac{\pi(x) - \mathrm{Li}(x)}{\mathrm{Li}(x)}\right| < e^{-\sqrt{c \log x}}$$

for all sufficiently large x. The next two sections are devoted to the proof of this fact. Section 5.4 contains an application of this theorem to the question of comparing $\mathrm{Li}(x)$ to other possible approximations; it is shown, in essence, that Dirichlet, Gauss, Chebyshev, and Riemann were correct in preferring $\mathrm{Li}(x)$ to another approximation suggested earlier by Legendre. The next section, 5.5, shows that de la Vallée Poussin's theorem can be improved considerably if the Riemann hypothesis is true and that in fact the Riemann hypothesis is *equivalent* to the statement that the relative error in $\pi(x) \sim \mathrm{Li}(x)$ goes to zero faster than $x^{-(1/2)+\epsilon}$ as $x \to \infty$ (for all $\epsilon > 0$). Finally, the last section is devoted to the very simple proof of von Mangoldt's theorem that Euler's formula $\sum \mu(n)/n = 0$ [$\mu(n)$ is the Möbius function] is true in the strong sense that the series $\sum \mu(n)/n$ is *convergent* to the sum zero; this proof makes very effective use of de la Vallée Poussin's theorem that the relative error in the prime number theorem approaches zero at least as fast as $\exp[-(c \log x)^{1/2}]$.

5.2 AN IMPROVEMENT OF Re $\rho < 1$

De la Vallée Poussin's estimate of the error in the prime number theorem is based on a strengthened version of the theorem that the roots $\rho = \beta + i\gamma$ satisfy $\beta < 1$, namely, the theorem that there exist constants† $c > 0$, $K > 1$, such that

$$\beta < 1 - \frac{c}{\log \gamma} \tag{1}$$

for all roots $\rho = \beta + i\gamma$ in the range $\gamma > K$. Since $\log \gamma > \log K > 0$, the inequality (1) is stronger than $\beta < 1$, but the amount by which it is stronger decreases as γ increases and (1) does not preclude the possibility that there are roots ρ arbitrarily near to the line Re $s = 1$; although (1) has been improved upon somewhat since 1899, no one has yet been able to prove that Re ρ has any upper bound less than 1.

De la Vallée Poussin's proof of (1) makes use of a technique by which Mertens [M6] simplified the proof that Re $\rho < 1$. This technique is based on the elementary inequality

$$4 \geq 2(1 - \cos\theta),$$
$$4(1 + \cos\theta) \geq 2(1 - \cos^2\theta) = 1 - \cos 2\theta,$$
$$3 + 4\cos\theta + \cos 2\theta \geq 0,$$

which holds for all θ. Combining this with the formula‡ $-\zeta'(s)/\zeta(s) = \int_0^\infty x^{-s}\, d\psi(x)$ for Re $s > 1$ gives

$$\mathrm{Re}\left\{-3\frac{\zeta'(\sigma)}{\zeta(\sigma)} - 4\frac{\zeta'(\sigma + it)}{\zeta(\sigma + it)} - \frac{\zeta'(\sigma + 2it)}{\zeta(\sigma + 2it)}\right\}$$
$$= \int_0^\infty x^{-\sigma}[3 + 4\cos(t\log x) + \cos(2t\log x)]\, d\psi(x) \geq 0.$$

Hence

$$\mathrm{Re}\left\{3\frac{\zeta'(\sigma)}{\zeta(\sigma)} + 4\frac{\zeta'(\sigma + it)}{\zeta(\sigma + it)} + \frac{\zeta'(\sigma + 2it)}{\zeta(\sigma + 2it)}\right\} \leq 0 \tag{2}$$

for all $\sigma > 1$ and for all real t.

This inequality can be used to prove $\beta < 1$ as follows. Assume there is a real value of t such that $\zeta(1 + it) = 0$ and let $f(s)$ be the function $[\zeta(s)]^3 \cdot [\zeta(s + it)]^4 \cdot [\zeta(s + 2it)]$. Then the first factor of $f(s)$ has a pole of order 3 at

†Specifically, de la Vallée Poussin proved that $\beta < 1 - c(\log\gamma - \log n)^{-1}$ for $\gamma \geq 705$, where $c = 0.034$ and $n = 47.8$. This requires, of course, much more careful estimates than those given here.

‡Mertens uses $\log \zeta(s) = \int_0^\infty x^{-s}\, dJ(x)$ instead.

$s = 1$, the second has a zero of order at least 4, and the third has no pole; hence $f(s)$ has a zero of order at least 1 at $s = 1$, say $f(s) = (s - 1)^n g(s)$, where $n \geq 1$ and where $g(s)$ has no zero and no pole at $s = 1$. Then the logarithmic derivative of g approaches a limit as $s \longrightarrow 1$, so the logarithmic derivative of $f(s)$ differs from $n(s - 1)^{-1}$ by a bounded amount as $s \longrightarrow 1$. Thus the logarithmic derivative of $f(s)$ has large positive real part when $s = \sigma$ is real and greater than 1. But (2) shows that this is impossible and hence that no t satisfies $\zeta(1 + it) = 0$.

De la Vallée Poussin used the inequality (2) to prove (1) by using the formula

$$\frac{\zeta'(s)}{\zeta(s)} = -\frac{1}{s-1} + \sum_{\rho} \frac{1}{s-\rho} - \frac{1}{2}\frac{\Pi'(s/2)}{\Pi(s/2)} + \frac{1}{2}\log\pi$$

[see (6) of Section 3.2] to estimate the terms of (2). Specifically, this formula and the formula $\Pi'(x)/\Pi(x) \sim \log x$ of Section 6.3 show that for $1 \leq \sigma \leq 2$ and for t large, the real part of $\zeta'(s)/\zeta(s)$ is approximately

$$\sum_{\rho} \operatorname{Re}\frac{1}{s-\rho} - \frac{1}{2}\log\left|\frac{t}{2}\right| + \frac{1}{2}\log\pi$$

[because $(s - 1)^{-1}$ is small] and in particular that one can choose $K > 1$ such that

$$\operatorname{Re}\frac{\zeta'(\sigma + it)}{\zeta(\sigma + it)} \geq \sum_{\rho} \operatorname{Re}\frac{1}{\sigma + it - \rho} - \log t + \frac{1}{2}\log\pi$$

holds throughout the region $1 \leq \sigma \leq 2$, $t \geq K$. The terms of the sum over ρ are positive (because $\sigma \geq 1 > \operatorname{Re}\rho$), so the same inequality holds *a fortiori* if some or all terms of the sum over ρ are omitted. Now $(\sigma - 1)\zeta(\sigma)$ is nonsingular at $\sigma = 1$, so the logarithmic derivative of $\zeta(\sigma)$ differs from $-(\sigma - 1)^{-1}$ by a bounded amount and (2) gives

$$0 \geq -\frac{3}{\sigma - 1} + \frac{4}{\sigma + it - \rho} - 4\log t - \log(2t) + \text{const}$$

when all but one term of the sums over ρ are omitted. Thus for all roots ρ and $1 < \sigma \leq 2$, $t \geq K$

$$\frac{3}{4}\frac{1}{\sigma - 1} + C\log t \geq \frac{1}{\sigma + it - \rho},$$

where $C > 0$ is independent of σ, t, and ρ. If $\rho = \beta + i\gamma$, where $\gamma \geq K$, then one can set $t = \gamma$ to find

$$\frac{3}{4}\frac{1}{\sigma - 1} + C\log\gamma \geq \frac{1}{\sigma - \beta}$$

for all σ in the range $1 < \sigma \leq 2$. Thus

$$\left[\frac{3}{4(\sigma-1)} + C\log\gamma\right]^{-1} \leq \sigma - \beta = (\sigma - 1) - (\beta - 1),$$

$$\beta - 1 \leq (\sigma - 1) - \frac{4(\sigma-1)}{3 + 4(\sigma-1)C\log\gamma}$$

$$= \frac{-(\sigma-1) + 4(\sigma-1)^2 C\log\gamma}{3 + 4(\sigma-1)C\log\gamma},$$

$$\beta \leq 1 - \frac{y - 4y^2}{3 + 4y} \cdot \frac{1}{C\log\gamma},$$

where $y = (\sigma - 1)C\log\gamma$. Thus y can have any value between 0 and $C\log\gamma \geq C\log K$. Since $y - 4y^2 > 0$ for small values of y, one need only fix a small positive value of y such that $y < C\log K$, $y - 4y^2 > 0$ in order to draw the desired conclusion (1) for all roots $\rho = \beta + i\gamma$ which satisfy $\gamma \geq K$.

5.3 DE LA VALLÉE POUSSIN'S ESTIMATE OF THE ERROR

Since the main step in the proof of the prime number theorem is to use the estimate $\beta < 1$ to prove that $\sum x^{\rho-1}/\rho(\rho+1)$ approaches zero as $x \to \infty$, it is natural to try to use the improved estimate $\beta < 1 - c(\log\gamma)^{-1}$ (for $\gamma \geq K$) of the last section to prove an improved estimate of $\sum x^{\rho-1}/\rho(\rho+1)$. De la Vallée Poussin accomplished this by the following very simple argument. Note first that

$$\left|\sum_\rho \frac{x^{\rho-1}}{\rho(\rho+1)}\right| \leq \sum_\rho \frac{x^{\beta-1}}{\gamma^2} = \sum_{|\gamma|<K} \frac{x^{\beta-1}}{\gamma^2} + 2\sum_{\gamma\geq K} \frac{x^{\beta-1}}{\gamma^{1-\delta}} \cdot \frac{1}{\gamma^{1+\delta}}.$$

The first term on the right is the sum of a finite number of terms each of which is a constant times a negative power of x (namely, $x^{\beta-1}$); hence there are positive constants C, ϵ such that this term is less than $Cx^{-\epsilon}$ for all $x > 1$. If δ is any positive constant, then $2\sum \gamma^{-1-\delta}$ converges by the theorem of Section 2.5; so the second term on the right is less than a constant times the maximum value of $x^{\beta-1}/\gamma^{1-\delta}$ if this expression does have a maximum value for $\gamma \geq K$. But

$$\frac{x^{\beta-1}}{\gamma^{1-\delta}} \leq \frac{x^{-c/(\log\gamma)}}{\gamma^{1-\delta}}$$

and the logarithmic derivative of the quantity on the right with respect to γ (considered for the moment as a continuous variable) is

$$\frac{c\log x}{(\log\gamma)^2} \cdot \frac{1}{\gamma} - \frac{1-\delta}{\gamma}$$

which is positive, negative, or zero according to whether $c\log x$ is greater than, less than, or equal to $(1 - \delta)(\log \gamma)^2$. Thus if x is large enough that $c \log x > (1 - \delta)(\log K)^2$ and if $1 - \delta > 0$, then there is a maximum for $\gamma \geq K$ at the point where $c \log x = (1 - \delta)(\log \gamma)^2$. At this point

$$\frac{c \log x}{\log \gamma} = (1 - \delta) \log \gamma = [c(1 - \delta) \log x]^{1/2};$$

hence

$$\frac{x^{\beta-1}}{\gamma^{1-\delta}} \leq \frac{x^{-c/(\log \gamma)}}{\gamma^{1-\delta}} \leq \frac{\exp\{-[c(1 - \delta) \log x]^{1/2}\}}{\exp\{[c(1 - \delta) \log x]^{1/2}\}} = \exp\{-2[c(1 - \delta) \log x]^{1/2}\}.$$

Set $\delta = \frac{3}{4}$ and let C_1 denote $2 \sum \gamma^{-1-\delta}$. Then the above estimates combine to give

$$\left| \sum \frac{x^{\rho-1}}{\rho(\rho + 1)} \right| < Cx^{-\epsilon} + C_1 \exp[-(c \log x)^{1/2}]$$

for all sufficiently large x. Finally, since $x^{-\epsilon}$ goes to zero much faster than $\exp[-(c \log x)^{1/2}]$, since the constants C, C_1 can be absorbed by decreasing c slightly, and since $2 \sum x^{\rho-1}/\rho(\rho + 1)$ is the relative error in the approximation $\int_0^x \psi(t)\,dt \sim x^2/2$ except for terms which are like x^{-1} as $x \to \infty$ [see (4) of Section 4.1] this proves that *there is a constant* $c > 0$ *such that the relative error in the approximation* $\int_0^x \psi(t)\,dt \sim x^2/2$ *is less than* $\exp[-(c \log x)^{1/2}]$ *for all sufficiently large* x.

Now, by essentially the same arguments which were used to deduce the prime number theorem $\pi(x) \sim \mathrm{Li}(x)$ from $\int_0^x \psi(t)\,dt \sim x^2/2$, one can deduce an estimate of the relative error in the prime number theorem from the above estimate of the relative error in $\int_0^x \psi(t)\,dt \sim x^2/2$.

The first step is to derive an estimate of the relative error in $\psi(x) \sim x$. Let $\varepsilon(x) = \exp[-(c \log x)^{1/2}]$, where c is as above so that

$$[1 - \varepsilon(x)]\frac{x^2}{2} \leq \int_0^x \psi(t)\,dt \leq [1 + \varepsilon(x)]\frac{x^2}{2}$$

for all sufficiently large x. Then, as before, $\int_0^y \psi(t)\,dt - \int_0^x \psi(t)\,dt$ for $y > x$ is on the one hand at most

$$[1 + \varepsilon(y)]\frac{y^2}{2} - [1 - \varepsilon(x)]\frac{x^2}{2} \leq [1 + \varepsilon(x)]\left(\frac{y^2 - x^2}{2}\right) + 2\varepsilon(x)\frac{x^2}{2}$$

and at least

$$[1 - \varepsilon(x)]\left(\frac{y^2 - x^2}{2}\right) - 2\varepsilon(x)\frac{x^2}{2}$$

and on the other hand is at most

$$\int_x^y \psi(y)\,dt = (y - x)\psi(y)$$

and at least

$$\int_x^y \psi(x)\,dt = (y - x)\psi(x).$$

Thus

$$(y-x)\psi(x) \le [1+\varepsilon(x)]\left(\frac{y^2-x^2}{2}\right) + 2\varepsilon(x)\frac{x^2}{2},$$

$$(y-x)\psi(y) \ge [1-\varepsilon(x)]\left(\frac{y^2-x^2}{2}\right) - 2\varepsilon(x)\frac{x^2}{2},$$

and, therefore,

$$\psi(x) - x \le \frac{y-x}{2} + \varepsilon(x)\frac{y+x}{2} + \varepsilon(x)\frac{x^2}{y-x},$$

$$\psi(y) - y \ge -\frac{y-x}{2} - \varepsilon(x)\frac{y+x}{2} - \varepsilon(x)\frac{x^2}{y-x}.$$

In the first inequality set $y = \{1 + [\varepsilon(x)]^{1/2}\}x$ to obtain

$$\begin{aligned}\frac{\psi(x)-x}{x} &\le \frac{[\varepsilon(x)]^{1/2}}{2} + \varepsilon(x)\left[1 + \frac{1}{2}\varepsilon(x)\right] + [\varepsilon(x)]^{1/2}\\ &\le \text{const}\,[\varepsilon(x)]^{1/2},\end{aligned}$$

and in the second inequality set $x = \{1 - [\varepsilon(y)]^{1/2}\}y$ to obtain

$$\begin{aligned}\frac{\psi(y)-y}{y} &\ge -\frac{[\varepsilon(y)]^{1/2}}{2} - \varepsilon(x)\left\{1 - \frac{1}{2}[\varepsilon(y)]^{1/2}\right\} - \varepsilon(x)\frac{1}{[\varepsilon(y)]^{1/2}}\\ &\ge -\frac{[\varepsilon(x)]^{1/2}}{2} - \varepsilon(x) - \frac{\varepsilon(x)}{[\varepsilon(y)]^{1/2}} \ge \frac{-2\varepsilon(x)}{[\varepsilon(y)]^{1/2}}\\ &= -2\exp\{-[c\log y + c\log[1-(\varepsilon(y))^{1/2}]]^{1/2} + \tfrac{1}{2}(c\log y)^{1/2}\}\\ &\ge -2\exp\{-(c\log y)^{1/2}[(1-\text{small})^{1/2} - \tfrac{1}{2}]\}.\end{aligned}$$

Thus *there is a constant* $c_0 > 0$ *such that the relative error in the approximation* $\psi(x) \sim x$ *is less than* $\exp[-(c_0 \log x)^{1/2}]$ *for all sufficiently large* x.

The next step is to consider the approximation $\theta(x) \sim x$. But since, as before,

$$\psi(x) - \theta(x^{1/2})(\log x)/\log 2 < \theta(x) < \psi(x)$$

and since $\theta(x^{1/2})\log x \sim x^{1/2}\log x$ grows much more slowly than $x\exp[-(c_0\log x)^{1/2}]$, it follows immediately that the relative error in this approximation is also less than $\exp[-(c_0\log x)^{1/2}]$ for all sufficiently large x.

The final step is to consider the approximation $\pi(x) \sim \mathrm{Li}(x)$. Here the formula

$$\pi(y) - \pi(x) = \int_x^y \frac{d\theta(t)}{\log t} = \frac{\theta(y)}{\log y} - \frac{\theta(x)}{\log x} + \int_x^y \frac{\theta(t)\,dt}{(\log t)^2 t}$$

shows that if $y > x$ are in the range where the relative error in $\theta(x) \sim x$ is less than $\varepsilon(x) = \exp[-(c_0\log x)^{1/2}]$, then $\pi(y)$ is at most

$$\begin{aligned}&\pi(x) + \frac{y[1+\varepsilon(y)]}{\log y} - \frac{x[1-\varepsilon(x)]}{\log x} + \int_x^y \frac{t[1+\varepsilon(t)]}{(\log t)^2 t}\,dt\\ &\quad = \pi(x) + \frac{y\varepsilon(y)}{\log y} + \frac{x\varepsilon(x)}{\log x} + \int_x^y \frac{dt}{\log t} + \int_x^y \frac{\varepsilon(t)\,dt}{(\log t)^2}.\end{aligned}$$

Let x be fixed, let y be larger than x^2, and let the final integral be divided into an integral from x to $y^{1/2}$ and an integral from $y^{1/2}$ to y. Then

$$\begin{aligned}\pi(y) &\le \text{const} + \frac{y\varepsilon(y)}{\log y} + \operatorname{Li}(y) \\ &\quad + (y^{1/2} - x)\frac{\varepsilon(x)}{(\log x)^2} + (y - y^{1/2})\frac{4\varepsilon(y^{1/2})}{(\log y)^2} \\ &\le \text{const} + \operatorname{Li}(y) \\ &\quad + \frac{y}{\log y}\left[\varepsilon(y) + \frac{\text{const}\log y}{y^{1/2}} + \frac{4\varepsilon(y^{1/2})}{\log y}\right]\end{aligned}$$

and similarly

$$\pi(y) \ge \text{const} + \operatorname{Li}(y) - \frac{y}{\log y}\left[\varepsilon(y) + \frac{\text{const}\log y}{y^{1/2}} + \frac{4\varepsilon(y^{1/2})}{\log y}\right].$$

Since the quantity in square brackets is less than $\varepsilon(y^{1/2})$ for all sufficiently large y, it will suffice to prove that $y/\log y$ divided by $\operatorname{Li}(y)$ is bounded as $y \to \infty$ in order to prove that the relative error in $\pi(y) \sim \operatorname{Li}(y)$ is less than $\varepsilon(y^{1/2})$ for y sufficiently large. It will be shown in the next section that $\operatorname{Li}(y) \sim y/\log y$, and this will complete the proof that *there is a constant* $c_1 > 0$ *such that the relative error in the approximation* $\pi(y) \sim \operatorname{Li}(y)$ *is less than* $\exp[-(c_1 \log y)^{1/2}]$ *for all sufficiently large values of* y. This is de la Vallée Poussin's estimate† of the error in the prime number theorem.

5.4 OTHER FORMULAS FOR $\pi(x)$

The approximate formula for $\pi(x)$ which appears in Legendre's *Theorie des Nombres* [L4] is

$$\pi(x) \sim \frac{x}{\log x - A}, \tag{1}$$

where A is a constant whose value Legendre gives as 1.08366, apparently on empirical grounds. Legendre does not specify the sense in which the approximation (1) is to be understood, but if it is interpreted in the usual sense of "the relative error approaches zero as $x \to \infty$," then the value of A is irrelevant because

$$\frac{x}{\log x - A} \sim \frac{x}{\log x - B}$$

†De la Vallée Poussin wrote the estimate in the form $(c_1 \log y)^{1/2} \exp[-(c_1 \log y)^{1/2}]$ and gave the explicit value 0.032 of c_1. He did not, however, give any explicit estimate of how large y must be in order for this estimate of the relative error in $\pi(y) \sim \operatorname{Li}(y)$ to be valid.

for any two numbers A, B (because the ratio is $\log x - A$ over $\log x - B$, which approaches 1) and hence if (1) is true for one value of A, it must be true for all values of A (because "$\sim$" is transitive). Therefore if Legendre's value $A = 1.08366$ has any significance, it must lie in some other interpretation of the approximation (1).

The prime number theorem $\pi(x) \sim \mathrm{Li}(x)$ shows that (1) is true if and only if $\mathrm{Li}(x) \sim x/(\log x - A)$ for some—and hence all—values of A. But by integration by parts

$$\begin{aligned}\mathrm{Li}(x) &= \mathrm{Li}(2) + \int_2^x \frac{dt}{\log t}\\ &= \mathrm{Li}(2) + \frac{x}{\log x} - \frac{2}{\log 2} + \int_2^x \frac{dt}{(\log t)^2},\end{aligned}$$

$$\mathrm{Li}(x) - \frac{x}{\log x} = \text{const} + \int_2^x \frac{dt}{(\log t)^2},$$

and it suffices to show that the integral on the right divided by $x/\log x$ approaches zero as $x \to \infty$ in order to conclude that (1) is true with $A = 0$. [Intuitively this is the obvious statement that the average value of $(\log t)^{-2}$ for $2 \le t \le x$ is much less than $(\log x)^{-1}$.] This is easily accomplished by dividing the interval of integration at $x^{1/2}$ to find

$$\begin{aligned}\frac{\log x}{x}\int_2^x \frac{dt}{(\log t)^2} &= \frac{\log x}{x}\int_2^{x^{1/2}} \frac{dt}{(\log t)^2} + \frac{\log x}{x}\int_{x^{1/2}}^x \frac{dt}{(\log t)^2}\\ &\le \frac{\log x}{x}\cdot\frac{x^{1/2}-2}{(\log 2)^2} + \frac{\log x}{x}\cdot\frac{x - x^{1/2}}{(\log x^{1/2})^2}\\ &\le \frac{\log x}{(\log 2)^2 x^{1/2}} + \frac{4}{\log x} \to 0.\end{aligned}$$

Thus the prime number theorem implies (1) but it implies no particular value of A.

Chebyshev [C2] was able to show that if any value of A is any better than any other, then this value must be $A = 1$. The special property of the value $A = 1$ which is needed is the fact that the approximation

$$\mathrm{Li}(x) \sim \frac{x}{\log x - A} \tag{2}$$

is best when $A = 1$. To see this, note that

$$\frac{x}{\log x - A} = \frac{x}{\log x}\left[1 + \left(\frac{A}{\log x}\right) + \left(\frac{A}{\log x}\right)^2 + \cdots\right],$$

$$\frac{x}{\log x - A} \sim \frac{x}{\log x} + \frac{Ax}{(\log x)^2}, \tag{3}$$

where $\sim$ means that the error is much smaller than the last term $Ax(\log x)^{-2}$

in the sense that their ratio approaches zero as $x \to \infty$, while

$$\begin{aligned}\mathrm{Li}(x) &= \frac{x}{\log x} + \text{const} + \int_2^x \frac{dt}{(\log t)^2} \\ &= \frac{x}{\log x} + \text{const} + \frac{x}{(\log x)^2} - \frac{2}{(\log 2)^2} + 2\int_2^x \frac{dt}{(\log t)^3}.\end{aligned}$$

Hence

$$\mathrm{Li}(x) \sim \frac{x}{\log x} + \frac{x}{(\log x)^2}, \tag{4}$$

where the error is much smaller than the last term because

$$\frac{(\log x)^2}{x}\int_2^x \frac{dt}{(\log t)^3} \le \frac{(\log x)^2}{x}\frac{x^{1/2}-2}{(\log 2)^3} + \frac{(\log x)^2}{x}\frac{x-x^{1/2}}{(\log x^{1/2})^3} \to 0.$$

The combination of (3) and (4) shows that the error in (2) divided by $x(\log x)^{-2}$ is $A - 1$ plus a quantity which approaches zero as $x \to \infty$. This gives a precise sense in which $A = 1$ is the best value of A in (2). Chebyshev was able to prove enough about the approximation $\pi(x) \sim \mathrm{Li}(x)$ to prove that if A has a best value in (1), then that value must also be $A = 1$.

However, even the prime number theorem is not enough to prove that $A = 1$ is best in (1). To prove this, one would have to show that the approximation $\pi(x) \sim \mathrm{Li}(x)$, like the approximations (3) and (4), has the property that its error grows much less rapidly than $x(\log x)^{-2}$, and the prime number theorem says only that it grows much less rapidly than $\mathrm{Li}(x)$, which in turn grows like $x(\log x)^{-1}$ by (2). Thus a stronger estimate of the error in $\pi(x) \sim \mathrm{Li}(x)$ is needed.

Now since $\exp[(c \log x)^{1/2}]$ grows more rapidly than any power of $(c \log x)^{1/2}$, it follows from de la Vallée Poussin's estimate of the error in $\pi(x) \sim \mathrm{Li}(x)$ that

$$\left|\frac{\pi(x) - \mathrm{Li}(x)}{x(\log x)^{-2}}\right| \le \frac{\mathrm{Li}(x)}{x} \cdot \frac{(\log x)^2}{\exp[(c \log x)^{1/2}]} \to 0$$

and hence that the error in (1) divided by $x(\log x)^{-2}$ is $A - 1$ plus a quantity which approaches zero as $x \to \infty$. Thus de la Vallée Poussin's estimate proves that the value $A = 1$ in (1) is better than any other value.

More generally, successive integration by parts shows that (4) can be generalized to

$$\begin{aligned}\mathrm{Li}(x) \sim \frac{x}{\log x} &+ \frac{x}{(\log x)^2} + 2\frac{x}{(\log x)^3} \\ &+ 6\frac{x}{(\log x)^4} + \cdots + (n-1)!\,\frac{x}{(\log x)^n},\end{aligned}$$

where the error (for any fixed n) grows much less rapidly than the last term $x(\log x)^{-n}$ as $x \to \infty$. De la Vallée Poussin's estimate shows that the error in

$\pi(x) \sim \mathrm{Li}(x)$ also grows less rapidly than $x(\log x)^{-n}$ and hence proves that the approximation

$$\pi(x) \sim \frac{x}{\log x} + \frac{x}{(\log x)^2} + \cdots + (n-1)!\frac{x}{(\log x)^n} \tag{5}$$

is valid in the sense that (for any fixed n) the error in the approximation divided by the last term on the right approaches zero as $x \longrightarrow \infty$. The case $n = 1$ is essentially the prime number theorem and the case $n = 2$ is essentially the theorem that $A = 1$ is best in (1).

Thus de la Vallée Poussin's estimate of the error proves that the approximation $\pi(x) \sim \mathrm{Li}(x)$ is better than the approximation (5) for any value of n. This is the principal consequence which de la Vallée Poussin himself drew from his estimate of the error.

Formula (5) is an example of an *asymptotic expansion*, which is an expansion such as $\pi(x) = \sum (n-1)!\, x(\log x)^{-n}$ in which the error resulting from taking a finite number of terms is of a lower order of magnitude than the last term used. Another more familiar example of an asymptotic expansion is the Taylor series of an infinitely differentiable function $f(x) = \sum f^{(n)}(a)(x-a)^n/n!$. The fact that this is an asymptotic expansion—that is, the fact that the error resulting from using just n terms decreases more rapidly than $(x-a)^n$ as $x \longrightarrow a$—is Taylor's theorem. This in no way implies, of course, that for fixed $x \neq a$ the error approaches zero as $n \longrightarrow \infty$. For any fixed x formula (5) for $\pi(x)$ becomes worthless as $n \longrightarrow \infty$ because $(n-1)!$ grows much faster than $(\log x)^n$. Another example of an asymptotic expansion is Stirling's series (3) of Section 6.3. Although pure mathematicians shun asymptotic expansions which do not converge as $n \longrightarrow \infty$, mathematicians who engage in computation are well aware that asymptotic expansions (for example, Stirling's series) are often more practical than convergent expansions [for example, the product formula (3) of Section 1.3 for Π].

Recall that Riemann's approximate formula for $\pi(x)$ was

$$\pi(x) \sim \mathrm{Li}(x) + \sum_{n=2}^{N} \frac{\mu(n)}{n} \mathrm{Li}(x^{1/n}), \tag{6}$$

where $N > \log x/\log 2$, and that on empirical grounds this formula appeared to be much better than $\pi(x) \sim \mathrm{Li}(x)$. The second term in this formula is $-\frac{1}{2}\mathrm{Li}(x^{1/2}) \sim -\frac{1}{2}x^{1/2}(\log x^{1/2})^{-1} = -x^{1/2}(\log x)^{-1}$, whereas de la Vallée Poussin's estimate shows that the error in $\pi(x) \sim \mathrm{Li}(x)$ is less than $\mathrm{Li}(x) \cdot \exp[-(c \log x)^{1/2}] \sim x(\log x)^{-1} \exp[-(c \log x)]^{1/2}$ so that $-\frac{1}{2}\mathrm{Li}(x^{1/2})$ divided by the error estimate is about $-x^{-1/2} \exp[+(c \log x)^{1/2}] = -\exp[-\frac{1}{2} \log x + (c \log x)^{1/2}] \longrightarrow 0$. Thus de la Vallée Poussin's estimate is not strong enough to prove that even the second term of (6) has any validity, much less the remaining terms. It was, in fact, proved by Littlewood [L13] that Riemann's formula (6) is *not* better as $x \longrightarrow \infty$ than the simpler formula $\pi(x) \sim \mathrm{Li}(x)$. In other words, Littlewood proved that in the long run as $x \longrightarrow \infty$ the "periodic" terms $\mathrm{Li}(x^\rho)$ in the formula for $\pi(x)$ (see Section 1.17) are as significant as the monotone increasing term $-\frac{1}{2}\mathrm{Li}(x^{1/2})$ and *a fortiori* as significant as the following terms $\mathrm{Li}(x^{1/3})$, $\mathrm{Li}(x^{1/4})$, . . . of (6).

5.5 ERROR ESTIMATES AND THE RIEMANN HYPOTHESIS

In view of the strong relationship between de la Vallée Poussin's estimate of the error in the prime number theorem and his estimate $\beta < 1 - c(\log\gamma)^{-1}$ of $\beta = \mathrm{Re}\,\rho$, it is not surprising that the Riemann hypothesis $\mathrm{Re}\,\rho = \frac{1}{2}$ should imply much stronger estimates of the error. The best such estimate that has been found up to now is the estimate proved by von Koch [K1] in 1901, namely, that if the Riemann hypothesis is true then the relative errors in $\psi(x) \sim x$ and $\pi(x) \sim \mathrm{Li}(x)$ are both less than a constant times $(\log x)^2 x^{-1/2}$ for all sufficiently large x. This estimate implies that the relative errors are eventually less than $x^{-(1/2)+\epsilon}$ for all $\epsilon > 0$, whereas de la Vallée Poussin's estimate $\exp[-(c\log x)^{1/2}]$ fails to show that they are less than $x^{-\epsilon}$ for any $\epsilon > 0$.

If the Riemann hypothesis is true, then the magnitude of the relative error in the approximation $\int_0^x \psi(t)\,dt \sim \frac{1}{2}x^2$ is less than

$$\begin{aligned} 2\left|\sum_\rho \frac{x^{\rho-1}}{\rho(\rho+1)}\right| &+ \text{const}\, x^{-2} + x^{-1}\zeta'(0)/\zeta(0) \\ &\leq 2\cdot x^{-1/2}\cdot\sum_\rho\left|\frac{1}{\rho(\rho+1)}\right| + \text{const}\, x^{-1} \\ &\leq \text{const}\, x^{-1/2} \end{aligned}$$

for all sufficiently large x. However, the previous method of deducing from this an estimate of the relative error in $\psi(x) \sim x$ involves taking a square root and hence yields only the estimate $x^{-1/4}$ and not von Koch's estimate $x^{-1/2}(\log x)^2$. Therefore some other method of estimating the error in $\psi(x) \sim x$ is necessary. The following method is due to Holmgren [H10]. In the estimate (see Section 4.3)

$$\begin{aligned} \psi(x) &\leq \int_x^{x+1} \psi(t)\,dt = \int_0^{x+1} \psi(t)\,dt - \int_0^x \psi(t)\,dt \\ &= \left|\frac{(x+1)^2}{2} - \frac{x^2}{2} + \sum_\rho \frac{(x+1)^{\rho+1} - x^{\rho+1}}{\rho(\rho+1)} + \text{bounded}\right| \\ &\leq x + \sum_\rho \left|\frac{(x+1)^{\rho+1} - x^{\rho+1}}{\rho(\rho+1)}\right| + \text{const} \end{aligned}$$

let the terms corresponding to roots $\rho = \frac{1}{2} + i\gamma$ (assuming the Riemann hypothesis) for which $|\gamma| \leq x$ be estimated using

$$\left|\frac{(x+1)^{\rho+1} - x^{\rho+1}}{\rho(\rho+1)}\right| = \frac{1}{|\rho|}\left|\int_x^{x+1} t^\rho\, dt\right| \leq \frac{|x+1|^{1/2}}{|\rho|} \leq \frac{(2x)^{1/2}}{|\gamma|}$$

and let those corresponding to roots for which $|\gamma| > x$ be estimated using

$$\left|\frac{(x+1)^{\rho+1} - x^{\rho+1}}{\rho(\rho+1)}\right| \leq \frac{2|x+1|^{\operatorname{Re}\rho+1}}{|\rho(\rho+1)|} \leq \frac{2(2x)^{3/2}}{\gamma^2}.$$

Let H be as in the theorem of Section 3.4 so that the number of roots $\rho = \frac{1}{2} + i\gamma$ in the interval $t \leq \gamma \leq t+1$ is less than $2 \log t$ for $t \geq H$. Then the above estimates give (when $x > H$)

$$\begin{aligned}
\psi(x) &\leq x + x^{1/2} \sum_{|\gamma|<H} \frac{\sqrt{2}}{|\gamma|} + 2x^{1/2} \sum_{H<\gamma<x} \frac{\sqrt{2}}{\gamma} + 2x^{3/2} \sum_{x<\gamma} \frac{2^{5/2}}{\gamma^2} \\
&\leq x + \text{const}\, x^{1/2} + \text{const}\, x^{1/2} \int_H^x \frac{\log t\, dt}{t} + \text{const}\, x^{3/2} \int_x^\infty \frac{\log t\, dt}{t^2} \\
&\leq x + \text{const}\, x^{1/2} + \text{const}\, x^{1/2} \frac{(\log t)^2}{2}\bigg|_H^x \\
&\quad + \text{const}\, x^{3/2}\left\{-\frac{\log t}{t}\bigg|_x^\infty + \int_x^\infty \frac{dt}{t^2}\right\} \\
&\leq x + \text{const}\, x^{1/2} + \text{const}\, x^{1/2}(\log x)^2 + \text{const}\, x^{3/2}\left\{\frac{\log x}{x} + \frac{1}{x}\right\} \\
&\leq x + \text{const}\, x^{1/2}(\log x)^2.
\end{aligned}$$

The same technique applied to the estimate

$$\psi(x) \geq \int_{x-1}^x \psi(t)\, dt = \left|\frac{x^2}{2} - \frac{(x-1)^2}{2} + \sum_\rho \frac{x^{\rho+1} - (x-1)^{\rho+1}}{\rho(\rho+1)} + \text{bounded}\right|$$

gives

$$\begin{aligned}
\psi(x) &\geq x - \sum_{|\gamma|<H} \frac{x^{1/2}}{|\gamma|} - 2 \sum_{H<\gamma<x} \frac{x^{1/2}}{\gamma} - 2 \sum_{x>\gamma} \frac{2x^{3/2}}{\gamma^2} \\
&\geq x - \text{const}\, x^{1/2} - \text{const}\, x^{1/2} \int_H^x \frac{\log t\, dt}{t} - \text{const}\, x^{3/2} \int_x^\infty \frac{\log t\, dt}{t^2} \\
&\geq x - \text{const}\, x^{1/2}(\log x)^2
\end{aligned}$$

which completes the proof that the relative error in $\psi(x) \sim x$ is less than a constant times $(\log x)^2 x^{-1/2}$ (assuming the Riemann hypothesis). Since

$$\psi(x) - \theta(x^{1/2})\log x/\log 2 \leq \theta(x) \leq \psi(x)$$

and since $\theta(x^{1/2}) \log x \sim x^{1/2} \log x$ is much smaller than $x^{1/2}(\log x)^2$, the same is true of $\theta(x) \sim x$. Now

$$\begin{aligned}
\pi(x) - \operatorname{Li}(x) &= \int_c^x \frac{d[\theta(t) - t]}{\log t} + \text{const} \\
&= \frac{\theta(x) - x}{\log x} + \int_c^x \frac{\theta(t) - t}{t(\log t)^2}\, dt + \text{const}
\end{aligned}$$

for any constant $c > 1$; hence if $|\theta(t) - t| < Kt^{1/2}(\log t)^2$ for $t \geq c$, then $x > c$ implies

$$|\pi(x) - \mathrm{Li}(x)| \leq \frac{Kx^{1/2}(\log x)^2}{\log x} + \int_c^x \frac{K\,dt}{t^{1/2}} + \text{const}$$
$$\leq K\frac{x}{\log x}\frac{(\log x)^2}{x^{1/2}} + Kx^{1/2} + \text{const},$$

which proves, since $\mathrm{Li}(x) \sim (x/\log x)$ as $x \to \infty$, that the relative error in $\pi(x) - \mathrm{Li}(x)$ is eventually less than a constant times $(\log x)^2 x^{-1/2}$ if the Riemann hypothesis is true, as was to be shown.

On the other hand, if the Riemann hypothesis is false, then there is a root ρ with $\mathrm{Re}\,\rho > \frac{1}{2}$ and hence (see below) a "periodic" term in Riemann's formula for $\pi(x)$ which grows more rapidly in magnitude than $x^{1/2}$, so it is reasonable to assume that the error in the prime number theorem $\pi(x) \sim \mathrm{Li}(x)$ would not in that case grow less rapidly than $x^{(1/2)+\epsilon}$.

The rate of growth of $\mathrm{Li}(x^\rho)$ for ρ in the first quadrant $\mathrm{Re}\,\rho > 0$, $\mathrm{Im}\,\rho > 0$ is easily estimated using the formula $\mathrm{Li}(x^\rho) = \int_{C^+} (\log t)^{-1} t^{\rho-1}\,dt + i\pi$ of Section 1.15 and integration by parts as in Section 5.4 to find

$$\mathrm{Li}(x^\rho) = \int_{C^+} \frac{t^{\rho-1}\,dt}{\log t} + i\pi$$
$$= \int_{C^+} \frac{d}{dt}\left\{\frac{t^\rho}{\rho \log t}\right\} dt - \int_{C^+} \frac{t^\rho}{\rho}\frac{(-1)}{(\log t)^2}\frac{dt}{t} + i\pi$$
$$= \frac{x^\rho}{\rho \log x} + \int_{C^+} \frac{t^{\rho-1}\,dt}{\rho(\log t)^2} + i\pi$$
$$= \frac{x^\rho}{\rho \log x} + \int_2^x \frac{t^{\rho-1}\,dt}{\rho(\log t)^2} + \text{const.}$$

The first term, which has modulus $x^{\mathrm{Re}\,\rho}|\rho|^{-1}(\log x)^{-1}$, dominates as $x \to \infty$ because

$$\int_2^{x^{1/2}} \frac{t^{\rho-1}\,dt}{(\log t)^2} + \int_{x^{1/2}}^x \frac{t^{\rho-1}\,dt}{(\log t)^2}$$
$$\leq \int_2^{x^{1/2}} \frac{t^{\mathrm{Re}\,\rho-1}\,dt}{(\log 2)^2} + \int_{x^{1/2}}^x \frac{t^{\mathrm{Re}\,\rho-1}\,dt}{(\log x^{1/2})^2}$$
$$\leq \frac{x^{(1/2)\,\mathrm{Re}\,\rho}}{\mathrm{Re}\,\rho(\log 2)^2} + \frac{4x^{\mathrm{Re}\,\rho}}{\mathrm{Re}\,\rho(\log x)^2}$$

has modulus much less than a constant times $x^{\mathrm{Re}\,\rho}(\log x)^{-1}$.

Theorem The Riemann hypothesis is equivalent to the statement that for every $\epsilon > 0$ the relative error in the prime number theorem $\pi(x) \sim \mathrm{Li}(x)$ is less than $x^{-(1/2)+\epsilon}$ for all sufficiently large x. [If they are true, then the relative error in the prime number theorem is in fact less than a constant times $x^{-1/2}(\log x)^2$.]

Proof It remains only to show that if the relative error in $\pi(x) \sim \mathrm{Li}(x)$ is less than $x^{-(1/2)+\epsilon}$, then the Riemann hypothesis must be true. Assume therefore that for every $\epsilon > 0$ the relative error in $\pi(x) \sim \mathrm{Li}(x)$ is eventually less than $x^{-(1/2)+\epsilon}$. Since $\text{Li}(x) \sim (x/\log x)$, this implies that the absolute error is eventually less than $x^{(1/2)+\epsilon}$, say for $x \geq c$. Then since

$$\begin{aligned}\theta(x) - x &= \int_c^x \log t \, d[\pi(t) - \text{Li}(t)] + \text{const} \\ &= (\log x)[\pi(x) - \text{Li}(x)] - \int_c^x \frac{\pi(t) - \text{Li}(t)}{t} \, dt + \text{const},\end{aligned}$$

it follows that

$$|\theta(x) - x| \leq (\log x)x^{(1/2)+\epsilon} + \int_c^x t^{-(1/2)+\epsilon} \, dt + \text{const} \leq x^{(1/2)+2\epsilon}$$

for all sufficiently large x. Since

$$\theta(x) \leq \psi(x) \leq \theta(x) + \frac{\log x}{\log 2}\theta(x^{1/2}) \sim \theta(x) + \text{const } x^{1/2} \log x,$$

this implies immediately that $|\psi(x) - x| \leq x^{(1/2)+2\epsilon}$ for all sufficiently large x. But the formulas

$$-\frac{\zeta'(s)}{\zeta(s)} = \int_0^\infty x^{-s} \, d\psi(x) = s \int_1^\infty x^{-s-1}\psi(x) \, dx,$$

$$-\frac{1}{(1-s)} = \int_1^\infty x^{-s} \, dx = -1 + s \int_1^\infty x^{-s-1} \cdot x \, dx,$$

which hold for $\text{Re } s > 1$, combine to give

$$-\frac{d}{ds} \log[(s-1)\zeta(s)] = 1 + s \int_1^\infty x^{-s-1}[\psi(x) - x] \, dx$$

for $\text{Re } s > 1$. If $|\psi(x) - x| < x^{(1/2)+2\epsilon}$ for all large x, then the integral on the right converges throughout the halfplane $\text{Re } s > \frac{1}{2} + 2\epsilon$. By analytic continuation the right side (which is analytic by differentiation under the integral sign) must be equal to the left side throughout the halfplane $\text{Re } s > \frac{1}{2} + 2\epsilon$, which shows that $\zeta(s)$ could have no zeros in this halfplane. Thus if the relative error in $\pi(x) \sim \text{Li}(x)$ is less than $x^{-(1/2)+\epsilon}$ for all $\epsilon > 0$, the Riemann hypothesis must be true.

5.6 A POSTSCRIPT TO DE LA VALLÉE POUSSIN'S PROOF

The Euler product formula for $\zeta(s)$ implies that for $\text{Re } s > 1$

(1) $$\begin{aligned}\frac{1}{\zeta(s)} &= \prod_p \left(1 - \frac{1}{p^s}\right) = 1 - \frac{1}{2^s} - \frac{1}{3^s} - \frac{1}{5^s} + \frac{1}{6^s} - \cdots \\ &= \sum_{n=1}^\infty \frac{\mu(n)}{n^s},\end{aligned}$$

where, as in the Möbius inversion formula, $\mu(n)$ is 0 if n is divisible by a square, -1 if n is a product of an odd number of distinct prime factors, and

$+1$ if n is a product of an even number of distinct prime factors [$\mu(1) = 1$]. Since $\zeta(s)$ has a pole at $s = 1$, $[\zeta(s)]^{-1}$ has a zero at $s = 1$; so if (1) were valid for $s = 1$, it would say

$$(2)\quad 0 = 1 - \tfrac{1}{2} - \tfrac{1}{3} - \tfrac{1}{5} + \tfrac{1}{6} - \tfrac{1}{7} + \tfrac{1}{10} - \tfrac{1}{11} - \tfrac{1}{13} + \tfrac{1}{14} + \tfrac{1}{15} - \cdots.$$

This equation was stated by Euler [E5], but Euler quite often dealt with divergent series and his statement of this equation should not necessarily be understood to imply *convergence* of the series on the right but only *summability*. At any rate Euler did not give any proof that the series (2) was convergent, and the first proof of this fact was given by von Mangoldt [M2] in 1897. Von Mangoldt's proof is rather difficult and a much simpler proof of the formula (2), together with an estimate of the *rate* of convergence, was given by de la Vallée Poussin in 1899 in connection with his proof of the improved error estimate $\exp[-(c \log x)^{1/2}]$ in the prime number theorem.

Specifically, de la Vallée Poussin proved that there is a constant K such that

$$(3)\qquad \left| \sum_{n<x} \frac{\mu(n)}{n} \right| < \frac{K}{\log x}$$

for all sufficiently large x. As $x \to \infty$ this of course implies (2). De la Vallée Poussin's proof is based on two observations concerning the function $P(x)$ defined by

$$P(x) = \int_0^x t^{-1}\, d\psi(t) = \sum_{n<x} \frac{\Lambda(n)}{n},$$

namely, the observation that $P(x)$ is related to the series (2) by the formula

$$(4)\qquad \sum_{n<x} \frac{\mu(n)}{n}\left[P\left(\frac{x}{n}\right) + \log n\right] \equiv 0$$

and the observation that an estimate of $P(x)$ can be derived from the estimate of $\psi(x)$ obtained in Section 5.3.

Consider first the proof of (4). The essence of this identity is the identity

$$\frac{d}{ds}\left[\frac{1}{\zeta(s)}\right] = -\left[\frac{\zeta'(s)}{\zeta(s)}\right]\left[\frac{1}{\zeta(s)}\right]$$

which, since $-\zeta'(s)/\zeta(s) = \sum \Lambda(n)n^{-s}$ and $[\zeta(s)]^{-1} = \sum \mu(n)n^{-s}$ for $\operatorname{Re} s > 1$, gives

$$(5)\qquad \sum_{n=1}^{\infty} \frac{-\mu(n)\log n}{n^s} = \left[\sum_{j=1}^{\infty} \frac{\Lambda(j)}{j^s}\right]\left[\sum_{k=1}^{\infty} \frac{\mu(k)}{k^s}\right]$$
$$= \sum_{n=1}^{\infty} \frac{1}{n^s}\left[\sum_{jk=n} \Lambda(j)\mu(k)\right]$$

for $\operatorname{Re} s > 1$. It is natural to suppose that since these series have the same sum for all s they must be identical, and hence for all x

$$\sum_{n<x} \frac{-\mu(n) \log n}{n^s} \equiv \sum_{n<x} \frac{1}{n^s} \left[\sum_{jk=n} \Lambda(j)\mu(k) \right] \tag{6}$$

which for $s = 1$ gives

$$-\sum_{n<x} \frac{\mu(n) \log n}{n} = \sum_{n<x} \sum_{jk=n} \frac{\Lambda(j)}{j} \cdot \frac{\mu(k)}{k}. \tag{7}$$

Now for every fixed $k \leq n < x$ the sum on the right is over all integers j such that $jk < x$, that is, it is over all integers $j < x/k$, hence

$$-\sum_{n<x} \frac{\mu(n) \log n}{n} = \sum_{k<x} \frac{\mu(k)}{k} P\left(\frac{x}{k}\right)$$

from which (4) follows. The one step of the argument which requires further justification is the truncation (6) of the series (5). This is essentially a question of recovering the coefficients of a series $\sum A_n n^{-s}$ from a knowledge of its sum, which can be done by a technique similar to that used in Section 3.2 to recover $\Lambda(n)$ from $-\zeta'(s)/\zeta(s) = \sum \Lambda(n)n^{-s}$. Specifically, one can go directly from (5) to the desired equation (7) by using the identity

$$\begin{aligned} \frac{1}{2\pi i} \int_{a-i\infty}^{a+i\infty} \left[\sum_{n=1}^{\infty} \frac{A_n}{n^s} \right] \frac{x^s\, ds}{s-1} &= x \cdot \frac{1}{2\pi i} \int_{a-i\infty}^{a+i\infty} \left[\sum_{n=1}^{\infty} \frac{A_n}{n} \left(\frac{x}{n}\right)^{s-1} \right] \frac{ds}{s-1} \\ &= x \cdot \frac{1}{2\pi i} \int_{a-1-i\infty}^{a-1+i\infty} \left[\sum_{n=1}^{\infty} \frac{A_n}{n} \left(\frac{x}{n}\right)^{t} \right] \frac{dt}{t} \\ &= x \cdot \sum_{n<x} \frac{A_n}{n}, \end{aligned}$$

where the termwise integration in the last step is justified, by the same argument as before, whenever the A_n are such that $\sum (|A_n|/n)n^{-(a-1)} < \infty$, which is certainly the case for both series in (5) for large a. This completes the proof of (4).

Consider now the estimation of $P(x)$. The desired estimate is easily found heuristically by using

$$d\psi(x) = (1 - \sum x^{\rho-1} - \sum x^{-2n-1})\, dx$$

to find

$$dP(x) = x^{-1}\, d\psi(x) = (x^{-1} - \sum x^{\rho-2} - \sum x^{-2n-2})\, dx,$$

$$P(x) = \log x - \sum \frac{x^{\rho-1}}{\rho-1} + \sum \frac{x^{-2n-1}}{2n+1} + \text{const},$$

which indicates that $P(x) \sim \log x + \text{const}$ with an error which goes to zero faster than

$$\begin{aligned} \left| \sum \frac{x^{\rho-1}}{\rho-1} \right| &\leq \left| \sum \frac{x^{\rho-1}}{\rho} \right| + \left| \sum \frac{x^{\rho-1}}{\rho(\rho-1)} \right| \\ &\leq \exp[-(c \log x)^{1/2}] + \exp[-(c \log x)^{1/2}]. \end{aligned}$$

In order to prove this it suffices to write

$$\begin{aligned}
P(x) &= \int_0^x t^{-1}\,d\psi(t) = t^{-1}\psi(t)\Big|_0^x + \int_0^x t^{-2}\psi(t)\,dt \\
&= \frac{\psi(x)}{x} + \int_1^x t^{-2}\psi(t)\,dt \\
&= \frac{\psi(x)-x}{x} + \int_1^x t^{-1}\,dt + \int_1^x t^{-1}\frac{\psi(t)-t}{t}\,dt + 1 \\
&= \log x + \frac{\psi(x)-x}{x} + \int_1^x t^{-1}\frac{\psi(t)-t}{t}\,dt + 1.
\end{aligned}$$

The integral in the last expression converges as $x \longrightarrow \infty$ because if x increases to x', it increases for large x by less than

$$\begin{aligned}
\int_x^{x'} t^{-1}\exp[-(c\log t)^{1/2}]\,dt &= \frac{1}{c}\int_{c\log x}^{c\log x'} \exp(-u^{1/2})\,du \\
&= \frac{2}{c}\int_{(c\log x)^{1/2}}^{(c\log x')^{1/2}} e^{-v}v\,dv = \frac{2}{c}(-e^{-v}v - e^{-v})\Big|_{(c\log x)^{1/2}}^{(c\log x')^{1/2}} \\
&\le \frac{2}{c}(c\log x)^{1/2}\exp[-(c\log x)^{1/2}] + \frac{2}{c}\exp[-(c\log x)^{1/2}].
\end{aligned}$$

Thus the above formula can be rewritten

$$P(x) = \log x + 1 + \int_1^\infty \frac{\psi(t)-t}{t^2}\,dt + \frac{\psi(x)-x}{x} - \int_x^\infty \frac{\psi(t)-t}{t^2}\,dt,$$

$$P(x) = \log x + C + \eta(x), \tag{8}$$

where C is the constant† $1 + \int_1^\infty t^{-2}[\psi(t) - t]\,dt$ and where the error $\eta(x)$ is less than

$$\left(1 + \frac{2}{c}\right)\exp[-(c\log x)^{1/2}] + \frac{2}{c}(c\log x)^{1/2}\exp[-(c\log x)^{1/2}]$$

for all x large enough that $[\psi(t) - t]/t \le \exp[-(c\log t)^{1/2}]$ for $t \ge x$. Thus it is possible to choose $K' > 0$, $c > 0$ such that the error $\eta(x)$ is less than

†The constant C is in fact the negative of Euler's constant (see Section 3.8). To derive this fact, note first that C is $1 + \lim_{s\to 1}\int_1^\infty t^{-1-s}[\psi(t) - t]\,dt$ and then integrate by parts to find that it is the limit as $s \longrightarrow 1$ of $-s^{-1}\{[\zeta'(s)/\zeta(s)] + 1 - s + (s-1)^{-1}\}$ which is the limit as $s \longrightarrow 1$ of $-(d/ds)\log[(s-1)\zeta(s)]$. But the functional equation gives $(s-1)\zeta(s) = -\Pi(1-s)(2\pi)^{s-1}\zeta(1-s)2\sin(\pi s/2)$. Using the fact that the logarithmic derivative of ζ at 0 is $\log 2\pi$ while that of Π is $-\gamma$ (see Section 3.8), the result $C = -\gamma$ then follows. Since the left side of formula (2) of Section 4.1 is $P(x) - x^{-1}\psi(x) = \log x + C - 1 + \cdots$, this implies that the constant on the right is $-\gamma - 1$. The same fact can also be derived by setting $v = 0$ in the formula of the note of Section 4.3 and letting $u \longrightarrow 1$. Yet another proof that $C = -\gamma$ is given in Section 12.10.

$K' \exp[-(c \log x)^{1/2}]$ for all $x \geq 0$. Then (4) and (8) combine to give

$$\sum_{n<x} \frac{\mu(n)}{n}\left[\log\left(\frac{x}{n}\right) + C - \eta\left(\frac{x}{n}\right) + \log n\right] \equiv 0,$$

$$(\log x + C) \sum_{n<x} \frac{\mu(n)}{n} \equiv \sum_{n<x} \frac{\mu(n)}{n} \eta\left(\frac{x}{n}\right),$$

$$\left|\sum_{n<x} \frac{\mu(n)}{n}\right| \leq \frac{\sum_{n<x} (1/n) K' \exp[-(c \log (x/n))^{1/2}]}{\log x + C},$$

and the desired inequality (3) will follow if it can be shown that

$$\sum_{n<x} \frac{1}{n} \exp\left[-\left(c \log \frac{x}{n}\right)^{1/2}\right]$$

remains bounded as $x \to \infty$. Now the logarithmic derivative of the summand with respect to n (considered for the moment as a continuous variable) is

$$-\frac{1}{n} - \frac{1}{2}\left(c \log \frac{x}{n}\right)^{-1/2} \frac{cn}{x} \cdot \left(-\frac{x}{n^2}\right) = \frac{1}{n}\left[-1 + \frac{c}{2}\left(c \log \frac{x}{n}\right)^{-1/2}\right]$$

which is negative until

$$c/2 = [c \log (x/n)]^{1/2}, \qquad \log (x/n) = c/4, \qquad n = xe^{-c/4},$$

after which it is positive. Let N be the last integer before the sign change. Then

$$\begin{aligned}
&\sum_{n<x} \frac{1}{n} \exp\left[-\left(c \log \frac{x}{n}\right)^{1/2}\right] \\
&\quad < \exp[-(c \log x)^{1/2}] + \sum_{n=2}^{N} \frac{1}{n} \exp\left[-\left(c \log \frac{x}{n}\right)^{1/2}\right] \\
&\qquad + \sum_{N+1 \leq n < x-1} \frac{1}{n} \exp\left[-\left(c \log \frac{x}{n}\right)^{1/2}\right] \\
&\qquad + \frac{1}{x} \exp\left[-\left(c \log \frac{x}{x}\right)^{1/2}\right] \\
&\quad \leq 1 + \int_1^N \frac{1}{t} \exp\left[-\left(c \log \frac{x}{t}\right)^{1/2}\right] dt \\
&\qquad + \int_{N+1}^x \frac{1}{t} \exp\left[-\left(c \log \frac{x}{t}\right)^{1/2}\right] dt + \frac{1}{x} \\
&\quad \leq 1 + \int_1^x \frac{1}{t} \exp\left[-\left(c \log \frac{x}{t}\right)^{1/2}\right] dt + \frac{1}{x}
\end{aligned}$$

and it suffices to show that this integral remains bounded as $x \to \infty$. But $u = x/t$ gives

$$\int_1^x \frac{1}{t} \exp\left[-\left(c \log \frac{x}{t}\right)^{1/2}\right] dt = \int_1^x \frac{1}{u} \exp[-(c \log u)^{1/2}]\, du$$

and it was shown above that this integral converges as $x \to \infty$, so the proof of (3), and hence of (2), is complete.

Chapter 6

Numerical Analysis of the Roots by Euler–Maclaurin Summation

6.1 INTRODUCTION

The first substantial numerical information on the roots ρ was given by Gram [G5], who in 1903 published a list of 15 roots on the line $\operatorname{Re} s = \frac{1}{2}$. Gram computed the first 10 of these roots to about 6 decimal places and the remaining 5 to about 1 place. Specifically, the values he gave were $\rho = \frac{1}{2} + i\alpha$, where

$$\begin{aligned} \alpha_1 &= 14.134\,725, & \alpha_6 &= 37.586\,176, & \alpha_{11} &= 52.8, \\ \alpha_2 &= 21.022\,040, & \alpha_7 &= 40.918\,720, & \alpha_{12} &= 56.4, \\ \alpha_3 &= 25.010\,856, & \alpha_8 &= 43.327\,073, & \alpha_{13} &= 59.4, \\ \alpha_4 &= 30.424\,878, & \alpha_9 &= 48.005\,150, & \alpha_{14} &= 61.0, \\ \alpha_5 &= 32.935\,057, & \alpha_{10} &= 49.773\,832, & \alpha_{15} &= 65.0. \end{aligned}$$

Subsequent calculations have confirmed that these values are correct except, as Gram stated, for slight errors in the last place given. (For the correct values to 6 places see Haselgrove's tables [H8].) Gram was also able to prove that this list includes *all* of the roots ρ in the range $0 \leq \operatorname{Im} s \leq 50$ and thus to prove that the Riemann hypothesis is true in this range.

The basis of Gram's calculations was the straightforward method of Euler–Maclaurin summation to evaluate both the function ζ and the factorial function Π, and consequently to evaluate $\xi(s) = \Pi(\frac{1}{2}s)\pi^{-s/2}(s-1)\zeta(s)$. It is interesting from the point of view of the psychology of mathematical discovery to note that Gram had initially attempted more original and more complicated techniques but had met with very limited success. It was several years before he tried using the classical method of Euler–Maclaurin summation, and when he did he was surprised at the ease with which he was able to compute the numbers he had been searching for for so long.

Euler–Maclaurin summation is a technique for the numerical evaluation of sums which was developed in the carly part of the eighteenth century. The original impetus came from Bernoulli's success in generalizing $\sum_{n=1}^{N} n = N(N+1)/2$ to find an analogous formula for $\sum_{n=1}^{N} n^k$ which involved the "Bernoulli numbers." Forms of the technique were used by Stirling and De Moivre as early as 1730, but the definitive statement of the method together with a proof of sorts did not come until around 1740 when it was published by Euler and, independently (see Cantor [C1]), by Maclaurin. Euler in his well-known calculus book [E6] included examples of the use of Euler–Maclaurin summation to compute $\zeta(s)$ for $s = 2, 3, \ldots, 15, 16$ and to compute $\Pi(s)$ for large s (Stirling's series), so Gram's computations are direct descendants of Euler's.

Gram's work was carried farther by Backlund [B3, B4] around 1912–1915. Backlund's major contribution was a method of computing, for certain values of T, the number of roots in the range $0 \leq \text{Im}\, s \leq T$. This method enabled him to show that the Riemann hypothesis was true up to the level $T = 200$, that is, to prove that all of the roots in the range $0 \leq \text{Im}\, s \leq 200$ lie on the line $\text{Re}\, s = \frac{1}{2}$. It also enabled him to prove that Riemann's estimate [see (d) of Section 1.19] of the number of roots in $0 \leq \text{Im}\, s \leq T$ for large T was correct, a fact which von Mangoldt had already proved in 1905 by a method which was more complicated. Backlund's proof of this fact is included in this chapter, even though it does not contribute to the numerical analysis of the roots, because it is a natural outgrowth of techniques which Backlund developed for the numerical analysis of the roots.

Some ten years later the Riemann hypothesis was verified up to the level $T = 300$ by Hutchinson [H11], who contributed some improvements of Gram's and Backlund's methods. As far as the distribution of the roots on the line $\text{Re}\, s = \frac{1}{2}$ is concerned, Hutchinson showed that it is usually true in the range $0 \leq \text{Im}\, s \leq 300$ that there is exactly one root between two consecutive Gram points (see Section 6.5)—in other words the set of roots and the set of Gram points usually separate each other—but that there are two slight exceptions to this rule in this range, the first near $\text{Im}\, s = 282.5$ and the second near $\text{Im}\, s = 295.5$.

Broadly speaking, the computations of Gram, Backlund, and Hutchinson contributed substantially to the plausibility of the Riemann hypothesis, but they gave no insight into the question of why it might be true or into the question of why Riemann might have been led to make such a hypothesis.

6.2 EULER–MACLAURIN SUMMATION

Consider the problem of finding the numerical value of the sum S defined by

$$S = \left(\frac{1}{10}\right)^2 + \left(\frac{1}{11}\right)^2 + \left(\frac{1}{12}\right)^2 + \cdots + \left(\frac{1}{100}\right)^2. \tag{1}$$

As a first approximation to S one might note that if half the first term and half the last term are omitted from S, then the sum which remains is an approximation to $\int_{10}^{100} x^{-2}\,dx$; specifically, the trapezoidal rule

$$\int_a^b f(x)\,dx \sim \sum_{i=1}^{n} \frac{f(x_i) + f(x_{i-1})}{2}(x_i - x_{i-1})$$

$$(a = x_0 < x_1 < \cdots < x_n = b)$$

in the case $x_0 = 10$, $x_1 = 11, \ldots, x_n = 100$, $f(x) = x^{-2}$ gives

$$\int_{10}^{100} \frac{dx}{x^2} \sim \frac{1}{2}\left[\left(\frac{1}{10}\right)^2 + \left(\frac{1}{11}\right)^2\right] \cdot 1$$

$$+ \frac{1}{2}\left[\left(\frac{1}{11}\right)^2 + \left(\frac{1}{12}\right)^2\right] \cdot 1 + \cdots + \frac{1}{2}\left[\left(\frac{1}{99}\right)^2 + \left(\frac{1}{100}\right)^2\right] \cdot 1.$$

Hence

$$S \sim \int_{10}^{100} \frac{dx}{x^2} + \frac{1}{2}\left(\frac{1}{10}\right)^2 + \frac{1}{2}\left(\frac{1}{100}\right)^2$$

$$= -\frac{1}{x}\bigg|_{10}^{100} + 0.005 + 0.00005 = 0.09505.$$

Euler–Maclaurin summation is a method of computing the error in this approximation and in the analogous approximation

$$\sum_{n=M}^{N} f(n) \sim \int_M^N f(x)\,dx + \tfrac{1}{2}[f(M) + f(N)] \tag{2}$$

for other sums.

The first step is to develop a closed formula for the error. Let $[x]$ denote, as usual, the largest integer less than or equal to x. Then the function $[x]$ is a step function which has jumps of one at integers, so the Stieltjes measure $d([x])$ assigns the weight one to integers and is zero elsewhere. Hence

$$\int_M^N f(x)\,d([x]) = \tfrac{1}{2}f(M) + f(M+1) + f(M+2) + \cdots$$
$$+ f(N-1) + \tfrac{1}{2}f(N),$$

where the usual convention of counting half the weight at an end point is followed. Thus to make the approximation (2) correct, the right side should be increased by

$$-\int_M^N f(x)\,dx + \int_M^N f(x)\,d([x]) = \int_M^N f(x)\,d([x]-x).$$

It is more natural to describe the measure $d([x]-x)$ as $d([x]-x+\frac{1}{2})$ because the function $[x]-x+\frac{1}{2}$ is positive half the time and negative half the time, and because it is zero when x is an integer (by the usual convention that at discontinuities the value is the average of the left-hand limit and the right-hand limit). Integration by parts then expresses the error in the form

$$\begin{aligned}\int_M^N f(x)\,d([x]-x+\tfrac{1}{2}) &= -\int_M^N ([x]-x+\tfrac{1}{2})\,df(x)\\ &= \int_M^N (x-[x]-\tfrac{1}{2})f'(x)\,dx.\end{aligned}$$

Having arrived at this formula by sketchy arguments based on Stieltjes integration, one can easily justify it using ordinary Riemann integration to find

$$\begin{aligned}\int_M^N (x-[x]-\tfrac{1}{2})f'(x)\,dx &= \sum_{n=M}^{N-1}\int_n^{n+1}(x-[x]-\tfrac{1}{2})f'(x)\,dx\\ &= \sum_{n=M}^{N-1}\int_0^1 (t-\tfrac{1}{2})f'(n+t)\,dt\\ &= \sum_{n=M}^{N-1}\left[(t-\tfrac{1}{2})f(n+t)\Big|_0^1 - \int_0^1 f(n+t)\,dt\right]\\ &= \sum_{n=M}^{N-1}[\tfrac{1}{2}f(n+1)+\tfrac{1}{2}f(n)] - \sum_{n=M}^{N-1}\int_0^1 f(n+t)\,dt\\ &= \tfrac{1}{2}f(M)+f(M+1)+f(M+2)+\cdots\\ &\quad + f(N-1)+\tfrac{1}{2}f(N) - \int_M^N f(x)\,dx\end{aligned}$$

which proves the desired formula

(3)
$$\begin{aligned}\sum_{n=M}^{N} f(n) &= \int_M^N f(x)\,dx + \tfrac{1}{2}[f(M)+f(N)]\\ &\quad + \int_M^N (x-[x]-\tfrac{1}{2})f'(x)\,dx\end{aligned}$$

for continuously differentiable functions f on $[M, N]$.

In the case of the sum S in (1) this formula shows that its value is 0.09505 plus

(4)
$$\int_{10}^{100}\frac{(x-[x]-\frac{1}{2})(-2)}{x^3}\,dx.$$

The integrand in this integral is positive from 10 to $10\frac{1}{2}$, negative from $10\frac{1}{2}$ to 11, positive from 11 to $11\frac{1}{2}$, etc. Since x^{-3} decreases, the integral can thus be written as an alternating series of terms which decrease in absolute value, so its value is positive but less than

$$\int_{10}^{10.5} \frac{(x-[x]-\frac{1}{2})(-2)}{x^3}\,dx = \int_0^{1/2} \frac{1-2t}{(10+t)^3}\,dt$$
$$\leq \frac{1}{10^3}\int_0^{1/2}(1-2t)\,dt = 0.00025.$$

Thus S lies between 0.09505 and 0.09530, which gives its value to three places.

This much is quite elementary. The real substance of Euler–Maclaurin summation is *repeated integration by parts of the last term of* (3) which puts this term in a form that can in many cases be evaluated numerically with great accuracy. This integration by parts requires the use of *Bernoulli polynomials.* The nth Bernoulli polynomial is by definition the unique polynomial of degree n with the property that

$$\int_x^{x+1} B_n(t)\,dt = x^n. \tag{5}$$

Thus, for example, $B_3(x) = ax^3 + bx^2 + cx + d$ is determined by the equation

$$\begin{aligned} x^3 &= \int_x^{x+1} (at^3 + bt^2 + ct + d)\,dt \\ &= a\frac{(x+1)^4 - x^4}{4} + b\frac{(x+1)^3 - x^3}{3} + c\frac{(x+1)^2 - x^2}{2} \\ &\quad + d\frac{x+1-x}{1} \\ &= ax^3 + \left(\frac{3}{2}a + b\right)x^2 + (a+b+c)x \\ &\quad + \left(\frac{a}{4} + \frac{b}{3} + \frac{c}{2} + \frac{d}{1}\right) \end{aligned}$$

from which one finds successively $a = 1$, $b = -3/2$, $c = 1/2$, $d = 0$. It is easy to see that this process always yields a polynomial $B_n(x)$ satisfying (5) and, since a polynomial $p(x)$ which satisfies $\int_x^{x+1} p(t)\,dt \equiv 0$ must be identically zero, this suffices to prove that condition (5) defines a unique polynomial $B_n(x)$. Differentiation of (5) gives

$$B_n(x+1) - B_n(x) = nx^{n-1}, \tag{6}$$
$$\int_x^{x+1} \frac{1}{n} B_n'(t)\,dt = x^{n-1},$$

which shows that $B_n'(x)/n$ satisfies the definition of $B_{n-1}(x)$ and hence that

$$B_n'(x) = nB_{n-1}(x). \tag{7}$$

Thus, starting with $B_3(x)$, which was computed above, one has immediately

$$B_3(x) = x^3 - \tfrac{3}{2}x^2 + \tfrac{1}{2}x, \qquad B_2(x) = x^2 - x + \tfrac{1}{6},$$
$$B_1(x) = x - \tfrac{1}{2}, \qquad B_0(x) = 1.$$

These polynomials can be used to integrate the last term of (3) by parts to put it in the form

$$\begin{aligned}\int_M^N \left(x - [x] - \frac{1}{2}\right) f'(x)\,dx &= \sum_{n=M}^{N-1} \int_0^1 \left(t - \frac{1}{2}\right) f'(n+t)\,dt = \sum_{n=M}^{N-1} \int_0^1 B_1(t) f'(n+t)\,dt \\ &= \sum_{n=M}^{N-1} \left[\frac{B_2(t)}{2} f'(n+t)\Big|_0^1 - \frac{1}{2}\int_0^1 B_2(t) f''(n+t)\,dt \right] \\ &= -\frac{B_2(0)}{2} f'(M) + \frac{B_2(1)}{2} f'(M+1) - \frac{B_2(0)}{2} f'(M+1) \\ &\quad + \frac{B_2(1)}{2} f'(M+2) - \frac{B_2(0)}{2} f'(M+2) + \cdots + \frac{B_2(1)}{2} f'(N) \\ &\quad - \frac{1}{2}\int_M^N B_2(x - [x]) f''(x)\,dx.\end{aligned}$$

The long sum telescopes because (6) with $n = 2$, $x = 0$ gives $B_2(1) = B_2(0)$. Thus if $\bar{B}_2(x)$ is used to denote the periodic function $B_2(x - [x])$, the last term of (3) can be written in the form

$$\frac{B_2(0)}{2} f'(x)\Big|_M^N - \frac{1}{2}\int_M^N \bar{B}_2(x) f''(x)\,dx.$$

The second term in this formula can be integrated by parts by exactly the same sequence of steps to put the last term of (3) in the form

$$\frac{B_2(0)}{2} f'(x)\Big|_M^N - \frac{B_3(0)}{2\cdot 3} f''(x)\Big|_M^N + \frac{1}{2\cdot 3}\int_M^N \bar{B}_3(x) f'''(x)\,dx,$$

where, of course, $\bar{B}_3(x)$ denotes the periodic function $B_3(x - [x])$.

Applying this formula to the evaluation of the integral (4) gives

$$\begin{aligned}\frac{\frac{1}{6}}{2}\,\frac{-2}{x^3}\Big|_{10}^{100} - 0 + \frac{1}{2\cdot 3}\int_{10}^{100} \bar{B}_3(x) f'''(x)\,dx &= \frac{1}{6}\left[\frac{1}{10^3} - \frac{1}{100^3}\right] + \frac{(-2)(-3)(-4)}{2\cdot 3}\int_{10}^{100} \frac{\bar{B}_3(x)\,dx}{x^5} \\ &= \frac{999}{6\cdot 10^6} - 4\int_{10}^{100} \frac{\bar{B}_3(x)\,dx}{x^5}.\end{aligned}$$

The first term is 1.665×10^{-4} and the second term is much smaller, as can be seen by an alternating series technique similar to the one used to estimate (4). Note first that $B_3(x)$ is a polynomial of degree three which is zero at $x = 0$ and $x = \frac{1}{2}$ [by direct evaluation] and at $x = 1$ [because (6) shows that $B_n(1) = B_n(0)$ for $n \geq 2$]. This accounts for all its zeros and proves that $B_3(x) = x(x - \frac{1}{2})(x - 1)$. Thus it is positive for $0 < x < \frac{1}{2}$ and negative for $\frac{1}{2} < x < 1$. These positive and negative "bumps" are symmetrical because $B_3(1 - x) = (1 - x)(1 - x - \frac{1}{2})(1 - x - 1) = -B_3(x)$, so the graph of $\bar{B}_3(x)$

is a wave consisting of positive bumps on $(n, n + \frac{1}{2})$ and symmetrically negative bumps on $(n + \frac{1}{2}, n + 1)$. Since x^{-5} decreases, this shows immediately that

$$0 \le 4 \int_{10}^{100} \frac{\bar{B}_3(x)\,dx}{x^5} \le 4 \int_{10}^{10.5} \frac{\bar{B}_3(x)\,dx}{x^5}.$$

Now

$$\begin{aligned} 4 \int_{10}^{10.5} \frac{\bar{B}_3(x)\,dx}{x^5} &\le \frac{4}{10^5} \int_{10}^{10.5} \bar{B}_3(x)\,dx = \frac{4}{10^5} \int_{0}^{1/2} B_3(t)\,dt \\ &= \frac{4}{10^5}\left[\frac{1}{4}t^4 - \frac{1}{2}t^3 + \frac{1}{4}t^2\right]\Bigg|_0^{1/2} = \frac{4}{10^5}\left[\frac{1}{64} - \frac{1}{16} + \frac{1}{16}\right] \\ &= \frac{1}{10^5 \cdot 16} = 6.25 \times 10^{-7}. \end{aligned}$$

Thus

$$S = 0.09505 + 1.665 \times 10^{-4} - 4 \int_{10}^{100} \frac{\bar{B}_3(x)\,dx}{x^5}$$

lies in the range

$$0.095215875 \le S \le 0.0952165$$

which gives S to six places.

The integration by parts can of course be carried farther to put (3) in the form

$$\begin{aligned} (8) \qquad \sum_{n=M}^{N} f(n) = \int_M^N f(x)\,dx &+ \frac{1}{2}[f(M) + f(N)] + \frac{B_2(0)}{2} f'(x)\Big|_M^N \\ &- \frac{B_3(0)}{2 \cdot 3} f''(x)\Big|_M^N + \cdots + (-1)^k \frac{B_k(0)}{k!} f^{(k-1)}(x)\Big|_M^N \\ &+ (-1)^{k+1} \frac{1}{k!} \int_M^N \bar{B}_k(x) f^{(k)}(x)\,dx. \end{aligned}$$

If k is odd, then $\bar{B}_k(x)$ is an oscillating function similar to $\bar{B}_3(x)$ and the integral in this formula can be estimated, provided $f^{(k)}(x)$ is monotone, by an alternating series technique like the one used above. To prove that $\bar{B}_{2\nu+1}(x)$ has this oscillating character, note that the identities

$$\begin{aligned} \int_x^{x+1} B_n(1 - t)\,dt = -\int_{1-x}^{-x} B_n(u)\,du &= \int_{-x}^{-x+1} B_n(u)\,du \\ &= (-x)^n = (-1)^n \int_x^{x+1} B_n(t)\,dt \end{aligned}$$

and

$$\begin{aligned} \int_x^{x+(1/2)} B_n(2t)\,dt = \tfrac{1}{2} \int_{2x}^{2x+1} B_n(u)\,du = \tfrac{1}{2}(2x)^n &= 2^{n-1} \int_x^{x+1} B_n(t)\,dt \\ &= 2^{n-1} \int_x^{x+(1/2)} [B_n(t) + B_n(t + \tfrac{1}{2})]\,dt \end{aligned}$$

imply

$$(9)\qquad \begin{aligned} B_n(1-x) &= (-1)^n B_n(x), \\ B_n(2x) &= 2^{n-1}[B_n(x) + B_n(x+\tfrac{1}{2})]. \end{aligned}$$

Thus $B_{2\nu+1}(1-x) = -B_{2\nu+1}(x)$, $B_{2\nu+1}(\frac{1}{2}) = 0$, $B_{2\nu+1}(0) = 2^{2\nu}[B_{2\nu+1}(0) + 0]$, $B_{2\nu+1}(0) = 0$; that is,

$$(10)\qquad B_{2\nu+1}(0) = B_{2\nu+1}(\tfrac{1}{2}) = 0 \qquad (\nu = 1, 2, \ldots).$$

Now there are no zeros of $B_{2\nu+1}(x)$ between 0 and $\frac{1}{2}$ because such a zero would imply two zeros of the derivative $(2\nu+1)B_{2\nu}(x)$ between 0 and $\frac{1}{2}$, which would in turn imply a zero of $(2\nu+1)2\nu B_{2\nu-1}(x)$ and hence a zero of $B_{2\nu-1}(x)$; repeating this process one would ultimately arrive at the conclusion that $B_3(x)$ had a zero between 0 and $\frac{1}{2}$, which it does not; hence neither does $B_{2\nu+1}(x)$. Thus $B_{2\nu+1}(x)$ has one sign on $(0, \frac{1}{2})$; by $B_{2\nu+1}(1-x) = -B_{2\nu+1}(x)$ it has the opposite sign on $(\frac{1}{2}, 1)$. Therefore $\bar{B}_{2\nu+1}(x)$ oscillates as was to be shown.

This also shows, in passing, that many of the terms of (8) are zero, namely, the terms containing $B_{2\nu+1}(0)$. To apply formula (8), one must of course find the values of the constants $B_{2\nu}(0)$. This is accomplished by proving that *the constants $B_{2\nu}(0)$ coincide with the Bernoulli numbers $B_{2\nu}$* defined in Section 1.5. To see this, apply formula (8) in the case $f(x) = e^{-hx}$, $M = 0$, $N = \infty$ to find

$$\begin{aligned} 1 + e^{-h} + e^{-2h} + \cdots &= \int_0^\infty e^{-hx}\,dx + \frac{1}{2} - \frac{B_2(0)}{2}(-h) \\ &\quad + \frac{B_3(0)}{3!}(-h)^2 + \cdots - (-1)^k \frac{B_k(0)}{k!}(-h)^{k-1} \\ &\quad + (-1)^{k+1}(-h)^k \frac{1}{k!}\int_0^\infty \bar{B}_k(x)e^{-hx}\,dx, \end{aligned}$$

$$\begin{aligned} \frac{1}{1-e^{-h}} &= \frac{1}{h} + \frac{1}{2} + \frac{B_2(0)}{2}h + \frac{B_3(0)}{3!}h^2 \\ &\quad + \cdots + \frac{B_k(0)}{k!}h^{k-1} - \frac{h^k}{k!}\int_0^\infty \bar{B}_k(x)e^{-hx}\,dx, \end{aligned}$$

$$\begin{aligned} \sum_{n=0}^\infty \frac{B_n}{n!}(-h)^n &= \frac{-h}{e^{-h}-1} = h \cdot \frac{1}{1-e^{-h}} \\ &= 1 + \frac{1}{2}h + \frac{B_2(0)}{2}h^2 + \frac{B_3(0)}{3!}h^3 + \cdots \\ &\quad + \frac{B_k(0)}{k!}h^k - \frac{h^{k+1}}{k!}\int_0^\infty \bar{B}_k(x)e^{-hx}\,dx. \end{aligned}$$

This formula is valid for small positive h, and, provided k is odd, the absolute value of the last term is at most $(h^{k+1}/k!)\left|\int_0^{1/2} B_k(t)\,dt\right| \leq \text{const}\, h^{k+1}$. Thus the polynomial

$$\sum_{n=0}^k [(-1)^n B_n - B_n(0)]\frac{h^n}{n!} = p(h)$$

has the property that $p(h)/h^{k+1}$ is bounded as $h \downarrow 0$, which proves $p(h) \equiv 0$, $B_n(0) \equiv (-1)^n B_n$ for $n \leq k$. Since k was arbitrary and since only even† values of n are at issue, this proves the theorem.

Putting these facts together gives the *Euler–Maclaurin summation formula*

$$\sum_{n=M}^{N} f(n) = \int_M^N f(x)\,dx + \frac{1}{2}[f(M) + f(N)] + \frac{B_2}{2} f'(x)\Big|_M^N + \frac{B_4}{4!} f'''(x)\Big|_M^N + \cdots + \frac{B_{2\nu}}{(2\nu)!} f^{(2\nu-1)}(x)\Big|_M^N + R_{2\nu},$$

where the B_n are the Bernoulli numbers, where $f(x)$ is any function which has $2\nu + 1$ continuous derivatives on $[N, M]$, and where $R_{2\nu}$ is given by either of the formulas

$$R_{2\nu} = \frac{-1}{(2\nu)!} \int_M^N \bar{B}_{2\nu}(x) f^{(2\nu)}(x)\,dx$$

or

$$R_{2\nu} = \frac{1}{(2\nu + 1)!} \int_M^N \bar{B}_{2\nu+1}(x) f^{(2\nu+1)}(x)\,dx$$

in which $\bar{B}_k(x) = B_k(x - [x])$. [The two forms of $R_{2\nu}$ are obtained by setting $k = 2\nu$ and $k = 2\nu + 1$, respectively, in (8).]

Continuing with the example of the sum S in (1), the next few terms are

$$\frac{B_4}{4!} \cdot \frac{(-4)(-3)(-2)}{x^5}\Big|_{10}^{100} = -\frac{1}{30}\,\frac{10^5 - 1}{10^{10}} = -\frac{33333}{10^{11}} = -3.3333 \times 10^{-7}$$

$$\frac{B_6}{6!}\,\frac{(-6)(-5)\cdots(-2)}{x^7}\Big|_{10}^{100} = \frac{1}{42}\,\frac{10^7 - 1}{10^{14}} = (2.3809\cdots) \times 10^{-9}$$

$$\frac{B_8}{8!}\,\frac{(-1)8!}{x^9}\Big|_{10}^{100} = -\frac{1}{30}\,\frac{10^9 - 1}{10^{18}} = (-3.33\cdots) \times 10^{-11}$$

which gives the approximation 0.095 216 169 017 6 to the value of S. The

0.095 05	−0.000 000 333 33
0.000 166 5	−0. 33 3
0.000 000 002 380 9	−0.000 000 333 363 3
0.095 216 502 380 9	
−0.000 000 333 363 3	
$S \sim$ 0.095 216 169 017 6	

Computation of the approximation to S.

†For odd values of n this gives an alternative proof that $B_3, B_5, B_7, \ldots$ are all zero. See Section 1.6.

magnitude of the error in this approximation is

$$\begin{aligned}\left|\frac{1}{9!}\int_{10}^{100}\bar{B}_9(x)\frac{(-1)10!}{x^{11}}\,dx\right| &\leq \frac{1}{10^{11}}\left|\int_0^{1/2}10B_9(t)\,dt\right| \\ &= 10^{-11}\left|B_{10}\left(\frac{1}{2}\right)-B_{10}(0)\right| \\ &= 10^{-11}\left|B_{10}\left(\frac{1}{2}\right)+B_{10}(0)-2B_{10}(0)\right| \\ &= 10^{-11}\left|2^{-9}B_{10}(0)-2B_{10}(0)\right| \\ &\leq 2B_{10}\cdot 10^{-11}.\end{aligned}$$

Rather than use this error estimate directly, however, it is more effective to note that it says that the magnitude of the error cannot be more than about *twice the size of the first term omitted*, and that the same estimate would apply no matter where the series was truncated. Since the terms are still getting smaller—$B_{10} = 5/66$, $B_{12} = -691/2730$ so the next two are about 8 in the thirteenth place and 3 in the fourteenth place—this implies that *the error is of the order of magnitude of the first term omitted* because when this term is included, the error is reduced to an amount much smaller than this term (namely, to an amount comparable to the next term). Thus one can be confident that the error in the above approximation to S is *less than one in the twelfth place*, that is, $S = 0.095\,216\,169\,018 \pm 1$.

It is a general rule of thumb in applying the Euler–Maclaurin summation formula that *as long as the terms are decreasing rapidly in size, the bulk of the error is in the first term omitted.* In order to give a rigorous proof of this fact in specific cases it is not necessary to estimate $|R_{2\nu}|$ in any refined way but merely to give a crude estimate showing that it is of an order of magnitude comparable to the first term omitted at most. Then the error has absolute value at most

$$|R_{2\nu}| = \left|\frac{B_2}{2!}f^{(2\nu-1)}\Big|_N^M + R_{2\nu+2}\right| \leq |\text{first term omitted}| + |R_{2\nu+2}|$$

and $|R_{2\nu+2}|$ is comparable to the second term omitted, hence much smaller than the first term omitted.

However, it must be observed that the terms do *not* continue to decrease indefinitely [except for very special functions f such as $f(x) = e^{-hx}$ for small h] and that in fact they ultimately grow without bound. To see this, it suffices to combine Euler's formula

$$\zeta(2\nu) = (2\pi)^{2\nu}(-1)^{\nu+1}B_{2\nu}/2\cdot(2\nu)!$$

[see (2) of Section 1.5] with the trivial observation that $\zeta(2\nu) = 1 + 2^{-2\nu} + 3^{-2\nu} + \cdots$ is only slightly larger than one for ν large; hence

$$B_{2\nu} \sim \pm 2\cdot(2\nu)!(2\pi)^{-2\nu},$$

and in the example the term containing $B_{2\nu}$ is roughly $\pm 2\cdot(2\nu)!(2\pi)^{-2\nu}\cdot 10^{-2\nu-1}$ which decreases quite rapidly at first but which ultimately grows without bound. This shows that there is a limit to the degree of accuracy with which one can evaluate S by extending the above process, and that this limit is roughly determined by the minimum value of $\Pi(x)(20\pi)^{-x}\cdot\frac{1}{5}$ for $x \geq 0$. If for any reason greater accuracy is required, then the first few terms can be summed separately, say $10^{-2} + 11^{-2} + 12^{-2} + 13^{-2} + 14^{-2}$, and the remaining sum evaluated by Euler–Maclaurin summation, which is now much more accurate because the denominators $10^{2\nu+1}$ are replaced by $15^{2\nu+1}$. [This observation is of crucial importance in the evaluation of $\zeta(s)$ by Euler–Maclaurin summation in Section 6.4. After all, the sum S is essentially $\zeta(2)$ except that the first nine terms and all terms past the hundredth are missing.]

In summary, the Euler–Maclaurin summation formula says that

$$\sum_{n=M}^{N} f(n) \sim \int_M^N f(x)\,dx + \frac{1}{2}[f(M) + f(N)] + \sum_{j=1}^{\nu} \frac{B_{2j}}{(2j)!} f^{(2j-1)}(x)\Big|_M^N.$$

In many examples the terms of the series on the right are at first rapidly decreasing in size and the bulk of the error in the approximation is accounted for by the first term omitted. In any case the error is precisely equal to

$$R_{2\nu} = \frac{1}{(2\nu+1)!}\int_M^N \bar{B}_{2\nu+1}(x) f^{(2\nu+1)}(x)\,dx,$$

and when $f^{(2\nu+1)}$ is monotone this leads to a simple estimate of $|R_{2\nu}|$ using the fact that $\bar{B}_{2\nu+1}(x)$ alternates in sign.

6.3 EVALUATION OF Π BY EULER–MACLAURIN SUMMATION. STIRLING'S SERIES

To evaluate $\Pi(s)$, it of course suffices to evaluate $\log \Pi(s)$. Now if s is a positive integer, say $s = N$, then $\log \Pi(N) = \sum_1^N \log n$ which, by Euler–Maclaurin summation (with $\nu = 0$), is

$$\int_1^N \log x\,dx + \frac{1}{2}[\log 1 + \log N] + \int_1^N \frac{\bar{B}_1(x)\,dx}{x}.$$

The first integral can be evaluated using $\int \log x\,dx = x\log x - x$. The second integral approaches a limit as $N \to \infty$ (by the alternating series test), so it can be written in the form

$$\int_1^N \frac{\bar{B}_1(x)\,dx}{x} = \int_1^\infty \frac{\bar{B}_1(x)\,dx}{x} - \int_N^\infty \frac{\bar{B}_1(x)\,dx}{x}$$

in which the first term is constant and the second term approaches zero as $N \to \infty$. This gives

$$\log \Pi(N) = \left(N + \frac{1}{2}\right) \log N - N + A - \int_N^\infty \frac{\bar{B}_1(x)\, dx}{x}, \tag{1}$$

where A is the constant

$$A = 1 + \int_1^\infty \frac{\bar{B}_1(x)\, dx}{x}.$$

Except for the fact that the constant A must still be evaluated, this gives a simple approximate formula for $\log \Pi(N)$, and it is natural to ask whether there is a similar formula for $\log \Pi(s)$ for other values of s.

As it stands, formula (1) cannot possibly be valid for all real numbers N because the derivative of its right side with respect to N is discontinuous at all integers [because $\bar{B}_1(x)$ is discontinuous]. However, if (1) is rewritten in the form

$$\log \Pi(s) = \left(s + \frac{1}{2}\right) \log s - s + A - \int_0^\infty \frac{\bar{B}_1(t)\, dt}{t + s}, \tag{2}$$

then both sides are well-behaved functions of s for all positive real numbers and, since equality holds whenever s is an integer, it is reasonable to expect that equality will hold for all real $s > 0$. The fact that it does hold follows quite easily from the application of Euler–Maclaurin summation (with $\nu = 0$) to the definition of $\Pi(s)$. Explicitly, from the definition

$$\Pi(s) = \lim_{N \to \infty} \frac{1 \cdot 2 \cdot 3 \cdots N}{(s+1)(s+2)\cdots(s+N)} (N+1)^s$$

of $\Pi(s)$ [see (3) of Section 1.3] it follows that

$$\begin{aligned}
\log \Pi(s) &= \lim_{N \to \infty} \left\{ s \log(N+1) + \sum_{n=1}^{N} \log n - \sum_{n=1}^{N} \log(s+n) \right\} \\
&= \lim_{N \to \infty} \left\{ s \log(N+1) + \int_1^N \log x\, dx + \frac{1}{2} \log N + \int_1^N \frac{\bar{B}_1(x)\, dx}{x} \right. \\
&\quad - \int_1^N \log(s+x)\, dx - \frac{1}{2}[\log(s+1) + \log(s+N)] \\
&\quad \left. - \int_1^N \frac{\bar{B}_1(x)\, dx}{s+x} \right\} \\
&= \lim_{N \to \infty} \left\{ s \log(N+1) + N \log N - N + 1 + \frac{1}{2} \log N + \int_1^N \frac{\bar{B}_1(x)\, dx}{x} \right. \\
&\quad - (s+N) \log(s+N) + (s+N) + (s+1) \log(s+1) \\
&\quad \left. - (s+1) - \frac{1}{2} \log(s+1) - \frac{1}{2} \log(s+N) - \int_1^N \frac{\bar{B}_1(x)\, dx}{s+x} \right\}
\end{aligned}$$

$$= \left(s+\frac{1}{2}\right)\log(s+1) + \int_1^\infty \frac{\bar{B}_1(x)\,dx}{x} - \int_1^\infty \frac{\bar{B}_1(x)\,dx}{s+x}$$
$$+ \lim_{N\to\infty}\left\{s\log(N+1) + \left(N+\frac{1}{2}\right)\log N\right.$$
$$\left. - \left(s+N+\frac{1}{2}\right)\log(s+N)\right\}$$
$$= \left(s+\frac{1}{2}\right)\log(s+1) + (A-1) - \int_1^\infty \frac{\bar{B}_1(x)\,dx}{s+x}$$
$$+ \lim_{N\to\infty}\left\{s\log\frac{N+1}{N+s} - \left(N+\frac{1}{2}\right)\log\left(1+\frac{s}{N}\right)\right\}$$
$$= \left(s+\frac{1}{2}\right)\log(s+1) + A - 1 - \int_1^\infty \frac{\bar{B}_1(x)\,dx}{s+x} - s.$$

This differs from the right side of (2) by

$$\left(s+\frac{1}{2}\right)\log\left(\frac{s+1}{s}\right) - 1 + \int_0^1 \frac{\bar{B}_1(t)\,dt}{s+t}$$
$$= \left(s+\frac{1}{2}\right)\log\left(\frac{s+1}{s}\right) - 1 + \int_0^1 \frac{t-\frac{1}{2}}{s+t}\,dt$$
$$= \left(s+\frac{1}{2}\right)\log\left(\frac{s+1}{s}\right) - 1 + \int_0^1 \frac{t+s-s-\frac{1}{2}}{s+t}\,dt$$
$$= \left(s+\frac{1}{2}\right)\log\left(\frac{s+1}{s}\right) - 1 + \int_0^1 dt - \left(s+\frac{1}{2}\right)\int_0^1 \frac{dt}{s+t}$$
$$= 0$$

which shows that (2) is indeed true for all positive real numbers s.

Formula (2) can be combined with the Legendre relation

$$\Pi(2s) = \frac{1}{\pi^{1/2}} 2^{2s}\Pi(s)\Pi\left(s-\frac{1}{2}\right)$$

[see (7) of Section 1.3] to give† the value of the constant A. To this end let (2) be rewritten in the form

$$\Pi(s) = s^{s+(1/2)}e^{-s}e^{A}r(s),$$

where

$$r(s) = \exp\left[-\int_0^\infty \frac{\bar{B}_1(t)\,dt}{s+t}\right].$$

†An alternative method of proving $A = \frac{1}{2}\log 2\pi$, not relying on the Legendre relation, is given later in this section.

Then $r(s) \longrightarrow 1$ as $s \longrightarrow \infty$, and the Legendre relation says

$$(2s)^{2s+(1/2)}e^{-2s}e^{A}r(2s)$$
$$= \pi^{-1/2}2^{2s}s^{s+(1/2)}e^{-s}e^{A}r(s)\left(s-\frac{1}{2}\right)^{s}e^{-s+(1/2)}e^{A}r\left(s-\frac{1}{2}\right)$$
$$2^{1/2}\left(1-\frac{1}{2s}\right)^{-s}e^{-1/2}\pi^{1/2}\frac{r(2s)}{r(s)r(s-\frac{1}{2})} = e^{A}.$$

As $s \longrightarrow \infty$ the left side approaches $2^{1/2}e^{1/2}e^{-1/2}\pi^{1/2}\cdot 1 = (2\pi)^{1/2}$, so $A = \frac{1}{2}\log 2\pi$. Thus (2) says

$$\log \Pi(s) = \left(s+\frac{1}{2}\right)\log s - s + \frac{1}{2}\log 2\pi - \int_0^\infty \frac{\bar{B}_1(x)\,dx}{s+x}$$

or, if the last term is integrated by parts a number of times,

$$\log \Pi(s) = \left(s+\frac{1}{2}\right)\log s - s + \frac{1}{2}\log 2\pi + \frac{B_2}{2s} \tag{3}$$
$$+ \frac{B_4}{4\cdot 3\cdot s^3} + \cdots + \frac{B_{2\nu}}{2\nu(2\nu-1)s^{2\nu-1}} + R_{2\nu},$$

where

$$R_{2\nu} = -\int_0^\infty \frac{\bar{B}_{2\nu}(x)\,dx}{2\nu(s+x)^{2\nu}} = -\int_0^\infty \frac{\bar{B}_{2\nu+1}(x)\,dx}{(2\nu+1)(s+x)^{2\nu+1}}.$$

This formula, which is known as *Stirling's series*†, is very effective for finding the approximate numerical value of $\log \Pi(s)$.

For example, consider the case $s = 10$. Then

$$\frac{B_2}{2s} = \frac{1}{6\cdot 2\cdot 10} = \frac{1}{120} = 8.3333\cdots \times 10^{-3},$$
$$\frac{B_4}{4\cdot 3\cdot s^3} = \frac{-1}{30\cdot 4\cdot 3\cdot 10^3} = \frac{-1}{36\cdot 10^4} = -2.7777\cdots \times 10^{-6},$$
$$\frac{B_6}{6\cdot 5\cdot s^5} = \frac{1}{42\cdot 6\cdot 5\cdot 10^5} = \frac{1}{126\cdot 10^6} = 7.936507\cdots \times 10^{-9},$$
$$\frac{B_8}{8\cdot 7\cdot s^7} = \frac{-1}{30\cdot 8\cdot 7\cdot 10^7} = \frac{-1}{168\cdot 10^8} = -5.9523\cdots \times 10^{-11},$$
$$\frac{B_{10}}{10\cdot 9\cdot s^9} = \frac{5}{66\cdot 10\cdot 9\cdot 10^9} = \frac{1}{1188\cdot 10^9} \sim 9 \times 10^{-13}.$$

The terms are still decreasing rapidly and the "rule of thumb" of the previous section would lead one to believe that if the B_{10} term is the first one

†It is named for James Stirling, who published it [S7] in 1730. The integral formula for the remainder $R_{2\nu}$ was published by Stieltjes [S5] in 1889 and does not seem to have been known, or at least used, earlier than that.

omitted, then the error will not be much larger than 9×10^{-13}, so the answer will be correct to 12 places. Now in fact

$$
\begin{aligned}
|R_{10}| &= \left| \int_0^\infty \frac{\bar{B}_{11}(x)\,dx}{11\cdot(10+x)^{11}} \right| \le \left| \int_0^{1/2} \frac{B_{11}(x)\,dx}{11\cdot(10+x)^{11}} \right| \\
&\le \frac{1}{12\cdot 11\cdot 10^{11}} \left| \int_0^{1/2} 12B_{11}(x)\,dx \right| \\
&= \frac{1}{12\cdot 11\cdot 10^{11}} \left| B_{12}\left(\frac{1}{2}\right) - B_{12}(0) \right| = \frac{|B_{12}|(2-2^{-11})}{12\cdot 11\cdot 10^{11}} \\
&\le \frac{2\cdot 691}{12\cdot 11\cdot 10^{11}\cdot 2730},
\end{aligned}
$$

so $|R_{10}|$ is very much smaller than the B_{10} term, and hence $|R_8| \sim 9 \times 10^{-13}$ as expected. Thus $\log \Pi(10) \sim 15.10441\ 25730\ 7470$ is accurate to 12 places.

$(10.5)\log 10 =$	$24.17714\ 34764\ 3748$
$-10 =$	$-10.$
$+\frac{1}{2}\log \pi =$	$+0.57236\ 49429\ 2470$
$+\frac{1}{2}\log 2 =$	$+0.34657\ 35902\ 7997$
B_2 term $=$	$+0.00833\ 33333\ 3333$
B_4 term $=$	$-0.00000\ 27777\ 7777$
B_6 term $=$	$+0.00000\ 00079\ 3651$
B_8 term $=$	$-0.00000\ 00000\ 5952$
$\log \Pi(10) \sim$	$15.10441\ 25730\ 7470$

Computation of the approximation to $\log \Pi(10)$.

Of course, if $s > 10$, then the terms of Stirling's series decrease even faster at first and it is even easier to compute $\log \Pi(s)$ with 10- or 12-place accuracy. In fact, Stirling's series is so effective a means of computation that one tends to forget that the series does not converge and that there is a limit to the accuracy with which $\log \Pi(s)$ can be computed using it. Nonetheless, the terms of Stirling's series do ultimately grow without bound as $\nu \to \infty$, and one cannot expect to reduce the error in the approximation to less than the size of the smallest term of the series. If it is desired to compute $\log \Pi(s)$ with greater accuracy than is possible with Stirling's series, then the formula

$$
\begin{aligned}
\log \Pi(s) = \log \Pi(s+N) &- \log(s+1) \\
&- \log(s+2) - \cdots - \log(s+N)
\end{aligned}
$$

can be used. Given $\epsilon > 0$ there is an N such that Stirling's series can be used to compute $\log \Pi(s+N)$ with an error of less than ϵ; since the other logarithms can then be computed by elementary means, this makes possible the evaluation of $\log \Pi(s)$ with any prescribed degree of accuracy. In the numerical analysis of the roots ρ it will not be necessary to use this technique, however, because Stirling's series itself gives the needed values with sufficient accuracy.

Once $\log \Pi(s)$ is found, $\Pi(s)$ can of course be found simply by exponentiating. Note that if $\log \Pi(s)$ is found to 12 decimal places, then $\Pi(s)$ is known to within a *factor* of $\exp(\pm 10^{-12}) \sim 1 \pm 10^{-12}$; so it is the *relative* error which is small, that is, the error divided by the value is less than 10^{-12}. Since $\Pi(s)$ is very large for large s, it is well to keep in mind in evaluating $\Pi(s)$ using Stirling's series that a small relative error may still mean a large absolute error. However, in the numerical analysis of the roots ρ only $\log \Pi(s)$—and in fact only $\operatorname{Im} \log \Pi(s)$—will need to be evaluated, so these considerations will not be necessary.

What *will* be necessary in the numerical analysis of the roots ρ is the use of Stirling's series for complex values of s. Let the "slit plane" be the set of all complex numbers other than the negative reals and zero. Then all terms of Stirling's series (3) are defined throughout the slit plane and are analytic functions of s. (This is obvious for all terms except the integral for R_0. However, as will be shown below, this integral too is convergent for all s in the slit plane.) Since (3) is true for positive real s, the theory of analytic continuation implies that it is true throughout the slit plane—in Riemann's terminology formula (3) for $\log \Pi(s)$ remains valid throughout the slit plane. However, the alternating series method of estimating $|R_{2\nu}|$ cannot be used when s is not real, so some other method of estimating $|R_{2\nu}|$ is needed if Stirling's series is to be used to compute $\log \Pi(s)$ for complex s. The following estimate was given by Stieltjes [S5]:

The objective is to show that the error at any stage is comparable in magnitude to the first term omitted. To this end, let the $B_{2\nu}$ term be the first term omitted and consider the resulting error

$$R_{2\nu-2} = -\int_0^\infty \frac{\bar{B}_{2\nu-1}(x)\, dx}{(2\nu-1)(s+x)^{2\nu-1}}.$$

The function $B_{2\nu} - \bar{B}_{2\nu}(x)$ is an antiderivative of $-2\nu \bar{B}_{2\nu-1}(x)$ which is zero at $x = 0$; hence by integration by parts

$$R_{2\nu-2} = \int_0^\infty \frac{[B_{2\nu} - \bar{B}_{2\nu}(x)]\, dx}{2\nu(s+x)^{2\nu}}.$$

[This integration by parts can be justified, even in the case $\nu = 1$ where $\bar{B}_{2\nu-1}(x)$ is discontinuous, by writing the integral for $R_{2\nu-2}$ as a sum over n of

Riemann integrals, as in the previous section.] Thus

$$|R_{2\nu-2}| \le \int_0^\infty \frac{|B_{2\nu} - \bar{B}_{2\nu}(x)|\,dx}{2\nu|s+x|^{2\nu}}.$$

Now $B_{2\nu} - B_{2\nu}(x)$ is zero at $x = 0$ and has only one extremum in the interval $0 < x < 1$, namely, at $x = \frac{1}{2}$ where its derivative $-2\nu B_{2\nu-1}(x)$ is zero. This implies that $B_{2\nu} - \bar{B}_{2\nu}(x)$ never changes sign. Since $\bar{B}_{2\nu}(x)$ has zeros—namely, at the extrema of $\bar{B}_{2\nu+1}$—the sign of $B_{2\nu} - \bar{B}_{2\nu}(x)$ is always the same as the sign of $B_{2\nu}$, which by Euler's formula for $\zeta(2\nu)$ [(2) of Section 1.5] is $(-1)^{\nu+1}$. Therefore the numerator of the above integral can be rewritten

$$|B_{2\nu} - \bar{B}_{2\nu}(x)| = (-1)^{\nu+1}[B_{2\nu} - \bar{B}_{2\nu}(x)].$$

On the other hand, it is a simple calculus problem to show that $|s+x|(|s|+x)^{-1}$ for $s = re^{i\theta}$ in the slit plane $\{-\pi < \theta < \pi\}$ assumes its minimum value for $x \ge 0$ at $x = |s|$ where it is $\cos(\theta/2)$. Thus

$$\frac{1}{|s+x|} = \frac{1}{|s|+x} \cdot \frac{|s|+x}{|s+x|} \le \frac{1}{(|s|+x)\cos(\theta/2)}$$

and

$$\begin{aligned}
|R_{2\nu-2}| &\le \int_0^\infty \frac{(-1)^{\nu+1}[B_{2\nu} - \bar{B}_{2\nu}(x)]\,dx}{2\nu\cos^{2\nu}(\theta/2)(|s|+x)^{2\nu}} \\
&= \int_0^\infty \frac{(-1)^{\nu+1}B_{2\nu}\,dx}{2\nu\cos^{2\nu}(\theta/2)(|s|+x)^{2\nu}} \\
&\quad - \int_0^\infty \frac{(-1)^{\nu+1}\bar{B}_{2\nu}(x)\,dx}{2\nu\cos^{2\nu}(\theta/2)(|s|+x)^{2\nu}} \\
&= \frac{1}{\cos^{2\nu}(\theta/2)} \cdot \frac{|B_{2\nu}|}{2\nu(2\nu-1)|s|^{2\nu-1}} \\
&\quad - \frac{1}{\cos^{2\nu}(\theta/2)} \int_0^\infty \frac{(-1)^{\nu+1}\bar{B}_{2\nu+1}(x)\,dx}{(2\nu+1)(|s|+x)^{2\nu+1}}.
\end{aligned}$$

By the alternating series method, the second integral is easily seen to have the same sign as $(-1)^{\nu+1}\bar{B}_{2\nu+1}(x)$ on $\{0 < x < \frac{1}{2}\}$. Since this function starts at $x = 0$ with the value zero and the derivative $(-1)^{\nu+1}(2\nu+1)B_{2\nu} > 0$, this sign is $+$ and the inequality becomes stronger if the last term is deleted

$$|R_{2\nu-2}| \le \left(\frac{1}{\cos(\theta/2)}\right)^{2\nu} \left|\frac{B_{2\nu}}{(2\nu)(2\nu-1)s^{2\nu-1}}\right|$$

which is the desired inequality. In words, *if the $B_{2\nu}$ term is the first term omitted in Stirling's series, then the magnitude of the error is at most* $\cos^{-2\nu}(\theta/2)$ *times the magnitude of the first term omitted.*

In the special case $\nu = 1$, $\operatorname{Re} s \ge 0$, $\cos(\theta/2) \ge \sqrt{2}/2$, this gives the estimate used in Section 3.4, namely, that the magnitude of the error in the approximation $\log \Pi(s/2) \sim (s + \frac{1}{2})\log s - s + \frac{1}{2}\log 2\pi$ is at most $(6|s|)^{-1}$ in the halfplane $\operatorname{Re} s \ge 0$. More generally, it shows that if s is a real number $s > 0$, the magnitude of the error is at most the magnitude of the first term omitted (each new term overshoots the mark and the true value lies between any two consecutive partial sums of Stirling's series) and if s is any complex number in the slit plane, then the magnitude of the error is *comparable* to the magnitude of the first term omitted unless s is quite near the negative real axis. This implies that, unless s is near the negative real axis, one is reasonably safe in using the rule of thumb that when the terms are rapidly decreasing in magnitude, the first term omitted accounts for the bulk of the error (because when it is included, the error is reduced to the next lower order of magnitude).

The constant A in Stirling's formula can be evaluated quite easily by applying Stirling's formula on the imaginary axis and combining the result with the estimate of Re log Π on the imaginary axis which follows from

$$\sin \pi s = \pi s/\Pi(s)\Pi(-s)$$

[see (6) of Section 1.3]. Since Π is real on the real axis, the reflection principle implies $\Pi(\bar{s}) = \overline{\Pi(s)}$, and with $s = it$, the above gives

$$|\Pi(it)|^2 = \frac{\pi it}{\sin \pi it} = \frac{2\pi t}{e^{\pi t} - e^{-\pi t}},$$

$$2 \log |\Pi(it)| = \log 2\pi + \log t - \log e^{\pi t} - \log(1 - e^{-2\pi t}),$$

$$\operatorname{Re} \log \Pi(it) = \frac{1}{2} \log 2\pi + \frac{1}{2} \log t - \frac{\pi t}{2} - \frac{1}{2} \log(1 - e^{-2\pi t}).$$

On the other hand by Stirling's formula (2)

$$\begin{aligned}\operatorname{Re} \log \Pi(it) &= \operatorname{Re}\left\{\left(it + \frac{1}{2}\right) \log it - it + A - \int_0^\infty \frac{\bar{B}_1(u)\,du}{u + it}\right\}\\ &= \frac{1}{2} \log t - t\frac{\pi}{2} + A - \int_0^\infty \frac{u\bar{B}_1(u)\,du}{u^2 + t^2};\end{aligned}$$

hence

$$A = \frac{1}{2} \log 2\pi - \frac{1}{2} \log(1 - e^{-2\pi t}) + \int_0^\infty \frac{u\bar{B}_1(u)\,du}{u^2 + t^2}$$

and the desired result $A = \frac{1}{2} \log 2\pi$ follows by taking the limit as $t \to \infty$.

The derivative of Stirling's series is the series

$$\text{(4)} \qquad \frac{\Pi'(s)}{\Pi(s)} = \log s + \frac{1}{2s} - \frac{B_2}{2s^2} - \frac{B_4}{4s^4} - \cdots - \frac{B_{2\nu}}{2\nu s^{2\nu}} + R'_{2\nu},$$

where

$$R'_{2\nu} = \int_0^\infty \frac{\bar{B}_{2\nu}(x)\,dx}{(s + x)^{2\nu+1}} = \int_0^\infty \frac{\bar{B}_{2\nu+1}(x)\,dx}{(s + x)^{2\nu+2}}.$$

This gives a precise form of the estimate $\Pi'(s)/\Pi(s) \sim \log s$ which was used in de la Vallée Poussin's proof in Section 5.2. Specifically, to prove that there is a $K > 0$ such that

$$\operatorname{Re} \frac{\Pi'[(\sigma + it)/2]}{\Pi[(\sigma + it)/2]} \leq 2 \log t$$

in the region $1 \le \sigma \le 2$, $t \ge K$ as was claimed in Section 5.2, it suffices to estimate R_0' to find

$$|R_0'| \le \frac{1}{\cos^3(\theta/2)} \cdot \left|\frac{B_2}{2s^2}\right|,$$

$$\left|\frac{\Pi'(s)}{\Pi(s)} - \log s\right| \le \frac{1}{2|s|} + \frac{B_2}{\cos^3(\theta/2)2|s|^2}.$$

For s in the halfplane $\operatorname{Re} s \ge 0$ and for $|s|$ sufficiently large, this gives

$$\left|\operatorname{Re}\frac{\Pi'(s)}{\Pi(s)} - \operatorname{Re}\log s\right| \le \frac{1}{|s|},$$

$$\operatorname{Re}\frac{\Pi'[(\sigma+it)/2]}{\Pi[(\sigma+it)/2]} \le \log\left|\frac{\sigma+it}{2}\right| + \frac{1}{|s|}$$

from which the desired inequality follows.

6.4 EVALUATION OF ζ BY EULER–MACLAURIN SUMMATION

Euler–Maclaurin summation applied directly to the series $\zeta(s) = \sum_1^\infty n^{-s}$ ($\operatorname{Re} s \ge 1$) does not give a workable method of evaluating $\zeta(s)$ because the remainders are not at all small. [This is the case $N = 1$ of formula (1) below.] However, if Euler–Maclaurin is applied instead to the series $\sum_{n=N}^\infty n^{-s}$, it gives a quite workable means of approximating numerically the sum of this series and hence, since the terms $\sum_{n=1}^{N-1} n^{-s}$ can be summed directly, a workable means of evaluating $\zeta(s)$. Explicitly, if $\operatorname{Re} s > 1$, then

$$\sum_{n=1}^{\infty} n^{-s} - \sum_{n=1}^{N-1} n^{-s} = \sum_{n=N}^{\infty} n^{-s},$$

$$\zeta(s) - \sum_{n=1}^{N-1} n^{-s} = \int_N^\infty x^{-s}\,dx + \tfrac{1}{2}N^{-s} + \int_N^\infty \bar{B}_1(x)(-s)x^{-s-1}\,dx,$$

(1) $$\zeta(s) = \sum_{n=1}^{N-1} n^{-s} + \frac{N^{1-s}}{s-1} + \frac{1}{2}N^{-s} + \frac{B_2}{2}sN^{-s-1} + \cdots + \frac{B_{2\nu}}{(2\nu)!}s(s+1)\cdots(s+2\nu-2)N^{-s-2\nu+1} + R_{2\nu},$$

where

$$R_{2\nu} = -\frac{s(s+1)\cdots(s+2\nu-1)}{(2\nu)!}\int_N^\infty \bar{B}_{2\nu}(x)x^{-s-2\nu}\,dx$$
$$= -\frac{s(s+1)\cdots(s+2\nu)}{(2\nu+1)!}\int_N^\infty \bar{B}_{2\nu+1}(x)x^{-s-2\nu-1}\,dx.$$

If N is at all large, say N is about the same size as $|s|$, then the terms of the series (1) decrease quite rapidly at first and it is natural to expect that the remainder $R_{2\nu}$ will be quite small. The same method by which Stieltjes estimated the remainder in Stirling's series can be applied to estimate the remainder $R_{2\nu}$ above. It gives

$$\begin{aligned}
|R_{2\nu-2}| &= \left|\frac{s(s+1)\cdots(s+2\nu-2)}{(2\nu-1)!}\int_N^\infty \bar{B}_{2\nu-1}(x)x^{-s-2\nu+1}\,dx\right| \\
&= \left|\frac{s(s+1)\cdots(s+2\nu-1)}{(2\nu)!}\int_N^\infty [B_{2\nu}-\bar{B}_{2\nu}(x)]x^{-s-2\nu}\,dx\right| \\
&\le \left|\frac{s(s+1)\cdots(s+2\nu-1)}{(2\nu)!}\right|\int_N^\infty (-1)^{\nu+1}[B_{2\nu}-\bar{B}_{2\nu}(x)]x^{-\sigma-2\nu}\,dx \\
&= \left|\frac{s(s+1)\cdots(s+2\nu-1)}{(2\nu)!}\right| |B_{2\nu}| \int_N^\infty x^{-\sigma-2\nu}\,dx \\
&\quad - \left|\frac{s(s+1)\cdots(s+2\nu-1)}{(2\nu)!}\right| \int_N^\infty \frac{(-1)^{\nu+1}\bar{B}_{2\nu+1}(x)}{(2\nu+1)} \\
&\quad \times (\sigma+2\nu)x^{-\sigma-2\nu-1}\,dx \\
&\le \left|\frac{s(s+1)\cdots(s+2\nu-1)B_{2\nu}N^{-\sigma-2\nu+1}}{(2\nu)!(\sigma+2\nu-1)}\right| \\
&= \left|\frac{s+2\nu-1}{\sigma+2\nu-1}\right| |B_{2\nu} \text{ term of (1)}|,
\end{aligned}$$

where $\sigma = \operatorname{Re} s$. In words, *if the $B_{2\nu}$ term is the first term omitted, then the magnitude of the remainder in the series* (1) *is at most* $|s+2\nu-1|(\sigma+2\nu-1)^{-1}$ *times the magnitude of the first term omitted.* This estimate is due to Backlund [B2]. In particular, if s is real, the remainder is less than the first term omitted, every term overshoots, and the actual value always lies between two consecutive partial sums of the series (1).

Although formula (1) is derived from the formula $\zeta(s) = \Sigma n^{-s}$ which is valid only for $\operatorname{Re} s > 1$, it obviously "remains valid," in Riemann's terminology, as long as the integral for $R_{2\nu}$ converges, which is true throughout the halfplane $\operatorname{Re}(s+2\nu+1) > 1$. Since ν is arbitrary, this gives an alternate proof of the fact that $\zeta(s)$ can be continued analytically throughout the s-plane with just one simple pole at $s = 1$ and no other singularities.

For some idea of how formula (1) works in actual practice, consider the case $s = 2$. A good way to proceed is to compute the first several numbers in the sequence $s(s+1)\cdots(s+2\nu-2)B_{2\nu}/(2\nu)!$ and to see how large N must be in order to make the terms of the series (1) decrease rapidly in size. Now

$$\frac{B_2}{2}\cdot 2 = \frac{1}{6} = 0.16666\ldots,$$

$$\frac{B_4}{4!}\cdot 2\cdot 3\cdot 4 = -\frac{1}{30} = -0.03333\ldots,$$

$$\frac{B_6}{6!}\cdot 2\cdot 3\cdot 4\cdot 5\cdot 6 = \frac{1}{42} = 0.0238095\ldots.$$

$1^{-2} =$	$1.000\,000\,00$
$2^{-2} =$	$0.250\,000\,00$
$3^{-2} =$	$0.111\,111\,11$
$4^{-2} =$	$0.062\,500\,00$
$5^{-1}(2-1)^{-1} =$	$0.200\,000\,00$
$\frac{1}{2}5^{-2} =$	$0.020\,000\,00$
B_2 term $=$	$0.001\,333\,33$
B_4 term $=$	$-0.000\,010\,66$
$\zeta(2) \sim$	$1.644\,933\,78$

Computation of approximation to $\zeta(2)$.

With $N = 5$ the last term is divided by $5^7 > 70{,}000$ so it is less than about 3 in the seventh place. Thus the approximation $\zeta(2) \sim 1.644\,933\,78$ is correct to six decimal places. For $s = \frac{1}{2}$ the terms decrease even more rapidly and with N only equal to 4 the B_6 term is just

$$\frac{B_6}{6!}\cdot\left(\frac{1}{2}\right)\left(\frac{3}{2}\right)\left(\frac{5}{2}\right)\left(\frac{7}{2}\right)\left(\frac{9}{2}\right)4^{-11/2} = \frac{1}{2^{21}} = \frac{1}{2}(2^{-10})^2 < \frac{1}{2}(10^{-3})^2,$$

so it does not affect the sixth decimal place and the approximation† $\zeta(\frac{1}{2}) \sim 1.460\,354\,96$ is correct to six decimal places.

$1^{-1/2} =$	$1.000\,000\,00$
$2^{-1/2} =$	$0.707\,106\,78$
$3^{-1/2} =$	$0.577\,350\,27$
$4^{1/2}(\frac{1}{2}-1)^{-1} =$	$-4.000\,000\,00$
$\frac{1}{2}4^{-1/2} =$	$0.250\,000\,00$
B_2 term $=$	$0.005\,208\,33$
B_4 term $=$	$-0.000\,020\,34$
$\zeta(\frac{1}{2}) \sim$	$-1.460\,354\,96$

Computation of approximation to $\zeta(\frac{1}{2})$.

Consider finally a case in which s is not real, the case $s = \frac{1}{2} + 18i$. Since $|s|$ is considerably larger in this case than in the two previous ones, it is clear that it will be necessary to use a much larger value of N in order to achieve comparable accuracy. On the other hand, the numbers $2^{-s}, 3^{-s}, \ldots$ are quite a bit more difficult to compute when s is complex (since $n^{-s} = n^{-1/2}e^{-i18\log n} = n^{-1/2}[\cos(18\log n) - i\sin(18\log n)]$ each one involves computing a square root, a logarithm, and two trigonometric functions), so this involves a great deal of computation. Rather than carry through all this computation, it will

†The square roots $(0.5)^{1/2}$, $(0.333\ldots)^{1/2}$ can be computed very easily using the usual iteration $x_{n+1} = \frac{1}{2}(x_n + Ax_n^{-1})$ for $A^{1/2}$.

be more illustrative of the technique to settle for less accuracy and to use a smaller value of N. Since $4^{-s} = (2^{-s})^2$ and $6^{-s} = 2^{-s} \cdot 3^{-s}$, the three values 2^{-s}, 3^{-s}, and 5^{-s} are the only values of n^{-s} which need to be computed in order to use (1) with $N = 6$. For this reason, the value $N = 6$ will be used in the computation below.

The first step is to estimate the size of the $B_{2\nu}$ terms in order to determine the degree of accuracy with which the calculations should be carried out. The B_2 term has modulus about $\frac{1}{2}|B_2| \cdot 18 \cdot 6^{-3/2} = (2 \cdot 2 \cdot \sqrt{6})^{-1} \sim 0.1$, the B_4 term modulus about $18 \cdot 18 \cdot 18/24 \cdot 30 \cdot 6^{7/2} = 3/8 \cdot 10\sqrt{6} \sim 3/196 \sim 0.015$, and the B_6 term modulus about $18 \cdot 18 \cdot 18 \cdot 18 \cdot 20/6! \cdot 42 \cdot 6^{11/2} = (2^3 \cdot 14 \cdot \sqrt{6})^{-1} \sim 1/270 \sim 0.0037$. The modulus of the B_8 term is about 0.0037 times

$$\left|\frac{B_8}{B_6 \cdot 8 \cdot 7} \cdot \left(\frac{11}{2} + 18i\right)\left(\frac{13}{2} + 18i\right)6^{-2}\right| \sim \frac{42}{30 \cdot 8 \cdot 7 \cdot 6 \cdot 6}|6 + 18i|^2 = \frac{1}{4},$$

so it is about 9 in the fourth place. Then the modulus of the B_{10} term is about 0.0009 times

$$\begin{aligned}\left|\frac{B_{10}}{B_8 \cdot 10 \cdot 9} \cdot \left(\frac{15}{2} + 18i\right)\left(\frac{17}{2} + 18i\right)6^{-2}\right| &\sim \frac{5 \cdot 30(8^2 + 18^2)}{66 \cdot 10 \cdot 9 \cdot 6^2} \\ &\sim \frac{5^2 \cdot 2^2(4^2 + 9^2)}{11 \cdot 10 \cdot 9 \cdot 6^2} \sim \frac{100 \cdot 97}{1{,}000 \cdot 36} \\ &\sim 0.27,\end{aligned}$$

so it is less than 3 in the fourth place. Thus, by Backlund's estimate of the remainder, if the B_{10} term is the first one omitted, the error is less than $|s + 9|$ $(\frac{1}{2} + 9)^{-1} \cdot 3 \cdot 10^{-4} \sim |1 + 2i| \cdot 3 \cdot 10^{-4} < 7 \cdot 10^{-4}$ so the answer is correct to three decimal places at least. The terms are now decreasing fairly slowly; two more terms would be required to obtain one more place, so it is reasonable to quit with the B_8 term. [If more than three-place accuracy were required, the easiest way to achieve it would probably be to compute 7^{-s}, after which $8^{-s} = 2^{-s} \cdot 4^{-s}$, $9^{-s} = (3^{-s})^2$, $10^{-s} = 2^{-s} \cdot 5^{-s}$ are easy and one can use $N = 10$ in (1).] Therefore the calculations will be carried out to five places with the intention of retaining three places in the final answer.

To compute 2^{-s} ($s = \frac{1}{2} + 18i$) with five-place accuracy, one must have accurate values of log 2 and π, say $\log 2 = 0.693\,147\,18$ and $\pi = 3.141\,592\,65$. Then $18 \log 2 = 4\pi - \theta$, where θ is 0.08972 to five places. The power series for $\cos\theta$ and $\sin\theta$ then give $\cos\theta = 0.99598$, $\sin\theta = 0.08960$ which, combined with the value $2^{-1/2} = 0.707\,11$ found above, gives $2^{-s} = (0.704\,27) + (0.063\,36)i$. Similarly, using accurate values of log 3 and π, one can find $18 \log 3 = 6\pi + (\pi/3) - \theta$ where $\theta = 0.12174$ to five places, from which

$$\begin{aligned}3^{-s} = 3^{-1/2}e^{-i\pi/3}e^{i\theta} &= \frac{1}{\sqrt{3}}\left(\frac{1}{2} - \frac{\sqrt{3}}{2}i\right)(\cos\theta + i\sin\theta) \\ &= 0.34726 - 0.46124i\end{aligned}$$

can be computed. Finally, $18 \log 5 = 9\pi + (\pi/4) - \theta$, where $\theta = 0.08985$ so that

$$5^{-s} = -\frac{1}{\sqrt{5}}\left(\frac{1}{\sqrt{2}} - \frac{i}{\sqrt{2}}\right)(\cos\theta + i\sin\theta) = -0.34333 + 0.28657i.$$

Using these values one can then compute $4^{-s} = 0.49199 + 0.08924i$ and $6^{-s} = 0.27379 - 0.30284i$ with (almost) five-place accuracy. Now

$$\frac{B_2}{2}s \cdot 6^{-s-1} = \left(\frac{1}{144} + \frac{1}{4}i\right)6^{-s} = 0.07761 + 0.06634i,$$

$$\begin{aligned}\frac{B_4}{4!}s(s+1)(s+2)6^{-s-3} &= -\frac{(\frac{1}{2} + 18i)(\frac{3}{2} + 18i)(\frac{5}{2} + 18i)}{30 \cdot 24 \cdot 6^3}6^{-s}\\ &= \frac{1456\frac{1}{8} + 5728\frac{1}{2}i}{720 \cdot 6^3} \cdot 6^{-s}\\ &= 0.01372 + 0.00725i.\end{aligned}$$

The B_6 term is the B_4 term times

$$\frac{B_6(\frac{7}{2} + 18i)(\frac{9}{2} + 18i)}{B_4 \cdot 6 \cdot 5 \cdot 6^2} = \frac{308\frac{1}{4} - 144i}{1512} = 0.20387 - 0.09524i$$

so it is $0.00349 + 0.00017i$. Similarly the B_8 term is the B_6 term times

$$\frac{B_8(\frac{11}{2} + 18i)(\frac{13}{2} + 18i)}{B_6 \cdot 8 \cdot 7 \cdot 6^2} = \frac{288\frac{1}{4} - 216i}{1440}$$

which gives $0.00072 - 0.00053i$ as the value of the B_8 term. Adding these values up then gives $\zeta(\frac{1}{2} + 18i) \sim 2.329\,22 - 0.188\,65i$ as the value of $\zeta(\frac{1}{2} + 18i)$ to three decimal places.

1^{-s}	=	$1.000\,00$
2^{-s}	=	$+0.704\,27 + 0.063\,36i$
3^{-s}	=	$+0.347\,26 - 0.461\,24i$
4^{-s}	=	$+0.491\,99 + 0.089\,24i$
5^{-s}	=	$-0.343\,33 + 0.286\,57i$
$6^{1-s}(s-1)^{-1}$	=	$-0.103\,40 - 0.088\,39i$
$\frac{1}{2}6^{-s}$	=	$+0.136\,89 - 0.151\,42i$
B_2 term	=	$+0.077\,61 + 0.066\,34i$
B_4 term	=	$+0.013\,72 + 0.007\,25i$
B_6 term	=	$+0.003\,49 + 0.000\,17i$
B_8 term	=	$+0.000\,72 - 0.000\,53i$
$\zeta(\frac{1}{2} + 18i)$	$\sim$	$2.329\,22 - 0.188\,65i$

Computation of approximation to $\zeta(\frac{1}{2} + 18i)$.

6.5 TECHNIQUES FOR LOCATING ROOTS ON THE LINE

The roots ρ are by definition the zeros of the function $\xi(s) = \Pi(s/2)\pi^{-s/2}(s-1)\zeta(s)$. For any given s, the value of $\xi(s)$ can be computed to any prescribed degree of accuracy by combining the techniques of the preceding two sections. Since $\xi(s)$ is *real valued* on the line $\operatorname{Re} s = \frac{1}{2}$, it can be shown to have zeros on the line by showing that it *changes sign*. This, in a nutshell, is the method by which roots ρ on the line $\operatorname{Re} s = \frac{1}{2}$ will be located in this section.

Consider, then, the problem of determining the sign of $\xi(\frac{1}{2}+it)$. If this function of t is rewritten in the form

$$\begin{aligned}\xi\left(\frac{1}{2}+it\right) &= \frac{s}{2}\Pi\left(\frac{s}{2}-1\right)\pi^{-s/2}(s-1)\zeta(s)\\ &= e^{\log \Pi[(s/2)-1]}\pi^{-s/2}\cdot\frac{s(s-1)}{2}\cdot\zeta(s)\\ &= \left[e^{\operatorname{Re}\log \Pi[(s/2)-1]}\pi^{-1/4}\cdot\frac{-t^2-\frac{1}{4}}{2}\right]\\ &\quad\times\left[e^{i\operatorname{Im}\log \Pi[(s/2)-1]}\pi^{-it/2}\zeta\left(\frac{1}{2}+it\right)\right]\end{aligned}$$

(where $s = \frac{1}{2}+it$), then the determination of its sign can be simplified by the observation that the factor in the first set of brackets is a negative real number; hence that the sign of $\xi(\frac{1}{2}+it)$ is opposite to the sign of the factor in the second set of brackets. The standard notation for this second factor is $Z(t)$, that is,

$$Z(t) = e^{i\vartheta(t)}\zeta(\tfrac{1}{2}+it)$$

where $\vartheta(t)$ is defined by

$$\vartheta(t) = \operatorname{Im}\log \Pi\left(\frac{it}{2}-\frac{3}{4}\right) - \frac{t}{2}\log\pi.$$

If $\vartheta(t)$ and $Z(t)$ are so defined, then *the sign of* $\xi(\frac{1}{2}+it)$ *is opposite to the sign*† *of* $Z(t)$. Thus, to determine the sign of $\xi(\frac{1}{2}+it)$ it suffices to compute $\vartheta(t)$, $\zeta(\frac{1}{2}+it)$ by the methods of the preceding two sections and to combine them to find $Z(t)$.

The computation of $\vartheta(t)$ can be simplified as follows:

$$\begin{aligned}\vartheta(t) &= \operatorname{Im}\left[\log \Pi\left(\frac{it}{2}+\frac{1}{4}\right) - \log\left(\frac{it}{2}+\frac{1}{4}\right)\right] - \frac{t}{2}\log\pi\\ &= \operatorname{Im}\left[\left(\frac{it}{2}-\frac{1}{4}\right)\log\left(\frac{it}{2}+\frac{1}{4}\right) - \left(\frac{it}{2}+\frac{1}{4}\right) + \frac{1}{2}\log 2\pi\right.\\ &\quad\left.+\frac{1}{12\left(\frac{it}{2}+\frac{1}{4}\right)} - \frac{1}{360\left(\frac{it}{2}+\frac{1}{4}\right)^3} + \cdots\right] - \frac{t}{2}\log\pi\end{aligned}$$

†In particular, $Z(t)$ is real when, of course, t is real.

$$= \frac{t}{2}\,\mathrm{Re}\log\left(\frac{it}{2}+\frac{1}{4}\right) - \frac{1}{4}\,\mathrm{Im}\log\left(\frac{it}{2}+\frac{1}{4}\right) - \frac{t}{2}$$

$$+ \frac{-\frac{t}{2}}{12\left(\frac{t^2}{4}+\frac{1}{16}\right)} - \frac{\mathrm{Im}\left(-\frac{it}{2}+\frac{1}{4}\right)^3}{360\left(\frac{t^2}{4}+\frac{1}{16}\right)^3} + \cdots - \frac{t}{2}\log\pi$$

$$= \frac{t}{2}\log\left[\left(\frac{t}{2}\right)^2\left(1+\frac{1}{4t^2}\right)\right]^{1/2} - \frac{1}{4}\left[\frac{\pi}{2} - \mathrm{Arctan}\left(\frac{1}{4}\Big/\frac{t}{2}\right)\right] - \frac{t}{2}$$

$$- \frac{1}{6t\left(1+\frac{1}{4t^2}\right)} - \frac{\frac{t^3}{8} + 3\left(-\frac{t}{2}\right)\left(\frac{1}{4}\right)^2}{360\left(\frac{t^2}{4}\right)^3\left(1+\frac{1}{4t^2}\right)^3} + \cdots - \frac{t}{2}\log\pi$$

$$= \frac{t}{2}\log\frac{t}{2} + \frac{t}{4}\log\left(1+\frac{1}{4t^2}\right) - \frac{\pi}{8} + \frac{1}{4}\,\mathrm{Arctan}\left(\frac{1}{2t}\right)$$

$$- \frac{t}{2} - \frac{1}{6t}\left(1+\frac{1}{4t^2}\right)^{-1} - \frac{1}{45t^3}\left(1+\frac{1}{4t^2}\right)^{-3}$$

$$+ \frac{1}{60t^5}\left(1+\frac{1}{4t^2}\right)^{-3} + \cdots - \frac{t}{2}\log\pi$$

$$= \frac{t}{2}\log\frac{t}{2\pi} + \frac{t}{4}\left[\frac{1}{4t^2} - \frac{1}{2}\left(\frac{1}{4t^2}\right)^2 + \cdots\right]$$

$$- \frac{\pi}{8} + \frac{1}{4}\left[\left(\frac{1}{2t}\right) - \frac{1}{3}\left(\frac{1}{2t}\right)^3 + \cdots\right]$$

$$- \frac{t}{2} - \frac{1}{6t}\left[1 - \frac{1}{4t^2} + \cdots\right] - \frac{1}{45t^3}\left[1 - \frac{3}{4t^2} + \cdots\right]$$

$$+ \frac{1}{60t^5}\left(1+\frac{1}{4t^2}\right)^{-3} + \cdots.$$

This gives finally

$$\vartheta(t) = \frac{t}{2}\log\frac{t}{2\pi} - \frac{t}{2} - \frac{\pi}{8} + \frac{1}{48t} + \frac{7}{5760t^3} + \cdots. \tag{1}$$

Since the terms decrease very rapidly for t at all large and since the error is comparable to the first term omitted, it is clear that the error in the approximation

$$\vartheta(t) \sim \frac{t}{2}\log\frac{t}{2\pi} - \frac{t}{2} - \frac{\pi}{8} + \frac{1}{48t} \tag{2}$$

is very slight. Specifically, the error involved in the above use of Stirling's series is less than

$$\frac{1}{\cos^6\frac{\theta}{2}}\cdot\left|\frac{1}{42\cdot 6\cdot 5\left(\frac{it}{2}+\frac{1}{4}\right)^5}\right| \le \frac{1}{t^5}\,\frac{2^3}{42\cdot 6\cdot\left(\frac{1}{2}\right)^5\left(1+\frac{1}{4t^2}\right)^{5/2}}$$

$$\le \frac{1}{t^5}\cdot\frac{64}{63},$$

and the errors resulting from truncating the alternating series are less than the first term omitted, which gives a total error of less than

$$\frac{t}{4}\cdot\frac{1}{3}\left(\frac{1}{4t^2}\right)^3+\frac{1}{4}\cdot\frac{1}{5}\left(\frac{1}{2t}\right)^5+\frac{1}{6t}\left(\frac{1}{4t^2}\right)^2+\frac{1}{45t^3}\cdot\frac{1}{60t^5}<\frac{1}{t^5},$$

so the total error in the approximation (2) is comfortably less than

$$\frac{7}{5760t^3}+\frac{2}{t^5}.$$

The second term in this estimate is very crude and could be much improved by using more terms of Stirling's series, finding further terms of the series (1), and estimating the error in terms of t^{-7} or t^{-9} to show that the first term $7/5760t^3$ contains the bulk of the error when t is at all large. However, this form of the estimate is quite adequate for the numerical analysis of the roots ρ.

Thus to find the sign of $\xi(\frac{1}{2}+18i)$ one can simply compute

$$\begin{aligned}\vartheta(18) &\sim 9\log\frac{9}{\pi}-9-\frac{\pi}{8}+\frac{1}{48\cdot 18}\\ &= 9.472452-9-0.392699+0.001158\\ &= 0.080911\end{aligned}$$

and combine it with the value of $\zeta(\frac{1}{2}+18i)$ computed in the previous section to find that

$$Z(18)=e^{0.0809i}(2.329-0.189i)=2.337+0.000i$$

is positive and that $\xi(\frac{1}{2}+18i)$ is therefore *negative*. Since $\xi(\frac{1}{2})$ is positive (ξ is positive on the entire real axis), it follows that *there is at least one root ρ on the line segment from $\frac{1}{2}$ to $\frac{1}{2}+18i$.* By computing further values of Z one could obtain more detailed information on the sign of $\xi(\frac{1}{2}+it)$ and therefore more precise information on the location of roots ρ on $\operatorname{Re} s=\frac{1}{2}$. However, the evaluation of Z requires *both* an evaluation of ζ *and* an evaluation of ϑ, and in order to locate roots ρ on $\operatorname{Re} s=\frac{1}{2}$, it suffices to evaluate just ζ—and in fact just $\operatorname{Im}\zeta$—provided one analyzes the result with a certain amount of ingenuity. Consider, for example, in Table IV the values of $\zeta(\frac{1}{2}+it)$ computed with two-place accuracy at intervals of 0.2 from $t=0$ to $t=50$.

Although these values were taken from Haselgrove's tables [H8], it would not be too lengthy to compute them from scratch using Euler–Maclaurin, particularly in view of the fact that there are economies of scale in computing many values of n^{it} ($n=2, 3, 4, \ldots$) which is the major operation in the Euler–Maclaurin evaluation of ζ. Now examination of this table leads to a few very elementary but very useful observations.

In the first place, the real part of ζ has a strong tendency to be positive. There are brief intervals, 11 in all, where $\operatorname{Re}\zeta$ is negative, but apart from the first one, which is clearly atypical, the longest of them is from 47.2 to 48.0

TABLE IV[a]

t	$\zeta(\frac{1}{2}+it)$	t	$\zeta(\frac{1}{2}+it)$	t	$\zeta(\frac{1}{2}+it)$
0.0	-1.46	9.0	$+1.45+0.19i$	18.0	$+2.33-0.19i$
0.2	$-1.18-0.67i$	9.2	$+1.48+0.14i$	18.2	$+2.27-0.43i$
0.4	$-0.68-0.94i$	9.4	$+1.51+0.08i$	18.4	$+2.17-0.66i$
0.6	$-0.28-0.94i$	9.6	$+1.53+0.02i$	18.6	$+2.02-0.86i$
0.8	$-0.02-0.84i$	9.8	$+1.54-0.04i$	18.8	$+1.84-1.03i$
1.0	$+0.14-0.72i$	10.0	$+1.54-0.12i$	19.0	$+1.62-1.16i$
1.2	$+0.25-0.62i$	10.2	$+1.54-0.19i$	19.2	$+1.38-1.24i$
1.4	$+0.32-0.52i$	10.4	$+1.53-0.26i$	19.4	$+1.13-1.28i$
1.6	$+0.37-0.44i$	10.6	$+1.50-0.34i$	19.6	$+0.88-1.26i$
1.8	$+0.41-0.37i$	10.8	$+1.47-0.42i$	19.8	$+0.65-1.18i$
2.0	$+0.44-0.31i$	11.0	$+1.42-0.49i$	20.0	$+0.43-1.06i$
2.2	$+0.46-0.26i$	11.2	$+1.36-0.56i$	20.2	$+0.25-0.90i$
2.4	$+0.48-0.21i$	11.4	$+1.29-0.62i$	20.4	$+0.11-0.70i$
2.6	$+0.50-0.16i$	11.6	$+1.21-0.67i$	20.6	$+0.02-0.48i$
2.8	$+0.52-0.12i$	11.8	$+1.12-0.71i$	20.8	$-0.02-0.25i$
3.0	$+0.53-0.08i$	12.0	$+1.02-0.75i$	21.0	$-0.01-0.02i$
3.2	$+0.55-0.04i$	12.2	$+0.91-0.76i$	21.2	$+0.06+0.19i$
3.4	$+0.56-0.01i$	12.4	$+0.79-0.76i$	21.4	$+0.18+0.38i$
3.6	$+0.58+0.03i$	12.6	$+0.68-0.75i$	21.6	$+0.34+0.52i$
3.8	$+0.59+0.06i$	12.8	$+0.56-0.71i$	21.8	$+0.52+0.62i$
4.0	$+0.61+0.09i$	13.0	$+0.44-0.66i$	22.0	$+0.72+0.67i$
4.2	$+0.62+0.12i$	13.2	$+0.33-0.58i$	22.2	$+0.92+0.66i$
4.4	$+0.64+0.15i$	13.4	$+0.23-0.49i$	22.4	$+1.11+0.60i$
4.6	$+0.66+0.18i$	13.6	$+0.15-0.38i$	22.6	$+1.26+0.49i$
4.8	$+0.68+0.21i$	13.8	$+0.07-0.25i$	22.8	$+1.38+0.34i$
5.0	$+0.70+0.23i$	14.0	$+0.02-0.10i$	23.0	$+1.45+0.16i$
5.2	$+0.73+0.26i$	14.2	$-0.01+0.05i$	23.2	$+1.46-0.03i$
5.4	$+0.75+0.28i$	14.4	$-0.01+0.21i$	23.4	$+1.41-0.21i$
5.6	$+0.78+0.30i$	14.6	$+0.01+0.38i$	23.6	$+1.30-0.38i$
5.8	$+0.81+0.32i$	14.8	$+0.07+0.55i$	23.8	$+1.14-0.50i$
6.0	$+0.84+0.34i$	15.0	$+0.15+0.70i$	24.0	$+0.95-0.58i$
6.2	$+0.87+0.36i$	15.2	$+0.26+0.85i$	24.2	$+0.73-0.60i$
6.4	$+0.91+0.37i$	15.4	$+0.39+0.98i$	24.4	$+0.51-0.55i$
6.6	$+0.94+0.38i$	15.6	$+0.56+1.09i$	24.6	$+0.30-0.43i$
6.8	$+0.98+0.39i$	15.8	$+0.74+1.17i$	24.8	$+0.13-0.25i$
7.0	$+1.02+0.40i$	16.0	$+0.94+1.22i$	25.0	$\pm 0.00-0.01i$
7.2	$+1.06+0.40i$	16.2	$+1.15+1.23i$	25.2	$-0.05+0.26i$
7.4	$+1.11+0.40i$	16.4	$+1.36+1.20i$	25.4	$-0.04+0.55i$
7.6	$+1.15+0.39i$	16.6	$+1.57+1.14i$	25.6	$+0.06+0.85i$
7.8	$+1.20+0.38i$	16.8	$+1.77+1.04i$	25.8	$+0.25+1.11i$
8.0	$+1.24+0.36i$	17.0	$+1.95+0.90i$	26.0	$+0.50+1.34i$
8.2	$+1.29+0.34i$	17.2	$+2.10+0.72i$	26.2	$+0.82+1.49i$
8.4	$+1.33+0.31i$	17.4	$+2.22+0.52i$	26.4	$+1.17+1.56i$
8.6	$+1.37+0.28i$	17.6	$+2.30+0.29i$	26.6	$+1.55+1.54i$
8.8	$+1.41+0.24i$	17.8	$+2.34+0.06i$	26.8	$+1.92+1.42i$

[a]Values from Haselgrove [H8].

TABLE IV *continued*[a]

t	$\zeta(\frac{1}{2} + it)$	t	$\zeta(\frac{1}{2} + it)$	t	$\zeta(\frac{1}{2} + it)$
27.0	$+2.25 + 1.21i$	35.0	$+2.60 + 1.11i$	43.0	$+0.44 - 0.31i$
27.2	$+2.53 + 0.91i$	35.2	$+2.84 + 0.67i$	43.2	$+0.16 - 0.16i$
27.4	$+2.73 + 0.55i$	35.4	$+2.94 + 0.17i$	43.4	$-0.07 + 0.11i$
27.6	$+2.83 + 0.15i$	35.6	$+2.89 - 0.33i$	43.6	$-0.20 + 0.50i$
27.8	$+2.83 - 0.27i$	35.8	$+2.70 - 0.80i$	43.8	$-0.18 + 0.94i$
28.0	$+2.72 - 0.68i$	36.0	$+2.38 - 1.19i$	44.0	$+0.01 + 1.40i$
28.2	$+2.52 - 1.05i$	36.2	$+1.97 - 1.46i$	44.2	$+0.37 + 1.80i$
28.4	$+2.23 - 1.35i$	36.4	$+1.50 - 1.59i$	44.4	$+0.87 + 2.08i$
28.6	$+1.87 - 1.57i$	36.6	$+1.03 - 1.57i$	44.6	$+1.47 + 2.19i$
28.8	$+1.48 - 1.69i$	36.8	$+0.60 - 1.40i$	44.8	$+2.11 + 2.10i$
29.0	$+1.09 - 1.70i$	37.0	$+0.26 - 1.12i$	45.0	$+2.71 + 1.80i$
29.2	$+0.71 - 1.61i$	37.2	$+0.04 - 0.76i$	45.2	$+3.21 + 1.31i$
29.4	$+0.38 - 1.43i$	37.4	$-0.05 - 0.36i$	45.4	$+3.54 + 0.69i$
29.6	$+0.13 - 1.18i$	37.6	$+0.01 + 0.03i$	45.6	$+3.66 - 0.03i$
29.8	$-0.04 - 0.89i$	37.8	$+0.19 + 0.36i$	45.8	$+3.56 - 0.74i$
30.0	$-0.12 - 0.58i$	38.0	$+0.46 + 0.59i$	46.0	$+3.24 - 1.39i$
30.2	$-0.11 - 0.29i$	38.2	$+0.80 + 0.71i$	46.2	$+2.75 - 1.90i$
30.4	$-0.02 - 0.03i$	38.4	$+1.14 + 0.69i$	46.4	$+2.14 - 2.22i$
30.6	$+0.14 + 0.17i$	38.6	$+1.44 + 0.55i$	46.6	$+1.49 - 2.33i$
30.8	$+0.33 + 0.30i$	38.8	$+1.67 + 0.31i$	46.8	$+0.86 - 2.24i$
31.0	$+0.52 + 0.34i$	39.0	$+1.79 \pm 0.00i$	47.0	$+0.33 - 1.97i$
31.2	$+0.70 + 0.31i$	39.2	$+1.78 - 0.33i$	47.2	$-0.06 - 1.57i$
31.4	$+0.84 + 0.22i$	39.4	$+1.66 - 0.64i$	47.4	$-0.27 - 1.11i$
31.6	$+0.92 + 0.09i$	39.6	$+1.43 - 0.88i$	47.6	$-0.31 - 0.66i$
31.8	$+0.92 - 0.06i$	39.8	$+1.12 - 1.02i$	47.8	$-0.21 - 0.28i$
32.0	$+0.84 - 0.20i$	40.0	$+0.79 - 1.04i$	48.0	$-0.01 - 0.01i$
32.2	$+0.71 - 0.29i$	40.2	$+0.48 - 0.95i$	48.2	$+0.24 + 0.14i$
32.4	$+0.52 - 0.32i$	40.4	$+0.22 - 0.75i$	48.4	$+0.47 + 0.15i$
32.6	$+0.31 - 0.27i$	40.6	$+0.05 - 0.48i$	48.6	$+0.64 + 0.07i$
32.8	$+0.11 - 0.14i$	40.8	$-0.02 - 0.18i$	48.8	$+0.71 - 0.06i$
33.0	$-0.05 + 0.08i$	41.0	$+0.03 + 0.12i$	49.0	$+0.67 - 0.20i$
33.2	$-0.13 + 0.36i$	41.2	$+0.18 + 0.36i$	49.2	$+0.53 - 0.29i$
33.4	$-0.13 + 0.69i$	41.4	$+0.40 + 0.51i$	49.4	$+0.34 - 0.29i$
33.6	$-0.02 + 1.03i$	41.6	$+0.64 + 0.57i$	49.6	$+0.14 - 0.18i$
33.8	$+0.20 + 1.35i$	41.8	$+0.87 + 0.52i$	49.8	$-0.02 + 0.03i$
34.0	$+0.52 + 1.60i$	42.0	$+1.04 + 0.37i$	50.0	$-0.08 + 0.33i$
34.2	$+0.92 + 1.75i$	42.2	$+1.12 + 0.18i$		
34.4	$+1.37 + 1.79i$	42.4	$+1.08 - 0.04i$		
34.6	$+1.83 + 1.69i$	42.6	$+0.95 - 0.22i$		
34.8	$+2.25 + 1.46i$	42.8	$+0.72 - 0.32i$		

[a]Values from Haselgrove [H8].

and the value of Re ζ does not go much below -0.3 anywhere in the range $1 \le t \le 50$.

The sign of the imaginary part of ζ, on the other hand, oscillates fairly regularly between plus and minus. In fact, these oscillations are smooth enough and the passages through the value zero pronounced enough that one can be fairly certain that $\operatorname{Im}\zeta(\frac{1}{2}+it)$ has exactly 21 zeros in the range $0 < t \le 50$, one in each of the 21 intervals where the table shows a sign change in $\operatorname{Im}\zeta$ (between 3.4 and 3.6, between 9.6 and 9.8, etc.). Since $\xi(\frac{1}{2}+it) = 0$ if and only if $\operatorname{Re}\zeta(\frac{1}{2}+it) = 0$ *and* $\operatorname{Im}\zeta(\frac{1}{2}+it) = 0$, the problem of finding all roots ρ on the line segment from $\frac{1}{2}$ to $\frac{1}{2}+50i$ is reduced to the problem of determining which, if any, of these 21 zeros of $\operatorname{Im}\zeta$ are also zeros of $\operatorname{Re}\zeta$. Now 11 of them, namely, the first, second, fourth, sixth, eighth, . . . , eighteenth, and twentieth lie between points where Re ζ is strongly positive and clearly do not merit serious consideration as possible zeros. Judging from Table IV, however, it appears quite possible that any of the remaining 10—namely, the third, fifth, seventh, etc.—might be zeros of Re ζ. Now when $Z(t)$, $\vartheta(t)$ are defined as above, they are real-valued functions of the real variable t and

$$\begin{aligned}\zeta(\tfrac{1}{2}+it) &= e^{-i\vartheta(t)}Z(t)\\ &= Z(t)\cos\vartheta(t) - iZ(t)\sin\vartheta(t).\end{aligned}$$

Hence, $\operatorname{Im}\zeta(\frac{1}{2}+it) = -Z(t)\sin\vartheta(t)$ and a change of sign of $\operatorname{Im}\zeta$ implies a change of sign of either $Z(t)$ or $\sin\vartheta(t)$. But even very rough calculations of $\vartheta(t)$ on the 10 intervals in question suffice to show that on these intervals $\vartheta(t)$ is nowhere near a multiple of π; therefore $\sin\vartheta(t)$ definitely does not change sign; therefore $Z(t)$ definitely does change sign; therefore there is definitely a zero of Z in the interval and a root ρ on the corresponding interval of $\operatorname{Re} s = \frac{1}{2}$.

This proves the existence of 10 roots ρ on the line segment from $\frac{1}{2}$ to $\frac{1}{2} + 50i$ and locates them between $\frac{1}{2} + i14.0$ and $\frac{1}{2} + i14.2$, between $\frac{1}{2} + i21.0$ and $\frac{1}{2} + i21.2$, etc. To locate them more exactly, it suffices to estimate their position by linear interpolation (*regula falsi*) and calculate $\operatorname{Im}\zeta$ more precisely. For example, linear interpolation suggests, since $\operatorname{Im}\zeta$ goes from -0.10 to $+0.05$ as t goes from 14.0 to 14.2, that the zero lies two thirds of the way through the interval at 14.1333, and it is at this point that the value of $\operatorname{Im}\zeta$ should be computed more exactly. This is precisely the method by which Gram computed the 15 roots given in Section 6.1.

The above calculations prove conclusively the existence of *at least* 10 roots ρ on the line segment from $\frac{1}{2}$ to $\frac{1}{2} + 50i$, and they strongly indicate—but do not prove—that there are no others on this line segment, but they give no information at all about possible roots *not* on the line $\operatorname{Re} s = \frac{1}{2}$ in the range $0 \le \operatorname{Im} s \le 50$. However, Gram did prove, as will be explained in the next section, that the 10 roots he had found are the *only* roots in the range $0 \le \operatorname{Im} s \le 50$, and therefore that the Riemann hypothesis is true in this range.

Consider now the problem of locating more roots beyond $t = 50$. Since the computation of $\zeta(\frac{1}{2} + it)$ becomes increasingly long as t increases, it is desirable to use as few evaluations of $\zeta(\frac{1}{2} + it)$ as possible. Now in the range $10 \leq t \leq 50$ the zeros of $\operatorname{Im} \zeta$ follow a very simple pattern, namely, *in this range the zeros of* $\operatorname{Im} \zeta = Z \sin \vartheta$ *are alternately zeros of* Z *and zeros of* $\sin \vartheta$. Gram showed that it is not unlikely that this pattern persists, at least for a while, past $t = 50$ and he also showed that the zeros of $\sin \vartheta$ are quite easy to find. These two observations simplify considerably the search for further zeros.

Gram gave the following reasons for believing that the alternation of zeros of Z with zeros of $\sin \vartheta$ will persist beyond $t = 50$. In the first place, this phenomenon is closely related to the fact that $\operatorname{Re} \zeta$ has a strong tendency to be positive. To see this relation note first that the approximation

$$\frac{d}{ds} \log \Pi(s) \sim \log s$$

[see (4) of Section 6.3] shows that the derivative of $\vartheta(t)$ is

$$\operatorname{Im} \frac{i}{2} \frac{\Pi'\left(\frac{it}{2} - \frac{3}{4}\right)}{\Pi\left(\frac{it}{2} - \frac{3}{4}\right)} - \frac{1}{2} \log \pi \sim \frac{1}{2} \log \left| \frac{it}{2} - \frac{3}{4} \right| - \frac{1}{2} \log \pi$$

$$\sim \frac{1}{2} \log \frac{t}{2\pi}$$

from which it can easily be shown that $\vartheta'(t) > 0$ for $t \geq 10$. Thus ϑ is an increasing function of t, between consecutive zeros of $\sin \vartheta$ there is exactly one zero of $\cos \vartheta$, and $\cos \vartheta$ changes sign. Therefore if $\operatorname{Re} \zeta = Z \cos \vartheta$ is positive at two consecutive zeros of $\sin \vartheta$—which is likely in view of the preponderance of positive values of $\operatorname{Re} \zeta$—it follows that Z changes sign and hence that there is at least one zero of Z between these two consecutive zeros of $\sin \vartheta$. Since the first failure, if there is one, in the pattern of alternation would naturally be expected to be a pair of consecutive zeros of $\sin \vartheta$ between which there would be either 2 or 0 zeros of Z, the first failure, if there is one, should be indicated by a zero of $\sin \vartheta$, where $\operatorname{Re} \zeta$ is negative instead of positive. The zeros of $\sin \vartheta$ are called *Gram points*, and the conclusion is that the persistence of the pattern of alternation of zeros of Z with Gram points is closely related to the persistence of the positivity of $\operatorname{Re} \zeta$ at Gram points. But Gram argued that the predominance of positive values of $\operatorname{Re} \zeta$ is due to the fact that the Euler–Maclaurin formula for $\zeta(\frac{1}{2} + it)$ *begins* with a $+1$ in the real part and to the fact that, as long as it is not necessary to use too large a value of N, it will be unusual for the smaller† terms which follow to combine to over-

†The negative values of $\operatorname{Re} \zeta$ which occur for small values of t result, as is seen immediately from the computation of $\zeta(\frac{1}{2})$ in the preceding section, from the fact that the term $N^{1-s}/(s - 1)$ then has a large negative real part. This term becomes small as t increases.

whelm this advantage on the plus side. As Gram puts it, equilibrium between plus and minus values of $\operatorname{Re}\zeta$ will be achieved† only very slowly as t increases. Thus the positivity of $\operatorname{Re}\zeta$ at Gram points, and consequently the alternation of Gram points with zeros of Z, is likely to persist for some distance beyond $t = 50$.

To locate the Gram points computationally is quite easy. In the first place, 11 of them can be read off from Table IV, namely, the 11 sign changes of $\operatorname{Im}\zeta$ which are not roots ρ. Since these points are the points where $\vartheta(t)$ is a multiple of π, since $\vartheta(t)$ is increasing for $t \geq 10$ (at least), and since the Gram point near 18 must surely be the solution of $\vartheta(t) = 0$ [because $\vartheta(18) = 0.08$], it is clear that the last 10 of these 11 Gram points are solutions of the equation

$$\vartheta(t) = n\pi$$

for $n = -1, 0, 1, 2, 3, 4, 5, 6, 7, 8$. [The first one, near 3.4, is also a solution for $n = -1$, as can be seen quite easily by setting $t = 3.4$ in (2).] Thus the first Gram point beyond the range of Table IV is a solution of $\vartheta(t) = 9\pi$. Since $\vartheta(t) = 8\pi$ occurs near $t = 48.7$ and since $\vartheta'(t) \sim \frac{1}{2}\log(t/2\pi) \sim \frac{1}{2}\log 8 = \frac{3}{2}\log 2 = 1.0$ in this region, $\vartheta(t) = 9\pi$ should occur near $48.7 + \pi \sim 51.8$. Now (2) gives $\vartheta(51.8) = 28.344$, whereas $9\pi = 28.274$; thus ϑ is too large by 0.070 at 51.8. In the course of computing $\vartheta(51.8)$ one finds that the derivative $\vartheta'(t) \sim \frac{1}{2}\log(t/2\pi)$ is about 1.05. Therefore t should be decreased to about $51.8 - (0.070/1.05) = 51.8 - 0.0666 \sim 51.734$. This value is correct to three places as can be checked by substituting it into (2) and observing that the result is 9π to three places. In this way any Gram point can be found with any desired degree of accuracy with relatively little computation.

In summary, if the nth Gram point g_n is defined to be the unique real number satisfying $\vartheta(g_n) = n\pi$, $g_n \geq 10$ $(n = 0, 1, 2, \ldots)$, then g_n can be computed and the above arguments give some reason to believe that $\operatorname{Re}\zeta(\frac{1}{2} + ig_n)$ will be positive for n well beyond the limit $n = 8$ of Table IV. As long as this remains true it follows that there is at least one root ρ on the line segment from $\frac{1}{2} + ig_{n-1}$ to $\frac{1}{2} + ig_n$ and, in all likelihood, exactly one root.

This program of Gram's was followed by Hutchinson [H11] who computed all the values g_n up to $g_{137} = 300.468$ and determined the sign of $\operatorname{Re}\zeta(\frac{1}{2} + ig_n)$ for each of them. He found that there were two exceptions‡ to the rule

†Gram states: "Si cela est juste, on peut inférer que l'équilibre ne s'établira que peu à peu, de sorte que la même régle sur la répartition des α [the roots on the line] par rapport aux γ [the Gram points] se maintiendra aussi pour les α suivantes les plus rapprochées de α_{15}." Actually Gram was wrong in believing that equilibrium would be achieved at all because as Titchmarsh proved [T4], the average value of $\zeta(\frac{1}{2} + ig_n)$ is 2. See the concluding remarks of Section 11.1.

‡According to Haselgrove [H8] this should not have been a surprise since he says that Bohr and Landau proved in 1913 that there are infinitely many exceptions. However, this seems to be an error on Haselgrove's part. Titchmarsh [T5] proved the existence of infinitely many exceptions in 1935 and he, with his extensive knowledge of the literature, believed this to be a new result.

Re $\zeta(\frac{1}{2} + ig_n) > 0$, namely, $n = 126$ and $n = 134$. He found, moreover, that these exceptions are slight in the sense that the corresponding values of Re ζ are only slightly less than zero and that if the points are shifted only a little bit—g_{126} from 282.455 up to 282.6 and g_{134} from 295.584 down to 295.4—then the sign of Re ζ becomes positive. It is easily checked that these two shifts do not change the sign of $\cos\vartheta(g_n)$ and therefore by the same argument as before [namely, $\cos\vartheta(g_n)$ alternates in sign, $Z(g_n)\cos\vartheta(g_n)$ is always positive, and therefore $Z(g_n)$ alternates in sign] it follows that *there is at least one root* ρ *on the line segment from* $\frac{1}{2} + ig_{n-1}$ *to* $\frac{1}{2} + ig_n$ $(n = 1, 2, \ldots, 137)$ *when* g_{126} *and* g_{134} *are shifted as above*. This locates at least 138 roots ρ (there is one between $\frac{1}{2}$ and $\frac{1}{2} + ig_0$); and it is at least plausible that each of these segments contains only one so that there are only 138 in all. By methods explained in the next section, Hutchinson was in fact able to show that there are exactly 138 roots ρ in the range $0 \le \operatorname{Im} s \le g_{137}$, counted with multiplicities, which proves then that *all roots* ρ *in the range* $0 \le \operatorname{Im} s \le 300$ *lie on the line* Re $s = \frac{1}{2}$ *and all of them are simple zeros of* ξ.

Hutchinson called the tendency of the zeros of Z to alternate with the Gram points g_n *Gram's law*. Since Gram stated only that this pattern would persist beyond $n = 8$ and seemed to doubt that it would persist indefinitely, this is a rather poor choice of terminology in that a "law" is usually something which is true without exception. Hutchinson knew that Gram's "law" was not a "law" in this sense when he proposed the name, and therefore he clearly did not use the word in the way it is usually used in mathematics today. Nonetheless the term "Gram's law" has won acceptance in the literature and it will be used in what follows. In the range covered by present-day calculations, extending up to the three-and-a-half-millionth Gram point, the exceptions to Gram's law are surprisingly slight (see Chapter 8).

6.6 TECHNIQUES FOR COMPUTING THE NUMBER OF ROOTS IN A GIVEN RANGE

Gram in the course of earlier work with $\xi(\frac{1}{2} + it)$ had succeeded in computing the Taylor series coefficients of its logarithm with great accuracy. Since

$$\log\xi\left(\frac{1}{2} + it\right) = \sum_{\operatorname{Re}\alpha>0} \log\left(1 - \frac{t^2}{\alpha^2}\right) + \log\xi\left(\frac{1}{2}\right)$$

$$= \log\xi\left(\frac{1}{2}\right) - \left(\sum_{\operatorname{Re}\alpha>0} \frac{1}{\alpha^2}\right)t^2 - \left(\frac{1}{2}\sum_{\operatorname{Re}\alpha>0}\frac{1}{\alpha^4}\right)t^4 - \cdots,$$

this gives the numerical values of $\Sigma\alpha^{-2n}$, where α has the meaning given it by Riemann, namely, $\rho = \frac{1}{2} + i\alpha$ is the generic root of ξ (see Section 1.18). Now α^{-2n} decreases very rapidly as α increases, so the series $\Sigma\alpha^{-2n}$ is dominated by its first few terms. By comparing $\Sigma\alpha^{-10}$ with the sum extended over 15 roots he had already located on the line, Gram was able to show that there are no other roots in the range $0 \leq \operatorname{Im} s \leq 50$. However, this method rapidly becomes unworkable as the number of roots considered increases, so to extend the computations beyond 10 or 15 roots a new method was required. Such a method was found by Backlund [B1] around 1912.

Backlund's method is based on Riemann's observation that if $N(T)$ denotes the number of roots ρ in the range $0 < \operatorname{Im} s < T$, then†

$$N(T) = \frac{1}{2\pi i} \int_{\partial R} \frac{\xi'(s)}{\xi(s)}\, ds,$$

where R is a rectangle of the form $\{-\epsilon \leq \operatorname{Re} s \leq 1 + \epsilon,\ 0 \leq \operatorname{Im} s \leq T\}$, where ∂R is the boundary of R oriented in the usual counterclockwise direction, where it is assumed that T is such that there are no roots ρ on the line $\operatorname{Im} s = T$, and where $N(T)$ counts the roots with multiplicities. By symmetry and the fact that ξ is real on the real axis, this can be rewritten as

$$N(T) = \frac{1}{2\pi} \cdot 2 \operatorname{Im}\left[\int_C \frac{\xi'(s)}{\xi(s)}\, ds\right],$$

where C is the portion of ∂R from $1 + \epsilon$ to $\frac{1}{2} + iT$. Using the definition $\xi(s) = \pi^{-s/2}\Pi(\frac{1}{2}s - 1)\frac{1}{2}s(s - 1)\zeta(s)$ and the fact that the logarithmic derivative of a product is the sum of the logarithmic derivatives puts this in the form

$$\frac{1}{\pi} \operatorname{Im}\left\{\int_C \frac{d}{ds}\left[\log \pi^{-s/2}\Pi\left(\frac{s}{2} - 1\right)\right] ds\right\}$$
$$+ \frac{1}{\pi} \operatorname{Im}\left\{\int_C \frac{d}{ds} [\log s(s - 1)]\, ds\right\} + \frac{1}{\pi} \operatorname{Im}\left\{\int_C \frac{\zeta'(s)}{\zeta(s)}\, ds\right\}.$$

The first two terms, being integrals of derivatives, can be evaluated using the fundamental theorem of calculus; the first is $\pi^{-1}\vartheta(T)$ because it is π^{-1} times the imaginary part of $\log \pi^{-s/2}\Pi(\frac{1}{2}s - 1)$ at $s = \frac{1}{2} + iT$ when this log is defined to be real on the positive real axis, and the second is 1 because it is π^{-1} times the imaginary part of the log of $(\frac{1}{2} + iT)(\frac{1}{2} + iT - 1) = -T^2 - \frac{1}{4}$ when $\log s(s - 1)$ is taken to be real for $s > 1$. Thus

$$N(T) = \frac{1}{\pi}\vartheta(T) + 1 + \frac{1}{\pi} \operatorname{Im} \int_C \frac{\zeta'(s)}{\zeta(s)}\, ds.$$

Backlund observed that *if it can be shown that* $\operatorname{Re} \zeta$ *is never zero on* C, *then this formula suffices to determine* $N(T)$ *as the integer nearest to* $\pi^{-1}\vartheta(T) + 1$; this follows simply from noting that if $\operatorname{Re} \zeta$ is never 0 on C, then the curve $\zeta(C)$ never leaves the right halfplane so that $\log \zeta$ is defined all along $\zeta(C)$

†This follows from the "argument principle" of complex analysis or, more directly, from termwise integration of the uniformly convergent (see Section 3.2) series $\xi'(s)/\xi(s) = \Sigma(s - \rho)^{-1}$ using the Cauchy integral fromula.

and gives an antiderivative of ζ'/ζ on C whose imaginary part lies between $-\pi/2$ and $\pi/2$, which by the fundamental theorem shows that the last term above has absolute value less than $\frac{1}{2}$.

Backlund was able to prove by this method that $N(200) = 79$. By methods similar to Gram's he was also able to locate 79 changes of sign of $Z(t)$ on $0 < t < 200$ and thus to prove that *all roots ρ in the range* $0 < \text{Im}\, s < 200$ *are on the line* $\text{Re}\, s = \frac{1}{2}$ *and all are simple zeros of* ξ. This extended Gram's result from 50 to 200. In 1925 Hutchinson [H11] extended the same result to 300; that is, he proved that all roots ρ in the range $0 < \text{Im}\, s < 300$ are simple zeros on $\text{Re}\, s = \frac{1}{2}$. It was explained in the previous section how Hutchinson was able to prove that ξ has at least 138 zeros on the line $\text{Re}\, s = \frac{1}{2}$ in the range $0 < \text{Im}\, s < g_{137} = 300.468$ so that, in view of the above observations and in view of the fact that $\pi^{-1}\vartheta(g_{137}) + 1 = 137 + 1 = 138$, the proof of Hutchinson's theorem is reduced to proving that $\text{Re}\, \zeta$ is never zero on the broken line segment from $1\frac{1}{2}$ to $1\frac{1}{2} + ig_{137}$ to $\frac{1}{2} + ig_{137}$ (which is the curve C when $\epsilon = \frac{1}{2}$, $T = g_{137}$). This Hutchinson did as follows.

In the first place, it is easily shown that $\text{Re}\, \zeta$ is not zero anywhere on the line $\text{Re}\, s = 1\frac{1}{2}$. One need only observe that for $\text{Re}\, s = \sigma > 1$

$$|\text{Im} \log \zeta(s)| \le |\log \zeta(s)| = \left| \int_0^\infty x^{-s}\, dJ(x) \right|$$
$$\le \int_0^\infty x^{-\sigma}\, dJ(x) = \log \zeta(\sigma)$$

and that the value of $\zeta(1\frac{1}{2})$, which can be found by Euler–Maclaurin summation, is less than $e^{\pi/2}$. Hence $\text{Im} \log \zeta(s)$ on $\text{Re}\, s = 1\frac{1}{2}$ lies between $-\pi/2$ and $\pi/2$, and $\zeta(s)$ cannot lie on the imaginary axis. Thus it remains to prove only that $\text{Re}\, \zeta$ is never zero on the line segment from $\frac{1}{2} + ig_{137}$ to $1\frac{1}{2} + ig_{137}$. Hutchinson's method of doing this is simply to examine the real parts of the individual terms of the expansion

$$(1) \qquad \zeta(s) = \sum_{n=1}^{N-1} n^{-s} + \frac{N^{1-s}}{s-1} + \frac{1}{2} N^{-s} - s \int_N^\infty \bar{B}_1(x) x^{-s-1}\, dx$$

where $s = \sigma + ig_{137}$ ($\frac{1}{2} \le \sigma \le 1\frac{1}{2}$) and $N = 51$. The real part of the first sum consists of 50 terms, say $h_n(\sigma)$, namely,

$$h_n(\sigma) = n^{-\sigma} \cos(g_{137} \log n) \qquad (\tfrac{1}{2} \le \sigma \le 1\tfrac{1}{2}, \quad n = 1, 2, \ldots, 50).$$

For fixed n the sign of $h_n(\sigma)$ is the same for all values of σ, and the objective is to prove that the positive terms dominate the negative terms throughout the interval $\frac{1}{2} \le \sigma \le 1\frac{1}{2}$ with enough left over to dominate the remaining terms as well. Hutchinson found that 21 of the terms $h_n(\sigma)$ are negative, namely, those with $n = 3, 4, 6, 8, 11, 13, 15, 16, 17, 20, 21, 27, 28, 30, 32, 34, 36, 37, 40, 41, 42$. Now the larger n is, the more rapidly the absolute value of $h_n(\sigma)$ decreases as σ increases, so any negative terms dominated by the first two positive terms h_1, h_2 at $\sigma = \frac{1}{2}$ will remain dominated by them for

$\sigma > \frac{1}{2}$. Simple computation shows that they dominate the first four negative terms, that is,

$$[h_1(\tfrac{1}{2}) + h_2(\tfrac{1}{2})] + [h_3(\tfrac{1}{2}) + h_4(\tfrac{1}{2}) + h_6(\tfrac{1}{2}) + h_8(\tfrac{1}{2})] > 0;$$

so the same is true for $\sigma > \frac{1}{2}$. The next negative term is h_{11}, and there are four more positive terms h_5, h_7, h_9, h_{10} which precede it. These four suffice to dominate the next seven negative terms, that is,

$$[h_5(\sigma) + h_7(\sigma) + h_9(\sigma) + h_{10}(\sigma)]$$
$$+ [h_{11}(\sigma) + h_{13}(\sigma) + h_{15}(\sigma) + h_{16}(\sigma) + h_{17}(\sigma) + h_{20}(\sigma) + h_{21}(\sigma)] > 0$$

for $\sigma \geq \frac{1}{2}$. This is proved as before, by proving the inequality computationally for $\sigma = \frac{1}{2}$ and observing that the positive terms, having lower indices, decrease in absolute value more slowly than the negative terms. The next negative term is h_{27}, which is preceded by nine more positive terms. These nine suffice to dominate *all* the remaining 10 negative terms, which leaves 14 positive terms with which to dominate the real parts of the remaining three terms of formula (1) for $\zeta(s)$. The first of these three terms is $(s-1)^{-1}N^{1-s}$, the real part of which is

$$\operatorname{Re}\left[\frac{\sigma - 1 - iT}{(\sigma-1)^2 + T^2} \cdot \frac{\exp(-iT\log N)}{N^{\sigma-1}}\right]$$
$$= \frac{(\sigma-1)\cos(T\log N) - T\sin(T\log N)}{[(\sigma-1)^2 + T^2]N^{\sigma-1}},$$

where $N = 51$, $T = g_{137}$, and $\frac{1}{2} \leq \sigma \leq 1\frac{1}{2}$. Thus, for σ in this range, the absolute value of this term is at most

$$N^{-\sigma}\frac{\frac{1}{2}|\cos(T\log N)| + T\,|\sin(T\log N)|}{T^2 N^{-1}}$$

from which it is easily shown that the next positive term h_{29} is more than enough to dominate it, that is,

$$h_{29}(\sigma) + \operatorname{Re}\frac{N^{1-s}}{s-1} > 0$$

for $\frac{1}{2} \leq \sigma \leq 1\frac{1}{2}$. The next term $\frac{1}{2}N^{-s}$ of (1) has a positive real part when $N = 51$, $s = \sigma + ig_{137}$, so only the last term of (1) remains and there are still 13 positive terms with which to dominate it. Integration by parts in the usual manner puts the last term of (1) in the form

$$\frac{B_2}{2!}sN^{-s-1} + \frac{B_4}{4!}s(s+1)(s+2)N^{-s-3} + \cdots$$
$$+ \frac{B_{2\nu}}{(2\nu)!}s(s+1)\cdots(s+2\nu-2)N^{-s-2\nu+1} + R,$$

where, by Backlund's estimate of the remainder (Section 6.4),

$$|R| \leq \frac{|s+2\nu+1|}{\sigma+2\nu+1}\left|\frac{B_{2\nu+2}}{(2\nu+2)!}s(s+1)\cdots(s+2\nu)N^{-s-2\nu-1}\right|.$$

In the case under consideration $s \sim \sigma + 300i$, $\frac{1}{2} \leq \sigma \leq 1\frac{1}{2}$, and $N = 51$. Hutchinson sets $2\nu = 50$ and notes that $|s + k|/N < 6$ for $k = 0, 1, 2, \ldots, 51$ so that the modulus of the last term of (1) is less than

$$\frac{|B_2|}{2!}6 \cdot 51^{-\sigma} + \frac{|B_4|}{4!}6^3 51^{-\sigma} + \cdots + \frac{|B_{50}|}{50!}6^{49}51^{-\sigma} + 6\frac{|B_{52}|}{52!}6^{51}51^{-\sigma}.$$

Now the first few values of $|B_{2\nu}| \cdot 6^{2\nu-1}/(2\nu)!$ are

$$\frac{|B_2|}{2} \cdot 6 = \frac{1}{2}, \qquad \frac{|B_6|}{6!}6^5 = \frac{6^5}{720 \cdot 42} = \frac{9}{35},$$

$$\frac{|B_4|}{4!} \cdot 6^3 = \frac{6^3}{24 \cdot 30} = \frac{3}{10}, \qquad \frac{|B_8|}{8!}6^7 = \frac{6^7}{8!30} = \frac{81}{350},$$

and the ratio of two successive values is, by Euler's formula for $\zeta(2k)$ [(2) of Section 1.5], equal to

$$\frac{|B_{2k+2}|}{(2k+2)!}6^{2k+1}\frac{(2k)!}{|B_{2k}|}6^{-2k+1} = \frac{\zeta(2k+2)}{2^{2k+1}\pi^{2k+2}} \cdot \frac{2^{2k-1}\pi^{2k}}{\zeta(2k)} \cdot 6^2$$
$$= \frac{6^2}{4\pi^2}\frac{\zeta(2k+2)}{\zeta(2k)}.$$

As $k \to \infty$ this ratio approaches $6^2/4\pi^2 = (3/\pi)^2$. The first few ratios 3/5, 6/7, 9/10 are less than $6^2/4\pi^2$ and increase as k increases, a phenomenon which persists for all k as can be seen from the fact that

$$\log \zeta(2k+2) - 2\log\zeta(2k) + \log\zeta(2k-2)$$
$$= \int_0^\infty [x^{-2k-2} - 2x^{-2k} + x^{-2k+2}]\, dJ(x)$$
$$= \int_0^\infty x^{-2k}(x^{-2} - 2 + x^2)\, dJ(x)$$
$$= \int_0^\infty x^{-2k}(x^{-1} - x)^2\, dJ(x) \geq 0,$$

$$\log\zeta(2k+2) - \log\zeta(2k) \geq \log\zeta(2k) - \log\zeta(2k-2),$$

$$\frac{\zeta(2k+2)}{\zeta(2k)} \geq \frac{\zeta(2k)}{\zeta(2k-2)}.$$

Thus the ratios are all less than $(3/\pi)^2$, which gives the bound

$$\left[\frac{1}{2} + \frac{3}{10} + \frac{9}{35} + \frac{9}{35}\left(\frac{3}{\pi}\right)^2 + \frac{9}{35}\left(\frac{3}{\pi}\right)^4 + \cdots + \frac{9}{35}\left(\frac{3}{\pi}\right)^{44}\right]51^{-\sigma}$$
$$+ 6 \cdot \frac{9}{35}\left(\frac{3}{\pi}\right)^{46} 51^{-\sigma}$$

for the modulus of the last term of (1). When the geometric series is summed and the resulting number is estimated for $\sigma = \frac{1}{2}$, it is found to be decidedly less than 1.4. On the other hand, Hutchinson found that the remaining 13 positive terms have the sum 1.492 when $\sigma = \frac{1}{2}$. Thus, since $51^{-\sigma}$ decreases as

σ increases more rapidly than any of the 13 positive terms do, it follows that these 13 positive terms dominate the last term of (1) on $\frac{1}{2} \leq \sigma \leq 1\frac{1}{2}$, and the proof that $\operatorname{Re} \zeta > 0$ on the line segment from $\frac{1}{2} + ig_{137}$ to $1\frac{1}{2} + ig_{137}$ is complete.

Hutchinson states that by using the same methods he was also able to show that $\operatorname{Re} \zeta$ does not vanish on the line segment from $\frac{1}{2} + ig_{268}$ to $1\frac{1}{2} + ig_{268}$. This implies of course that $N(g_{268}) = \pi^{-1}\vartheta(g_{268}) + 1 = 269$ so, since $g_{268} = 499.1575$, it constitutes one half of a proof that all roots ρ in the range $0 < \operatorname{Im} s < (500 - \epsilon)$ are simple zeros on the line $\operatorname{Re} s = \frac{1}{2}$, that is, it constitutes one half of an extension of the previous result from 300 to just below 500. However, Hutchinson was apparently unable to complete the proof by locating another $269 - 138 = 131$ changes of sign in $\xi(\frac{1}{2} + it)$ for $300 < t < 500$. The evaluation of $\zeta(\frac{1}{2} + it)$ by Euler–Maclaurin summation is very lengthy for t in this range, and the project of performing enough such evaluations to locate the required 131 sign changes was more than Hutchinson was willing, or perhaps able, to undertake. It is just as well that he did not undertake it because only a few years later a much shorter method of evaluating $\xi(\frac{1}{2} + it)$, and hence of finding the required 131 sign changes, was discovered. This method, which uses the Riemann–Siegel formula, is the subject of the next chapter.

6.7 BACKLUND'S ESTIMATE OF $N(T)$

As in the preceding section, let $N(T)$ denote the number of roots ρ in the range $0 < \operatorname{Im} s < T$. Then Riemann's estimate of $N(T)$ [see (d) of Section 1.19] is the statement that the relative error in the approximation

$$N(T) \sim \frac{T}{2\pi} \log \frac{T}{2\pi} - \frac{T}{2\pi} \tag{1}$$

is less than a constant times T^{-1} as $T \longrightarrow \infty$. Since the right side is greater than a constant times $T \log T$ as $T \longrightarrow \infty$, this statement will follow if it can be shown that the absolute error in (1) is less than a constant times $\log T$. But the formulas

$$N(T) = \pi^{-1}\vartheta(T) + 1 + \pi^{-1} \operatorname{Im} \int_C \frac{\zeta'(s)}{\zeta(s)}\, ds,$$

$$\vartheta(T) = \frac{T}{2} \log \frac{T}{2\pi} - \frac{T}{2} - \frac{\pi}{8} + \frac{1}{48T} + \frac{7}{5760T^3} + \cdots$$

of Sections 6.6 and 6.5, respectively, combine to give

$$N(T) - \left[\frac{T}{2\pi} \log \frac{T}{2\pi} - \frac{T}{2\pi}\right] = \left[-\frac{1}{8} + \frac{1}{48\pi T} + \cdots\right] + 1 + \pi^{-1} \operatorname{Im} \int_C \frac{\zeta'(s)}{\zeta(s)}\, ds$$

which shows that the magnitude of the absolute error in (1) is less than

$$1 + \pi^{-1}\left|\operatorname{Im}\int_C \frac{\zeta'(s)}{\zeta(s)}\,ds\right| \tag{2}$$

for large T. Here C denotes the broken line segment from $1\frac{1}{2}$ to $1\frac{1}{2} + iT$ to $\frac{1}{2} + iT$, and it is assumed that $\zeta(s)$ is not zero on C, which is the same as to say that T is not a discontinuity of the step function $N(T)$. Now it suffices to prove Riemann's estimate (1) for such T, and for such T the integral in (2) can be evaluated by the fundamental theorem

$$\operatorname{Im}\int_C \frac{\zeta'(s)}{\zeta(s)}\,ds = \operatorname{Im}\log\zeta\left(\frac{1}{2} + iT\right)$$

when the log on the left is defined by analytic continuation along C. The generalization of the statement that this integral lies between $-\pi/2$ and $\pi/2$ if $\operatorname{Re}\zeta$ has no zeros on C is the statement that if $\operatorname{Re}\zeta$ has n zeros on C, then the integral lies between $-(n + \frac{1}{2})\pi$ and $(n + \frac{1}{2})\pi$. Thus the error in (1) has absolute value less than

$$1 + n + \tfrac{1}{2},$$

and in order to prove Riemann's estimate (1), it suffices to prove that *there is a constant K such that the number n of zeros of* $\operatorname{Re}\zeta$ *on the line segment from $\frac{1}{2} + iT$ to $1\frac{1}{2} + iT$ is at most $K\log T$ for all sufficiently large T.* Backlund was able to prove this theorem by a simple application of Jensen's theorem and was thereby able to give a much simpler proof [B1] of Riemann's estimate (1) than von Mangoldt's original proof [M3]. His proof is as follows.

Let $f(z) = \frac{1}{2}[\zeta(z + 2 + iT) + \zeta(z + 2 - iT)]$. Since $\zeta(\bar{s}) = \overline{\zeta(s)}$, the function f is identical with $\operatorname{Re}\zeta(z + 2 + iT)$ for real z, so the number n in question is equal to the number of zeros of $f(z)$ on the interval $-1\frac{1}{2} \le z \le -\frac{1}{2}$ of the real axis. Now consider Jensen's formula

$$\log|f(0)| + \sum \log\left|\frac{R}{z_i}\right| = \frac{1}{2\pi}\int_0^{2\pi} \log|f(Re^{i\theta})|\,d\theta \tag{3}$$

(see Section 2.2). Since $f(z)$ is analytic in the entire z-plane except for poles at $z + 2 \pm iT = 1$, that is, poles at $z = -1 \mp iT$, Jensen's formula applies whenever $R \le T$. Consider the case $R = 2 - \epsilon$ where ϵ is chosen so that f has no zeros on the circle $|z| = R$. The second sum on the left side of Jensen's formula (3) is a sum of positive terms and the terms corresponding to the n zeros in question are all at least $\log|(2 - \epsilon)/1\frac{1}{2}| = \log\frac{1}{3}(4 - 2\epsilon)$; hence

$$\log|f(0)| + n\log\tfrac{1}{3}(4 - 2\epsilon) \le \log M$$

where M is the maximum value of $|f(z)|$ on $|z| = 2 - \epsilon$. As $\epsilon \downarrow 0$, this gives

$$n \le \frac{\log|M/f(0)|}{\log\frac{4}{3}}$$

as an upper bound for n where M is an upper bound for $|f(z)|$ on $|z| = 2$.

Now $|f(0)| = |\operatorname{Re}\zeta(2+iT)| \geq 1 - 2^{-2} - 3^{-2} - 4^{-2} - \cdots = 1 - [\zeta(2) - 1] = 2 - (\pi^2/6) > \frac{1}{4}$, so this gives

$$n \leq \frac{1}{\log 4 - \log 3} \cdot \log 4M \leq \text{const} \log M + \text{const},$$

and to prove the theorem, it suffices to show that $\log M$ grows no faster than a constant times $\log T$. Now

$$\begin{aligned} M &= \max |\tfrac{1}{2}\zeta(2e^{i\theta} + 2 + iT) + \tfrac{1}{2}\zeta(2e^{i\theta} + 2 - iT)| \\ &\leq \max |\zeta(2e^{i\theta} + 2 + iT)|, \end{aligned}$$

so it suffices to estimate the growth of $|\zeta(s)|$ in the strip $0 \leq \operatorname{Re} s \leq 4$. But Backlund's estimate of the remainder R in

$$\zeta(s) = 1/(s-1) + \tfrac{1}{2} + R$$

gives (see Section 6.4)

$$|\zeta(s)| \leq \left|\frac{1}{s-1}\right| + \frac{1}{2} + \frac{|s+1|}{\sigma+1}\frac{B_2}{2!}|s|.$$

Hence for $s = 2e^{i\theta} + 2 + iT$,

$$\begin{aligned} M &\leq \frac{1}{T-2} + \frac{1}{2} + \frac{T+2+5}{0+1}\frac{1}{12} \cdot (T+2+4) \\ &\leq \text{const } T^2 \end{aligned}$$

for large T and the theorem follows.

By refining this estimate carefully, Backlund was able to obtain the specific estimate

$$\begin{aligned} &\left| N(T) - \left(\frac{T}{2\pi}\log\frac{T}{2\pi} - \frac{T}{2\pi} - \frac{7}{8}\right)\right| \\ &\qquad < 0.137 \log T + 0.443 \log\log T + 4.350 \end{aligned}$$

for all $T \geq 2$. (See Backlund [B3].)

6.8 ALTERNATIVE EVALUATION OF $\zeta'(0)/\zeta(0)$

Euler–Maclaurin summation can be used to prove the formula $\zeta'(0)/\zeta(0) = \log 2\pi$ of Section 3.8 as follows:

$$\begin{aligned} \zeta(s) - \frac{1}{s-1} &= \sum_{n=1}^{\infty} n^{-s} - \int_1^{\infty} x^{-s}\,dx \\ &= \sum_{n=1}^{N-1} n^{-s} + \sum_{n=N}^{\infty} n^{-s} - \int_1^N x^{-s}\,dx - \int_N^{\infty} x^{-s}\,dx \\ &= \sum_{n=1}^{N-1} n^{-s} - \int_1^N x^{-s}\,dx + \tfrac{1}{2}N^{-s} - s\int_N^{\infty} \bar{B}_1(x)x^{-s-1}\,dx \end{aligned}$$

at first for Re $s > 1$ but then by analytic continuation for Re $s > 0$. For $s = 1$ the right side is

$$1 + \frac{1}{2} + \frac{1}{3} + \cdots + \frac{1}{N-1} - \log N + \frac{1}{2N} - \int_N^\infty \bar{B}_1(x)x^{-2}\,dx$$

which approaches Euler's constant γ (see Section 3.8) as $N \to \infty$. Thus the Taylor series expansion of $(s-1)\zeta(s)$ around $s = 1$ begins $(s-1)\zeta(s) = 1 + \gamma(s-1) + \cdots$ and γ is the logarithmic derivative of $(s-1)\zeta(s)$ at $s = 1$. But the functional equation in the form (4) of Section 1.6 gives

$$(s-1)\zeta(s) = -\Pi(1-s)(2\pi)^{s-1}2\sin(s\pi/2)\zeta(1-s)$$

so logarithmic differentiation of both sides at $s = 1$ gives

$$\begin{aligned}\gamma &= -\frac{\Pi'(0)}{\Pi(0)} + \log 2\pi + \frac{\pi}{2}\frac{\cos(\pi/2)}{\sin(\pi/2)} - \frac{\zeta'(0)}{\zeta(0)}\\ &= \gamma + \log 2\pi - \frac{\zeta'(0)}{\zeta(0)}\end{aligned}$$

and the result follows.

Chapter 7

The Riemann–Siegel Formula

7.1 INTRODUCTION

In 1932 Carl Ludwig Siegel published an account [S4] of the work relating to the zeta function and analytic number theory found in Riemann's private papers in the archives of the University Library at Göttingen [R1a]. This was an event of very great importance in the history of the study of the zeta function not only because the work contained new and important information, but also because it revealed the profundity and technical virtuosity of Riemann's researches. Anyone who has read Siegel's paper is unlikely to assert, as Hardy did in 1915 [H3a], that Riemann "could not prove" the statements he made about the zeta function, or to call them, as Landau [L3] did in 1908, "conjectures." Whereas the eight-page resumé *Ueber die Anzahl. . .* , the only work which Riemann published on this subject, could possibly be interpreted as a series of remarkable heuristic insights, Siegel's paper shows clearly that there lay behind it an extensive analysis which may have lacked detailed error estimates but which surely did not lack extremely powerful methods and which in all likelihood was based on a very sure grasp of the magnitudes of error terms even when they were not explicitly estimated.

The difficulty of Siegel's undertaking could scarcely be exaggerated. Several first-rate mathematicians before him had tried to decipher Riemann's disconnected jottings, but all had been discouraged either by the complete lack of any explanation of the formulas, or by the apparent chaos in their arrangement, or by the analytical skill needed to understand them. One wonders whether anyone else would ever have unearthed this treasure if Siegel had not. It is indeed fortunate that Siegel's concept of scholarship derived from the older tradition of respect for the past rather than the contemporary style of novelty.

There are two topics covered in the paper, the one an asymptotic formula for the computation of $Z(t)$ and the other a new representation of $\zeta(s)$ in terms of definite integrals. This chapter is devoted mainly to the asymptotic formula for $Z(t)$, which is known as the Riemann–Siegel formula. The majority of the chapter, Sections 7.2–7.5, consists of the derivation of this formula. Some computations using the formula are given in Section 7.6, error estimates are discussed in Section 7.7, and the relation of the formula to the Riemann hypothesis is discussed in Section 7.8. Finally, in Section 7.9, the new representation of $\zeta(s)$ is derived.

7.2 BASIC DERIVATION OF THE FORMULA

Recall Riemann's formula for $\zeta(s)$ which "remains valid for all s," namely, the formula

$$\zeta(s) = \frac{\Pi(-s)}{2\pi i} \int_{+\infty}^{+\infty} \frac{(-x)^s}{e^x - 1} \cdot \frac{dx}{x}, \tag{1}$$

where the limits of integration indicate a contour which begins at $+\infty$, descends the real axis, circles the singularity at the origin once in the positive direction, and returns up the positive real axis to $+\infty$ [see (3) of Section 1.4] and where $(-x)^s = \exp[s \log(-x)]$ is defined in the usual way for $-x$ not on the negative real axis. There are two ways that finite sums can be split off from (1), the first being to use

$$\frac{e^{-Nx}}{e^x - 1} = \sum_{n=N+1}^{\infty} e^{-nx}$$

in place of $(e^x - 1)^{-1} = \sum e^{-nx}$ in the derivation of (1) to find

$$\zeta(s) = \sum_{n=1}^{N} n^{-s} + \frac{\Pi(-s)}{2\pi i} \int_{+\infty}^{+\infty} \frac{e^{-Nx}(-x)^s}{e^x - 1} \cdot \frac{dx}{x} \tag{2}$$

and the second being to change the contour of integration to a curve C_M which circles the poles $\pm 2\pi i M, \pm 2\pi i(M-1), \ldots, \pm 2\pi i$ as well as the singularity 0 of the integrand (say C_M is the path which descends the real axis from $+\infty$ to $(2M+1)\pi$, circles the boundary of the disk $|s| \leq (2M+1)\pi$ once in the positive direction, and returns to $+\infty$) and, using the residue theorem, to find

$$\zeta(s) = \Pi(-s)(2\pi)^{s-1} 2 \sin\frac{\pi s}{2} \sum_{n=1}^{M} n^{-(1-s)} + \frac{\Pi(-s)}{2\pi i} \int_{C_M} \frac{(-x)^s}{e^x - 1} \cdot \frac{dx}{x} \tag{3}$$

as in the derivation of the functional equation of ζ in Section 1.6. The first proof of the functional equation (see Section 1.6) amounts simply to showing that if Re $s > 1$, then the integral in (2) approaches zero as $N \longrightarrow \infty$ (because it is $\sum_{N+1}^{\infty} n^{-s}$) whereas if Re $s < 0$, then the integral in (3) approaches zero as $M \longrightarrow \infty$ (by the estimate of Section 1.6). For s in the critical strip $0 <$ Re $s < 1$, however, neither the integral in (2) nor the integral in (3) can be neglected.

The techniques by which (2) and (3) were derived can easily be combined to give

$$\zeta(s) = \sum_{n=1}^{N} n^{-s} + \Pi(-s)(2\pi)^{s-1} 2 \sin\frac{\pi s}{2} \sum_{n=1}^{M} n^{-(1-s)} + \frac{\Pi(-s)}{2\pi i}\int_{C_M} \frac{(-x)^s e^{-Nx}}{e^x - 1} \cdot \frac{dx}{x} \tag{4}$$

which can be put in a more symmetrical form by multiplying by $\frac{1}{2}s(s-1)$ $\Pi(\frac{1}{2}s-1)\pi^{-s/2}$ and using the identities of the factorial function needed for the derivation of the symmetrical form of the functional equation [see (5) of Section 1.6] to find

$$\begin{aligned}\xi(s) = {} & (s-1)\Pi\left(\frac{s}{2}\right)\pi^{-s/2} \sum_{n=1}^{N} n^{-s} \\ & + (-s)\Pi\left(\frac{1-s}{2}\right)\pi^{-(1-s)/2} \sum_{n=1}^{M} n^{-(1-s)} \\ & + \frac{(-s)\Pi(1-s/2)\pi^{-(1-s)/2}}{(2\pi)^{s-1} 2\sin(\pi s/2)\cdot 2\pi i}\int_{C_M} \frac{(-x)^s e^{-Nx}}{e^x - 1} \cdot \frac{dx}{x}\end{aligned} \tag{5}$$

for all N, M, s. The case of greatest interest is the case $s = \frac{1}{2} + it$ where t is real. Then, in view of the symmetry between s and $1 - s$, it is natural to set $N = M$. If $f(t)$ denotes $(-\frac{1}{2} + it)\Pi((\frac{1}{2} + it)/2)\pi^{-(1/2+it)/2}$, the formula then becomes

$$\begin{aligned}\xi\left(\frac{1}{2} + it\right) = {} & f(t) \sum_{n=1}^{N} n^{-(1/2)-it} + f(-t) \sum_{n=1}^{N} n^{-(1/2)+it} \\ & + \frac{f(-t)}{(2\pi)^{(1/2)+it} 2i \sin[\frac{1}{2}\pi(\frac{1}{2} + it)]}\int_{C_N} \frac{-(-x)^{-(1/2)+it} e^{-Nx}\,dx}{e^x - 1}.\end{aligned}$$

Now by definition $Z(t)$ satisfies $\xi(\frac{1}{2} + it) = r(t)Z(t)$, where (see Section 6.5)

$$\begin{aligned}r(t) &= \exp\left[\operatorname{Re}\log\Pi\left(\frac{1}{2}s - 1\right)\right]\pi^{-(1/4)} \cdot \frac{-t^2 - \frac{1}{4}}{2} \\ &= \exp\left[\log\Pi\left(\frac{1}{2}s - 1\right)\right]\pi^{-(1/4)} \cdot \frac{s(s-1)}{2}\exp\left[-i\operatorname{Im}\log\Pi\left(\frac{1}{2}s - 1\right)\right] \\ &= \Pi\left(\frac{s}{2}\right)(s-1)\pi^{-(1/4)}\, e^{-i\vartheta(t)}\pi^{-(1/2)it} \\ &= f(t)e^{-i\vartheta(t)}, \\ f(t) &= r(t)e^{i\vartheta(t)}\end{aligned}$$

($s = \frac{1}{2} + it$), so a factor $r(t) = r(-t)$ can be canceled from all terms above.

Using $\vartheta(-t) = -\vartheta(t)$ and the simplification $2i\sin(\pi s/2) = e^{-i\pi s/2}(e^{i\pi s} - 1) = e^{-i\pi/4}e^{t\pi/2}(e^{i\pi/2}e^{-t\pi} - 1) = -e^{-i\pi/4}e^{t\pi/2}(1 - ie^{-t\pi})$ then puts the formula in the form

$$Z(t) = \sum_{n=1}^{N} n^{-1/2}\cdot 2\cos[\vartheta(t) - t\log n] + \frac{e^{-i\vartheta(t)}e^{-t\pi/2}}{(2\pi)^{1/2}(2\pi)^{it}e^{-i\pi/4}(1 - ie^{-t\pi})}\int_{C_N}\frac{(-x)^{-(1/2)+it}e^{-Nx}\,dx}{e^x - 1}$$

for all real t. Although the series on the right diverges as $N \to \infty$, the terms do decrease in size, which gives some reason to believe that if N is suitably chosen, the approximation

$$Z(t) \sim 2\sum_{n=1}^{N} n^{-1/2}\cos[\vartheta(t) - t\log n]$$

might have some merit. The study of this approximation is of course equivalent to the study of the remainder term

(6) $$\frac{e^{-i\vartheta(t)}e^{-t\pi/2}}{(2\pi)^{1/2}(2\pi)^{it}e^{-i\pi/4}(1 - ie^{-t\pi})}\int_{C_N}\frac{(-x)^{-(1/2)+it}e^{-Nx}\,dx}{e^x - 1}.$$

The Riemann–Siegel formula is† a technique for the approximate numerical evaluation of this integral and hence of $Z(t)$ and $\xi(\frac{1}{2} + it)$.

The essence of Riemann's technique for evaluating the integral (6) is a standard technique for the approximate evaluation of definite integrals known as the *saddle point method* or the *method of steepest descent* (see,‡ for example, Jeffreys and Jeffreys [J2]). Consider the modulus of the integrand

(7) $$\frac{(-x)^{-(1/2)+it}e^{-Nx}}{e^x - 1}.$$

As long as the contour of integration stays well away from the zeros of the denominator $x = 0, \pm 2\pi i, \pm 4\pi i, \ldots$, this modulus is at most a constant times the modulus of the numerator, so in looking for places where the modulus of the integrand is large, it suffices to consider places where the modulus of the numerator is large. Now this modulus is $e^{\phi(x)}$, where

$$\phi(x) = \operatorname{Re}\{(-\tfrac{1}{2} + it)\log(-x) - Nx\}.$$

†The assumption $\operatorname{Re} s = \frac{1}{2}$ is made only for the sake of convenience; the entire analysis of (6) applies, with only slight modifications, to the last term of (4). Siegel carries through this analysis and makes the assumption $\operatorname{Re} s = \frac{1}{2}$ only as the last step. The assumption $N = M$ is a natural concomitant of the assumption $\operatorname{Re} s = \frac{1}{2}$, but if, for example, $\operatorname{Re} s > \frac{1}{2}$, then s is nearer the range where the series with N converges, and it would presumably be better to take $N > M$. Siegel indicates the method for dealing with the case $N \neq M$, but he does not carry it through.

‡Jeffreys and Jeffreys attribute the method to Debye, although Debye himself acknowledges (see [D1]) that the method occurs in a posthumously published fragment of Riemann [R1, pp. 405–406]. Be that as it may, the widespread use of the method in the theory of Bessel functions and in theoretical physics dates from Debye's rediscovery of it in 1910.

Since ϕ is a harmonic function, it has no local maxima or minima, but it does have a saddle point at the unique point where the derivative of $(-\frac{1}{2} + it)\log(-x) - Nx$ is zero, namely, at the point $(-\frac{1}{2} + it)/N$. Let α denote this point. In the vicinity of α the function ϕ can be written in the form

$$\begin{aligned}
\phi(x) &= \operatorname{Re}\Big\{\Big(-\frac{1}{2} + it\Big)\log(-\alpha) \\
&\quad + \Big(-\frac{1}{2} + it\Big)\log\Big(1 + \frac{x-\alpha}{\alpha}\Big) - N\alpha - N(x-\alpha)\Big\} \\
&= \text{const} + \operatorname{Re}\Big\{\Big(-\frac{1}{2} + it\Big) \\
&\quad \times \Big[\frac{x-\alpha}{\alpha} - \frac{1}{2}\Big(\frac{x-\alpha}{\alpha}\Big)^2 + \frac{1}{3}\Big(\frac{x-\alpha}{\alpha}\Big)^3 - \cdots\Big] - N(x-\alpha)\Big\} \\
&= \text{const} + \operatorname{Re}\Big\{-\frac{1}{2}\Big(-\frac{1}{2} + it\Big)\Big(\frac{x-\alpha}{\alpha}\Big)^2 \\
&\quad + \frac{1}{3}\Big(\frac{x-\alpha}{\alpha}\Big)^3\Big(-\frac{1}{2} + it\Big) - \cdots\Big\} \\
&= \text{const} - \frac{1}{2}\operatorname{Re}\Big\{\frac{N^2(x-\alpha)^2}{-\frac{1}{2} + it}\Big\} \\
&\quad + \text{terms in } (x-\alpha)^3, (x-\alpha)^4, \ldots.
\end{aligned}$$

If x passes through α along the line $\operatorname{Im}\log(x-\alpha) = \frac{1}{2}\operatorname{Im}\log(-\frac{1}{2} + it)$ where $(x-\alpha)^2/(-\frac{1}{2} + it)$ is real and positive, then ϕ has a local maximum at α and, consequently, the modulus of the integrand (7) has a local maximum at α. (On the other hand, if x passes through α along the line perpendicular to this one, then ϕ has a local *minimum* at α; thus the method is to cross the saddle point at α along the line of steepest descent, which gives the method its name.) Thus, if it can be arranged that the path of integration passes through α in this way and never enters regions away from α where the integrand is large, then the integral will have been concentrated into a small part of the total path of integration and this short integral will be approximable by local methods.

Now if t is large, then the saddle point $\alpha = (-\frac{1}{2} + it)/N$ lies near the positive imaginary axis, whatever value of N is chosen. But the path of integration C_N (recall $M = N$) crosses the positive imaginary axis between $2\pi Ni$ and $2\pi(N+1)i$, so in order for C_N to pass near the saddle point it is necessary to have $(-\frac{1}{2} + it)/N \sim 2\pi Ni$ or $N^2 \sim (-\frac{1}{2} + it)/2\pi i \sim t/2\pi$. This motivates the choice of N as the integer part of $(t/2\pi)^{1/2}$, that is, $N = [(t/2\pi)^{1/2}]$ is the largest integer less than $(t/2\pi)^{1/2}$. Then the saddle point α lies near $it(t/2\pi)^{-1/2} = i(2\pi t)^{1/2}$, which is between $2\pi Ni$ and $2\pi(N+1)i$ as desired. Since $\operatorname{Im}\log(-\frac{1}{2} + it) \sim \pi/2$, the path of integration should pass through the saddle point along a line of slope approximately 1 and, because of the configuration of C_N, it should go from upper right to lower left.

Let a denote the approximate saddle point $a = i(2\pi t)^{1/2}$ and let L denote the line of slope 1 through a directed from upper right to lower left. Then, in summary, the saddle point method suggests that *the integral* (6) *is approximately equal to the integral of the same function over a segment of L containing a, and this latter integral can be approximated by local methods* using the fact that the modulus of the integrand has a saddle point near a.

7.3 ESTIMATION OF THE INTEGRAL AWAY FROM THE SADDLE POINT

Let $t > 0$ be given, let $a = i(2\pi t)^{1/2}$, and let L be the line of slope 1 through a directed from upper right to lower left. The objective of this section is to show that if L_1 is a suitable segment of L, then the remainder term (6) of Section 7.2 is accurately estimated by the approximation

$$(1)\qquad \frac{e^{-i\vartheta(t)}e^{-t\pi/2}}{(2\pi)^{1/2}(2\pi)^{it}e^{-i\pi/4}(1-ie^{-t\pi})}\int_{C_N}\frac{(-x)^{-(1/2)+it}e^{-Nx}\,dx}{e^x-1}$$
$$\sim \frac{e^{-i\vartheta(t)}e^{-t\pi/2}}{(2\pi)^{1/2}(2\pi)^{it}e^{-i\pi/4}(1-ie^{-t\pi})}\int_{L_1}\frac{(-x)^{-(1/2)+it}e^{-Nx}\,dx}{e^x-1},$$

where, as in Section 7.2, N is the integer part of $(t/2\pi)^{1/2}$ and C_N is a contour which begins at $+\infty$, circles the poles $\pm 2\pi iN, \pm 2\pi i(N-1), \ldots, \pm 2\pi i$, and the singularity 0 once in the positive direction and returns to $+\infty$. [The value of t must be assumed to be such that $(t/2\pi)^{1/2}$ is not an integer so that a is not a pole of the integrand.] More precisely, the objective is to show that the error in the approximation (1) is very small for t at all large and approaches zero as $t \to \infty$.

Approximations to the right side of (1) will be given in the following section, 7.4. These approximations will be derived using termwise integration of a power series in $(x - a)$ whose radius of convergence is $|a|$ and for this reason it will be advantageous to take L_1 to lie well within this radius of convergence, say the portion of L which lies within $\frac{1}{2}|a|$ of a. Thus L_1 *will be the directed line segment from* $a + \frac{1}{2}e^{i\pi/4}|a|$ *to* $a - \frac{1}{2}e^{i\pi/4}|a|$, where $a = i(2\pi t)^{1/2}$.

With this choice of L_1, the path of integration C_N can be taken to be $L_0 + L_1 + L_2 + L_3$, where L_0 is the (infinite) portion of L which lies above and to the right of L_1, where L_2 is the vertical line segment from the lower left end of L_1 to the line $\{\operatorname{Im} x = -(2N+1)\pi\}$, and where L_3 is the (infinite) portion of $\{\operatorname{Im} x = -(2N+1)\pi\}$ to the right of the lower end of L_2. If Re x

is very large, then the very small term e^{-Nx} dominates the integrand and it is easily seen that this is a valid choice of C_N even though L_0 and L_3 do not approach infinity along the positive real axis. With these definitions, then, the approximation (1) to be proved becomes

$$(2)\qquad \frac{e^{-i\vartheta(t)}e^{-t\pi/2}}{(2\pi)^{1/2}(2\pi)^{it}e^{-i\pi/4}(1-ie^{-t\pi})} \times \int_{L_j} \frac{(-x)^{-(1/2)+it}e^{-Nx}\,dx}{e^x-1} \sim 0 \qquad (j=0,2,3).$$

These three approximations will be considered in turn.

The case $j=0$ *of* (2): The modulus of the numerator of the integrand is, as before, $e^{\phi(x)}$, where $\phi(x)=\operatorname{Re}\{(-\frac{1}{2}+it)\log(-x)-Nx\}$. The presence of a saddle point of ϕ near a, where L nearly passes over a maximum of ϕ, suggests that ϕ increases as x descends L_0 toward its terminal point and that $e^{\phi(x)}$ has its maximum on L_0 at this terminal point. This is easily confirmed by differentiating $\phi=\operatorname{Re}[(-\frac{1}{2}+it)\log(-a-ke^{i\pi/4})-N(a+ke^{i\pi/4})]$ with respect to k for k real and greater than or equal to $\frac{1}{2}|a|$ to find

$$\begin{aligned}\frac{d\phi}{dk} &= \operatorname{Re}\left\{\left(-\frac{1}{2}+it\right)(a+ke^{i\pi/4})^{-1}e^{i\pi/4}-Ne^{i\pi/4}\right\}\\ &= -\frac{1}{2}\operatorname{Re}\{(a+ke^{i\pi/4})^{-1}e^{i\pi/4}\}\\ &\quad+\operatorname{Re}\left\{\left(\frac{it}{a}\right)\left(1+\frac{k}{a}e^{i\pi/4}\right)^{-1}e^{i\pi/4}\right\}-N\operatorname{Re}\{e^{i\pi/4}\}\\ &\le \frac{1}{2}|a+ke^{i\pi/4}|^{-1}+\left(\frac{t}{2\pi}\right)^{1/2}\operatorname{Re}\left\{\left(1+\frac{k}{|a|}e^{-i\pi/4}\right)^{-1}e^{i\pi/4}\right\}\\ &\quad-N\left(\frac{\sqrt{2}}{2}\right).\end{aligned}$$

With $u=k|a|^{-1}\ge\frac{1}{2}$ this can be written in the form

$$\begin{aligned}\frac{d\phi}{dk} &\le \frac{1}{2}|a|^{-1}+\left(\frac{t}{2\pi}\right)^{1/2}\operatorname{Re}\{(1+ue^{-i\pi/4})^{-1}e^{i\pi/4}\}\\ &\quad-\left[\left(\frac{t}{2\pi}\right)^{1/2}-1\right]\operatorname{Re}\{e^{i\pi/4}\}\\ &= \frac{1}{2}(2\pi t)^{-1/2}+\left(\frac{t}{2\pi}\right)^{1/2}\operatorname{Re}\{(1+ue^{-i\pi/4})^{-1}(e^{i\pi/4}-e^{i\pi/4}-u)\}+\frac{\sqrt{2}}{2}\\ &= \frac{1}{2}(2\pi t)^{-1/2}-\left(\frac{t}{2\pi}\right)^{1/2}\operatorname{Re}\{(u^{-1}+e^{-i\pi/4})^{-1}\}+\frac{\sqrt{2}}{2}.\end{aligned}$$

The middle term is at most a positive constant times $-(t/2\pi)^{1/2}$, hence $d\phi/dk$ is negative on all of L_0 whenever t is at all large, as was to be shown. Thus the modulus of the numerator is at most

$$\begin{aligned}
&\exp \operatorname{Re}\left\{\left(-\frac{1}{2}+it\right)\log\left(-a-\frac{1}{2}|a|e^{i\pi/4}\right)-N\left(a+\frac{1}{2}|a|e^{i\pi/4}\right)\right\}\\
&\quad=\exp\left[-\frac{1}{2}\log\left|a+\frac{1}{2}ae^{-i\pi/4}\right|\right.\\
&\qquad\left.-t\operatorname{Im}\log\left(-a-\frac{1}{2}ae^{-i\pi/4}\right)-N\cdot\frac{1}{2}(2\pi t)^{1/2}\frac{\sqrt{2}}{2}\right]\\
&\quad=|a|^{-1/2}\left|1+\frac{1}{2}e^{-i\pi/4}\right|^{-1/2}\\
&\qquad\times\exp\left[-t\operatorname{Im}\log\left(-i-\frac{1}{2}e^{i\pi/4}\right)-N\cdot\frac{1}{2}(\pi t)^{1/2}\right]\\
&\quad\le(2\pi t)^{-1/4}\exp\left\{-t\left[-\frac{\pi}{2}-\operatorname{Arctan}\left(\frac{\frac{1}{2}\cdot\frac{1}{2}\sqrt{2}}{1+\frac{1}{2}\cdot\frac{1}{2}\sqrt{2}}\right)\right]\right.\\
&\qquad\left.-\left[\left(\frac{t}{2\pi}\right)^{1/2}-1\right]\frac{1}{2}(\pi t)^{1/2}\right\}\\
&\quad=(2\pi t)^{-1/4}e^{t\pi/2}\exp\left\{t\operatorname{Arctan}\left(\frac{1}{2\sqrt{2}+1}\right)-\frac{t}{2\sqrt{2}}\right\}\exp\left[\frac{1}{2}(\pi t)^{1/2}\right]\\
&\quad\le e^{t\pi/2}e^{-t/11}\exp\left[\frac{1}{2}(\pi t)^{1/2}\right]
\end{aligned}$$

because

$$\operatorname{Arctan}\left(\frac{1}{2\sqrt{2}+1}\right)-\frac{1}{2\sqrt{2}}\le\frac{1}{2\sqrt{2}+1}-\frac{1}{2\sqrt{2}}<-\frac{1}{11}$$

and because it can be assumed that $2\pi t>1$. Thus the integral to be estimated has modulus at most

$$\begin{aligned}
&\left|\frac{e^{-i\vartheta(t)}e^{-t\pi/2}e^{t\pi/2}e^{-t/11}\exp[\frac{1}{2}(\pi t)^{1/2}]}{(2\pi)^{1/2}(2\pi)^{it}e^{-i\pi/4}(1-ie^{-t\pi})}\right|\int_{L_0}\frac{|dx|}{|e^x-1|}\\
&\quad\le\frac{e^{-t/11}\exp[\frac{1}{2}(\pi t)^{1/2}]}{(2\pi)^{1/2}(1-e^{-t\pi})}\int_{(1/2)(\pi t)^{1/2}}^{\infty}\frac{e^{-u}\sqrt{2}\,du}{1-e^{-u}}\\
&\quad\le\frac{e^{-t/11}\exp[\frac{1}{2}(\pi t)^{1/2}]}{(2\pi)^{1/2}(1-e^{-t\pi})}\cdot\frac{\sqrt{2}}{1-\exp[-\frac{1}{2}(\pi t)^{1/2}]}[-e^{-u}]\Big|_{(\pi t)^{1/2}/2}^{\infty}\\
&\quad=\frac{e^{-t/11}}{\pi^{1/2}}\cdot\frac{1}{(1-e^{-t\pi})\{1-\exp[-\frac{1}{2}(\pi t)^{1/2}]\}}.
\end{aligned}$$

The second term differs negligibly from 1 when t is at all large and the integral (2) in the case $j=0$ is comfortably less than $e^{-t/11}$.

The case $j=2$ of (2): On L_2 the real part of x is constant, say $-b$, where $b=-\operatorname{Re}(a-\frac{1}{2}e^{i\pi/4}|a|)=\frac{1}{2}\cdot\frac{1}{2}\sqrt{2}\cdot(2\pi t)^{1/2}=\frac{1}{2}(\pi t)^{1/2}$. The denominator of the integrand on L_2 then has modulus at least $1-e^{-b}$ which is greater than $\frac{1}{2}$ for t at all large. The numerator of the integrand has modulus at most

$$|(-x)^{-(1/2)+it}e^{-Nx}|$$
$$\leq (\max|x|^{-1/2})\{\max \exp[-t \operatorname{Im}\log(-x)]\}\exp\left[\left(\frac{t}{2\pi}\right)^{1/2}\cdot b\right].$$

The maximum value of $|x|^{-1/2}$ occurs at the point where L_2 crosses the real axis, at which point it is $b^{-1/2}$. The maximum value of $\exp[-t \operatorname{Im}\log(-x)]$ occurs at the minimum value of $\operatorname{Im}\log(-x)$ at the initial point of L_2, where

$$\operatorname{Im}\log(-x) = \operatorname{Im}\log\left(-i + \frac{1}{2}e^{i\pi/4}\right) = -\frac{\pi}{2} + \operatorname{Arctan}\left(\frac{\frac{1}{2}\frac{1}{2}\sqrt{2}}{1 - \frac{1}{2}\frac{1}{2}\sqrt{2}}\right),$$
$$\exp[-t \operatorname{Im}\log(-x)] = e^{t\pi/2}\exp\left[-t \operatorname{Arctan}\frac{1}{2\sqrt{2}-1}\right].$$

Finally, $(t/2\pi)^{1/2}b = t/2\sqrt{2}$, so the modulus of the numerator is at most $b^{-1/2}e^{t\pi/2}e^{-kt}$ where $k = \operatorname{Arctan}(1/2\sqrt{2}-1) - (1/2\sqrt{2})$. By direct numerical evaluation it is found that $k < \frac{1}{8}$ so the integrand has modulus at most $2b^{-1/2}e^{t\pi/2}e^{-t/8}$ on L_2. Since the length of the path of integration L_2 is less than $2|a| = 2(2\pi t)^{1/2} = 4\sqrt{2}\,b$, this shows that (2) is at most

$$\frac{e^{-t\pi/2}}{(2\pi)^{1/2}(1-e^{-t\pi})}\cdot 2b^{-1/2}e^{t\pi/2}e^{-t/8}\cdot 4\sqrt{2}\,b \leq ke^{-t/8}t^{1/4},$$

where the constant k is about $8\sqrt{2}(\frac{1}{2}\pi^{1/2})^{1/2}(2\pi)^{-1/2} < 5$. Thus for $t \geq 100$ the modulus of (2) in this case is much less than in the case $j = 0$.

The case $j = 3$ *of* (2): On L_3 the imaginary part of x is identically equal to $-(2N+1)\pi$, hence the denominator of the integrand is $-e^{\operatorname{Re} x} - 1$ which has modulus at least 1. The least value of $|x|$ on L_3 is $(2N+1)\pi$ so $(-x)^{-1/2}$ has modulus less than $(2N+1)^{-1/2}\pi^{-1/2}$ on L_3. The least value of $\operatorname{Im}\log(-x)$ on L_3 is greater than $\operatorname{Im}\log(1+i) = \pi/4$ so $(-x)^{it}$ has modulus at most $e^{-t\pi/4}$. Thus (2) has modulus at most

$$\frac{e^{-t\pi/2}}{(2\pi)^{1/2}(1-e^{-t\pi})}\frac{e^{-t\pi/4}}{(2N+1)^{1/2}\pi^{1/2}}\int_{-(1/2)(\pi t)^{1/2}}^{\infty} e^{-Nu}\,du$$
$$= \frac{e^{-3\pi t/4}}{\pi(1-e^{-t\pi})\sqrt{2}\,(2N+1)^{1/2}}\frac{\exp[\frac{1}{2}N(\pi t)^{1/2}]}{N}$$
$$\leq e^{-3\pi t/4}\exp\left[\frac{1}{2}\left(\frac{t}{2\pi}\right)^{1/2}(\pi t)^{1/2}\right] \leq e^{-t}$$

for $t > 2\pi$. Thus this term is entirely negligible compared to $e^{-t/11}$ when t is large.

In summary, the above very crude estimates suffice to show that the error in the approximation (1) is considerably less than $e^{-t/11}$ for $t \geq 100$.

7.4 FIRST APPROXIMATION TO THE MAIN INTEGRAL

The estimates of the preceding section show that, with an absolute error which is very small and which decreases very rapidly as t increases, the remainder term R in the formula

$$Z(t) = 2 \sum_{n^2<(t/2\pi)} n^{-1/2} \cos[\vartheta(t) - t \log n] + R$$

is approximately equal to a definite integral

$$(1) \qquad R \sim \frac{e^{-i\vartheta(t)}e^{-t\pi/2}}{(2\pi)^{1/2}(2\pi)^{it}e^{-i\pi/4}(1 - ie^{-\pi t})} \int_{L_1} \frac{(-x)^{-(1/2)+it}e^{-Nx}\,dx}{e^x - 1},$$

where N is the integer part of $(t/2\pi)^{1/2}$ and where L_1 is a line segment in the complex x-plane which has slope 1, length $(2\pi t)^{1/2}$, and midpoint on the imaginary axis at $i(2\pi t)^{1/2}$. [It is assumed that $(t/2\pi)^{1/2}$ is not an integer so the integrand is not singular on L_1.] The objective of this section is to develop a first approximation to the value of this definite integral.

The argument which led to the above integral, namely, the technique of ignoring the denominator $e^x - 1$ and applying the saddle point method to the numerator, suggests that the path of integration should pass through the saddle point $\alpha = (-\frac{1}{2} + it)/N$ and the numerator should be expanded in powers of $(x - \alpha)$. This has two disadvantages, the first being that α depends on the discrete variable N and the second being that α has a small real part which complicates the computations. Instead set $a = i(2\pi t)^{1/2} = 2\pi i(t/2\pi)^{1/2}$ and expand the numerator in terms of $(x - a)$. This gives

$$\begin{aligned}&\exp\{(-\tfrac{1}{2} + it)\log(-a) + (-\tfrac{1}{2} + it)\\ &\qquad \times \log[1+ (x - a)/a] - Na - N(x - a)\}\\ &= (-a)^{-(1/2)+it}e^{-Na}\\ &\qquad \times \exp\{[(-\tfrac{1}{2} + it)a^{-1} - N](x - a)\\ &\qquad - (-\tfrac{1}{2} + it)\cdot\tfrac{1}{2}(x - a)^2a^{-2} + \cdots\}.\end{aligned}$$

Now the coefficient of $(x - a)$ in the exponential is approximately $it/i(2\pi t)^{1/2} - N = (t/2\pi)^{1/2} - N = p$, where p is the fractional part of $(t/2\pi)^{1/2}$. The coefficient of $(x - a)^2$ is approximately $-it\cdot\frac{1}{2}/(-2\pi t) = i/4\pi$. The coefficients of $(x - a)^3, (x - a)^4, \ldots$ are approximately $\pm(1/n)(it)/[i(2\pi t)^{1/2}]^n = \text{const}\, t^{(-n+2)/2}$ and are therefore small for large t. Thus it is natural to write the numerator of the integrand in the form

$$(-a)^{-(1/2)+it}e^{-Na}e^{p(x-a)}e^{i(x-a)^2/4\pi}g(x - a)$$

because then $g(x - a)$ is the exponential of the power series

$$-i\frac{(x - a)^2}{4\pi} - p(x - a) - N(x - a) + \left(-\frac{1}{2} + it\right)\log\left(1 + \frac{x - a}{a}\right)$$

whose coefficients are all small when t is large; the expansion $g(x-a) = \sum_{n=0}^{\infty} b_n(x-a)^n$ has radius of convergence $|a|$ because $x=0$ is the only singularity of the function it defines, its constant term b_0 is 1, and its remaining coefficients $b_1, b_2, \ldots$ are small for large t. Thus the integral in (1) becomes

$$\text{(2)}\qquad \frac{e^{-i\vartheta(t)}e^{-t\pi/2}(-a)^{-(1/2)+it}e^{-Na}}{(2\pi)^{1/2}(2\pi)^{it}e^{-i\pi/4}(1-ie^{-t\pi})}\int_{L_1}\frac{e^{i(x-a)^2/4\pi}e^{p(x-a)}\sum_{n=0}^{\infty} b_n(x-a)^n\,dx}{e^x-1}.$$

The factor $\exp[i(x-a)^2/4\pi]$ is real on L (where, as before, L is the line of which L_1 is a segment), has a maximum of 1 at $x=a$, and decreases rapidly as x moves away from a (for example, at the ends of L_1 it is $\exp\{i[\pm\frac{1}{2}e^{i\pi/4}(2\pi t)^{1/2}]^2/4\pi\} = e^{-t/8}$ which is very small for large t), so this integral is highly concentrated near $x=a$ where the b_0 term dominates. Thus the integral above is approximately

$$\text{(3)}\qquad \int_L \frac{e^{i(x-a)^2/4\pi}e^{p(x-a)}\,dx}{e^x-1}.$$

Riemann was able to evaluate this definite integral in closed form, and hence, since the factors in front of the integral can be evaluated numerically, he was able to find a numerical approximation to the value of R in (1).

Before evaluating the integral (3) it is advantageous to simplify the expression (2) by taking the change of variable $x = u + 2\pi iN$, $x - a = u + 2\pi iN - 2\pi i(t/2\pi)^{1/2} = u - 2\pi ip$, where p is the fractional part of $(t/2\pi)^{1/2}$. Then (2) takes the form

$$\begin{aligned}
&\frac{e^{-i\vartheta(t)}(e^{i\pi/2})^{it}[-i(2\pi t)^{1/2}]^{-(1/2)+it}e^{-N2\pi i(N+p)}}{(2\pi)^{(1/2)+it}(e^{i\pi/2})^{-1/2}(1-ie^{-t\pi})}\\
&\qquad\times\int_{\Gamma_1}\frac{e^{i(u-2\pi ip)^2/4\pi}e^{p(u-2\pi ip)}\sum b_n(u-2\pi ip)^n\,du}{e^u-1}\\
&=\frac{e^{-i\vartheta(t)}[(t/2\pi)^{1/2}]^{-(1/2)+it}}{2\pi(-i)(1-ie^{-t\pi})}e^{-2\pi iN^2-2\pi iNp-\pi ip^2-2\pi ip^2}\\
&\qquad\times\int_{\Gamma_1}\frac{e^{iu^2/4\pi}e^{2pu}\sum b_n(u-2\pi ip)^n\,du}{e^u-1}\\
&=\left(\frac{t}{2\pi}\right)^{-1/4}\left(\frac{t}{2\pi}\right)^{it/2}e^{-i\vartheta(t)-i\pi(N+p)^2-i\pi N^2-2\pi ip^2}\\
&\qquad\times\frac{1}{(1-ie^{-t\pi})(-1)2\pi i}\int_{\Gamma_1}\frac{e^{iu^2/4\pi}e^{2pu}\sum b_n(u-2\pi ip)^n\,du}{e^u-1}
\end{aligned}$$

where $\boldsymbol{\Gamma}_1$ is the line of slope 1 and length $(2\pi t)^{1/2}$, whose midpoint is $2\pi ip$, directed from upper right to lower left. Set

$$U = \frac{\exp\{i[(t/2)\log(t/2\pi) - (t/2) - (\pi/8) - \vartheta(t)]\}}{1-ie^{-t\pi}}.$$

Then, by the formula for $\vartheta(t)$ [(1) of Section 6.5], U is very near 1 for t large and, since $(N + p)^2 = t/2\pi$ and $(-1)^{N^2} = (-1)^N$, (2) takes the form

$$\left(\frac{t}{2\pi}\right)^{-1/4} Ue^{i\pi/8}(-1)^{N-1}e^{-2\pi i p^2}\frac{1}{2\pi i}\int_{\Gamma_1}\frac{e^{iu^2/4\pi}e^{2pu}\sum b_n(u-2\pi ip)^n\,du}{e^u-1}. \tag{4}$$

Riemann proved that

$$e^{i\pi/8}e^{-2\pi i p^2}\cdot\frac{1}{2\pi i}\int_\Gamma\frac{e^{iu^2/4\pi}e^{2pu}\,du}{e^u-1} = \frac{\cos 2\pi(p^2-p-\frac{1}{16})}{\cos 2\pi p}, \tag{5}$$

where Γ is a line of slope 1, directed from upper right to lower left, which crosses the imaginary axis between 0 and $2\pi i$. This shows that to a first approximation the remainder R in (1) is

$$R \sim (-1)^{N-1}\left(\frac{t}{2\pi}\right)^{-1/4}\frac{\cos 2\pi(p^2-p-\frac{1}{16})}{\cos 2\pi p} \tag{6}$$

which can, of course, be evaluated when t is given (recall that N is the integer part of $(t/2\pi)^{1/2}$ and p the fractional part—the apparent singularities at $p = \frac{1}{4}, \frac{3}{4}$ are discussed below) so that (1) can be used to give a first approximation to $Z(t)$. This is the first term of the Riemann–Siegel formula, the later terms of which will be developed in the next section. The remainder of this section is devoted to the proof of Riemann's formula (5).

Let $\Psi(p)$ denote the left side of (5). Since $\exp(iu^2/4\pi)$ approaches zero very rapidly as $|u| \to \infty$ in either direction along Γ, this integral converges for all p and defines an entire† function of the complex variable p. Let D denote the domain of the u-plane bounded by Γ and the line parallel to Γ which crosses the imaginary axis at the point which lies $2\pi i$ below the point where Γ crosses the imaginary axis. Then by the Cauchy integral formula

$$\frac{1}{2\pi i}\int_{\partial D}\frac{e^{iu^2/4\pi}e^{2pu}\,du}{e^u-1} = \text{value at 0 of } e^{iu^2/4\pi}e^{2pu}\frac{u}{e^u-1} = 1,$$

while on the other hand this integral over ∂D is

$$\begin{aligned}\frac{1}{2\pi i}\int_\Gamma\frac{e^{iu^2/4\pi}e^{2pu}\,du}{e^u-1} &- \frac{1}{2\pi i}\int_\Gamma\frac{e^{i(u-2\pi i)^2/4\pi}e^{2p(u-2\pi i)}\,du}{e^{u-2\pi i}-1}\\ &= e^{-i\pi/8}e^{2\pi i p^2}\Psi(p) - \frac{e^{-i\pi}e^{-4\pi i p}}{2\pi i}\int_\Gamma\frac{e^{iu^2/4\pi}e^{2pu+u}\,du}{e^u-1}\\ &= e^{-i\pi/8}e^{2\pi i p^2}\Psi(p) + e^{-4\pi i p}e^{-i\pi/8}e^{2\pi i(p+1/2)^2}\Psi(p+\tfrac{1}{2})\end{aligned}$$

so that equating the two expressions for the integral over ∂D gives a relation between $\Psi(p)$ and $\Psi(p+\frac{1}{2})$, namely,

$$\begin{aligned}e^{i\pi/8}e^{-2\pi i p^2} &= \Psi(p) + e^{-4\pi i p}e^{2\pi i p}e^{i\pi/2}\Psi(p+\tfrac{1}{2})\\ &= \Psi(p) + ie^{-2\pi i p}\Psi(p+\tfrac{1}{2}).\end{aligned} \tag{7}$$

†Thus the zeros of the denominator $\cos 2\pi p$ in (5) must be canceled by zeros in the numerator, and, indeed, if p is of the form (odd/4), then $p^2 - \frac{1}{16}$ is of the form (multiple of 8)/16 so $p^2 - p - \frac{1}{16}$ is also (odd/4) and the numerator is zero.

A second relationship between $\Psi(p)$ and $\Psi(p+\frac{1}{2})$ can be found by noting that when the integrals they contain are subtracted, the denominator $e^u - 1$ cancels to give

$$\begin{aligned} e^{-i\pi/8}e^{2\pi i p^2}\Psi(p) &- e^{-i\pi/8}e^{2\pi i(p+1/2)^2}\Psi(p+\tfrac{1}{2}) \\ &= \frac{1}{2\pi i}\int_\Gamma \frac{e^{iu^2/4\pi}[e^{2pu} - e^{2pu+u}]\,du}{e^u - 1} = -\frac{1}{2\pi i}\int_\Gamma e^{iu^2/4\pi}e^{2pu}\,du \\ &= -\frac{1}{2\pi i}\int_\Gamma e^{i(u-4\pi i p)^2/4\pi}e^{i4\pi p^2}\,du = e^{4\pi i p^2}\cdot K, \end{aligned}$$

where K is a constant independent of p; hence

$$\Psi(p) - e^{2\pi i p}\cdot i\cdot\Psi(p+\tfrac{1}{2}) = e^{i\pi/8}e^{2\pi i p^2}\cdot K. \tag{8}$$

With $p = \frac{1}{4}$ the two expressions (7) and (8) for $\Psi(\frac{1}{4}) + \Psi(\frac{3}{4})$ give

$$e^{i\pi/8}e^{-2\pi i/16} = e^{i\pi/8}e^{2\pi i/16}K, \qquad K = e^{-\pi i/4}.$$

If this is used in (8) and if $\Psi(p+\frac{1}{2})$ is eliminated between (7) and (8), one finds

$$e^{2\pi i p}\Psi(p) + e^{-2\pi i p}\Psi(p) = e^{2\pi i p}e^{i\pi/8}e^{-2\pi i p^2} + e^{-2\pi i p}e^{-i\pi/8}e^{2\pi i p^2}$$

which gives the desired expression (5) of $\Psi(p)$ as a quotient of cosines.

In summary, it has been shown that the remainder R has approximately the value (6). To deduce this approximation from the previous approximation (1), the series $\sum b_n(x-a)^n$ in (2) was truncated after the first term $b_0 = 1$, the factor U was replaced by 1, and the domain of integration of the resulting integral was extended to be the entire line of which it is a segment.

7.5 HIGHER ORDER APPROXIMATIONS

The only source of substantial error in the approximations of the preceding section is the truncation of the series $\sum b_n(x-a)^n = 1 + \cdots$ at the first term. In this section Riemann's method for obtaining higher order approximations using the higher order terms of this series will be described.

The computation of the individual coefficients b_n is not difficult. Recall that $\sum b_n(x-a)^n$ is by definition the exponential of the series

$$\begin{aligned} -[i(x-a)^2/4\pi] - (p+N)(x-a) \\ + (-\tfrac{1}{2} + it)\log[1 + (x-a)/a], \end{aligned} \tag{1}$$

where $a = i(2\pi t)^{1/2}$, $p + N = (t/2\pi)^{1/2}$, $0 \le p < 1$, N an integer. Let $\omega = (2\pi/t)^{1/2}$. Then ω is small for t large and the coefficients of the series (1) can be expressed in terms of ω as

$$-\frac{i}{4\pi}(x-a)^2 - \omega^{-1}(x-a) + \left(-\frac{1}{2} + 2\pi i\omega^{-2}\right)\log\left[1 + \frac{\omega}{2\pi i}(x-a)\right]$$
$$= -\frac{1}{2}\frac{\omega}{2\pi i}(x-a) + \frac{1}{4}\left(\frac{\omega}{2\pi i}\right)^2(x-a)^2$$
$$+ \frac{1}{3}\left(\frac{\omega^2}{8\pi^2} + \frac{1}{2\pi i}\right)\left(\frac{\omega}{2\pi i}\right)(x-a)^3 + \cdots$$
$$+ (-1)^{n-1}\frac{1}{n}\left(\frac{\omega^2}{8\pi^2} + \frac{1}{2\pi i}\right)\left(\frac{\omega}{2\pi i}\right)^{n-2}(x-a)^n + \cdots.$$

Thus the coefficients of $(x - a)$, $(x - a)^2$ are monomials of degree 1,2, respectively, in ω, and the coefficients of the higher terms $(x - a)^n$ are binomials in ω whose terms are of degree $n - 2$ and n. Since $\sum b_n(x - a)^n$ is the exponential of this series it follows immediately that *b_n is a polynomial in ω of degree at most n in which all terms have degree at least equal to the integer part of* $(n/3)$. [For example, to find b_{14} one could compute the coefficient of $(x - a)^{14}$ in the first 14 powers of the above series, divide by the appropriate factorial, and add. Many of the terms in the coefficient of $(x - a)^{14}$ would have degree 14 in ω; the terms of smallest degree would be those in which the first degree term in front of $(x - a)^3$ is used the maximum number of times, which in the case $n = 14$ will give terms of degree 6 in front of $(x - a)^3(x - a)^3(x - a)^3(x - a)^3(x - a)^2 = (x - a)^{14}$.] The easiest way to compute the b_n explicitly is to make use of the fact that the derivative of the series (1) is

$$-\frac{i\cdot 2(x-a)}{4\pi} - \omega^{-1} + \left(-\frac{1}{2} + 2\pi i\omega^{-2}\right)\frac{1}{1 + [(x-a)/a]}\cdot\frac{1}{a}$$
$$= \frac{(x-a)}{2\pi i} - \omega^{-1} + \left(-\frac{1}{2} + 2\pi i\omega^{-2}\right)\frac{1}{2\pi i\omega^{-1} + (x-a)}$$
$$= \frac{2\pi i\omega^{-1}(x-a) + (x-a)^2 - 2\pi i\omega^{-1}[2\pi i\omega^{-1} + (x-a)] + 2\pi i(-\frac{1}{2} + 2\pi i\omega^{-2})}{2\pi i[(2\pi i\omega^{-1} + (x-a)]}$$
$$= \frac{(x-a)^2 - \pi i}{2\pi i[(x-a) + 2\pi i\omega^{-1}]}.$$

Thus the logarithmic derivative of $\sum b_n(x - a)^n$ is simply

$$\frac{(\sum b_n(x-a)^n)'}{\sum b_n(x-a)^n} = \frac{(x-a)^2 - \pi i}{2\pi i[(x-a) + 2\pi i\omega^{-1}]}.$$

Hence

$$2\pi i[(x-a) + 2\pi i\omega^{-1}][\textstyle\sum nb_n(x-a)^{n-1}] = [(x-a)^2 - \pi i][\textstyle\sum b_n(x-a)^n] \tag{2}$$

from which it is easy to derive a recursion relation among the b's which makes their computation possible. This computation is a bit long, however, and it will not be needed in what follows.

If the series $\sum b_n(x-a)^n$ is truncated at the Kth term rather than at the 0th term, then the argument of the preceding section leads to the higher order approximation

$$\frac{e^{-i\vartheta(t)}e^{-t\pi/2}}{(2\pi)^{1/2}(2\pi)^{it}e^{-i\pi/4}(1-ie^{-t\pi})}\int_{\Gamma_1}\frac{(-x)^{-(1/2)+it}e^{-Nx}\,dx}{e^x-1} \sim (-1)^{N-1}\left(\frac{t}{2\pi}\right)^{-1/4}\cdot U[b_0c_0+b_1c_1+\cdots+b_Kc_K] \tag{3}$$

where

$$c_n = e^{i\pi/8}e^{-2\pi i p^2}\frac{1}{2\pi i}\int_\Gamma \frac{e^{iu^2/4\pi}e^{2pu}}{e^u-1}(u-2\pi ip)^n\,du.$$

Here N, p depend on t as before and Γ is the line of slope 1 through $2\pi ip$ oriented from upper right to lower left. The numbers c_n can be computed by expanding $(u-2\pi ip)^n$ and integrating termwise using the formula

$$\begin{aligned}\frac{d^k}{dp^k}[e^{2\pi ip^2}\Psi(p)] &= \frac{d^k}{dp^k}\left[\frac{e^{i\pi/8}}{2\pi i}\int_\Gamma\frac{e^{iu^2/4\pi}e^{2pu}}{e^u-1}\,du\right]\\ &= \frac{e^{i\pi/8}}{2\pi i}\int_\Gamma\frac{e^{iu^2/4\pi}e^{2pu}}{e^u-1}(2u)^k\,du.\end{aligned}$$

In this way the c_n can be expressed as finite linear combinations of $\exp(2\pi ip^2)$ $\Psi^{(k)}(p)$ $(k=0,1,\ldots,n)$ whose coefficients are polynomials in p ($\Psi^{(k)}$ is the kth derivative of Ψ). The easiest way to compute the c_n is to make use of the relationship

$$\begin{aligned}\Psi(p+y) &= e^{i\pi/8}e^{-2\pi i(p+y)^2}\frac{1}{2\pi i}\int_\Gamma\frac{e^{iu^2/4\pi}e^{2(p+y)u}}{e^u-1}\,du\\ e^{2\pi iy^2}\Psi(p+y) &= e^{i\pi/8}e^{-2\pi ip^2}\frac{1}{2\pi i}\int_\Gamma\frac{e^{iu^2/4\pi}e^{2pu}}{e^u-1}e^{2y(u-2\pi ip)}\,du\\ &= \sum_{n=0}^{\infty}e^{i\pi/8}e^{-2\pi ip^2}\frac{1}{2\pi i}\int_\Gamma\frac{e^{iu^2/4\pi}e^{2pu}}{e^u-1}\frac{[2y(u-2\pi ip)]^n}{n!}\,du,\end{aligned}$$

$$e^{2\pi iy^2}\sum_{m=0}^{\infty}\frac{\Psi^{(m)}(p)}{m!}y^m = \sum_{n=0}^{\infty}\frac{(2y)^n}{n!}c_n, \tag{4}$$

and to equate powers of y. The explicit expressions of the c_n which this relationship yields will not be needed in what follows. All that will be needed is the fact that c_n can be expressed as a linear combination of $\Psi(p), \Psi'(p), \ldots, \Psi^{(n)}(p)$ with coefficients which are independent of t.

In order to use the higher order approximation (3), it is necessary to be able to evaluate $b_0c_0+b_1c_1+\cdots+b_Kc_K$ for given t and K. Riemann devised the following method of accomplishing this without going through the computation of the b_n and c_n. Note that since the b_n are polynomials in ω, one can also regard $b_0c_0 + b_1c_1 + \cdots + b_Kc_K$ as a polynomial in ω and arrange it according to powers of ω. It is natural to do this for two reasons: first because ω is small for t large, so the importance of a term depends on the power of ω it contains; second because any given power of ω no longer occurs in $b_{K+1}, b_{K+2}, \ldots$ once K is sufficiently large, and hence the coefficient of any power of ω is independent of K for K sufficiently large. Riemann's method is a method for finding the coefficient of ω^k in $b_0c_0 + b_1c_1 + \cdots + b_Kc_K$ for large K. By the above observation about the form of the c_n, it is clear that this coefficient can be expressed as a linear combination of $\Psi(p)$, $\Psi'(p), \ldots, \Psi^{(n)}(p)$ with coefficients independent of t.

Let the implicit relation (4) satisfied by the c_n be multiplied by $\sum_{n=0}^{K} n!b_n(2y)^{-n}$. The right side becomes a power series in y in which a finite number of terms contain negative powers of y and in which, by the choice of the multiplier, the constant term (term in y^0) is $b_0c_0 + b_1c_1 + \cdots + b_Kc_K$. The left side becomes a product of three power series in y which, by the commutativity and associativity of multiplication of formal power series, is equal to $\sum \Psi^{(m)}(p)y^m/m!$ times

$$G(y) = \exp 2\pi iy^2 \sum_{n=0}^{K} n!b_n(2y)^{-n}.$$

If the coefficients of the nonpositive powers of y in $G(y)$ can be computed, then it will be a simple matter to find the constant term of its product with $\sum \Psi^{(m)}(p)y^m/m!$ and hence the desired expression $b_0c_0 + \cdots + b_Kc_K$. The essence of the argument below is to use the recurrence relation satisfied by the b_n to find a recurrence relation satisfied by the coefficients of G which makes it possible to compute the terms of G with nonpositive powers of y. More specifically, let the b_n be written as polynomials in ω and let the terms of G be rearranged in the order of powers of ω

$$G(y) = \sum_{j=0}^{\infty} A_j\omega^j$$

in which each A_j is a power series in y with a finite number of terms containing negative powers and in which the sum is actually finite because the largest power of ω in $b_0, b_1, \ldots, b_K$ is ω^K. The A_j depend on K, but if a particular positive integer ν is chosen and if K is large enough that $b_{K+1}, b_{K+2}, \ldots$ contain no terms in $\omega^0, \omega^1, \ldots, \omega^{\nu-1}$, then $A_0, A_1, \ldots, A_{\nu-1}$ are independent of K. The objective is to find the terms with nonpositive powers of y in A_0, $A_1, \ldots, A_{\nu-1}$. Computations will be carried out mod ω^ν so that G can be taken as

$$G(y) = e^{2\pi iy^2} \sum_{n=0}^{\infty} n!b_n(2y)^{-n} \qquad (\text{mod } \omega^\nu)$$

and the main step of the argument is to find a recurrence relation satisfied (mod ω^ν) by G which makes it possible to compute the A_j.

Of course the desired relation must be deduced from the relation (2) satisfied by the b_n, which can be stated

$$2\pi i n b_n - 4\pi^2\omega^{-1}(n+1)b_{n+1} = b_{n-2} - \pi i b_n \tag{5}$$

for $n = 0, 1, 2, 3, \ldots$ (provided b_{-1}, b_{-2} are defined to be zero). The first step is to state these relations among the b_n as a differential equation satisfied by the formal power series

$$F(y) = \sum_{n=0}^{\infty} n!\, b_n (2y)^{-n}.$$

If the nth equation (5) is multiplied by $n!(2y)^{-n-1}$, then the first term on the left becomes the general term of the series $-\pi i DF$ (where D denotes formal differentiation with respect to y), the second term on the left becomes the general term of $-4\pi^2\omega^{-1}F$, the first term on the right becomes the general term of $\frac{1}{4}D^2(2y)^{-1}F = \frac{1}{8}D^2y^{-1}F$, and the second term on the right becomes the general term of $-(\pi i/2y)F$. More precisely

$$\pi i DF = -2\pi i \sum_0^{\infty} n!\, n b_n (2y)^{-n-1},$$

$$4\pi^2\omega^{-1}F = 4\pi^2\omega^{-1} + 4\pi^2\omega^{-1}\sum_0^{\infty} n!(n+1)b_{n+1}(2y)^{-n-1},$$

$$\tfrac{1}{8}D^2y^{-1}F = \sum_0^{\infty} n!\, b_{n-2}(2y)^{-n-1},$$

$$-\tfrac{1}{2}\pi i y^{-1}F = -\pi i \sum_0^{\infty} n!\, b_n(2y)^{-n-1},$$

so the relation (5) for $n = 0, 1, 2, \ldots$ is equivalent to the differential equation

$$\pi i DF + 4\pi^2\omega^{-1}F + \tfrac{1}{8}D^2y^{-1}F - \tfrac{1}{2}\pi i y^{-1}F = 4\pi^2\omega^{-1}$$

for F. Then integration by parts gives a differential equation satisfied by $G(y) = \exp(2\pi i y^2)F(y) \pmod{\omega^\nu}$ as follows.

$$\begin{aligned}
4\pi^2\omega^{-1}e^{2\pi i y^2} &= [\pi i DF + 4\pi^2\omega^{-1}F + \tfrac{1}{8}D^2y^{-1}F - \tfrac{1}{2}\pi i y^{-1}F]e^{2\pi i y^2} \\
&= \pi i DG - \pi i F(De^{2\pi i y^2}) + 4\pi^2\omega^{-1}G + \tfrac{1}{8}D^2y^{-1}G \\
&\quad -\tfrac{1}{8}(y^{-1}F)(D^2e^{2\pi i y^2}) - \tfrac{1}{4}(Dy^{-1}F)(De^{2\pi i y^2}) - \tfrac{1}{2}\pi i y^{-1}G \\
&= \pi i DG + 4\pi^2 yG + 4\pi^2\omega^{-1}G + \tfrac{1}{8}D^2y^{-1}G \\
&\quad -\tfrac{1}{8}(y^{-1}F)(-16\pi^2y^2e^{2\pi i y^2} + 4\pi i e^{2\pi i y^2}) \\
&\quad -\tfrac{1}{4}(-y^{-2}F + y^{-1}DF)(4\pi i y e^{2\pi i y^2}) - \tfrac{1}{2}\pi i y^{-1}G \\
&= \pi i DG + 4\pi^2 yG + 4\pi^2\omega^{-1}G + \tfrac{1}{8}D^2y^{-1}G + 2\pi^2 yG \\
&\quad -\tfrac{1}{2}\pi i y^{-1}G + \pi i y^{-1}G - \pi i DG + \pi i F(De^{2\pi i y^2}) \\
&\quad -\tfrac{1}{2}\pi i y^{-1}G \\
&= 6\pi^2 yG + 4\pi^2\omega^{-1}G + \tfrac{1}{8}D^2y^{-1}G - 4\pi^2 yG
\end{aligned}$$

and finally

$$e^{2\pi i y^2} = G + \tfrac{1}{2}\omega y G + \tfrac{1}{32}\omega\pi^{-2}D^2y^{-1}G$$

mod ω^ν. Since $G = \sum A_j\omega^j \pmod{\omega^\nu}$, this gives

$$e^{2\pi i y^2} = \sum_j A_j\omega^j + \sum_j \left(\frac{y}{2}A_j + \tfrac{1}{32}\pi^{-2}D^2y^{-1}A_j\right)\omega^{j+1};$$

so the terms in $\omega, \omega^2, \omega^3, \ldots$ cancel on the left, which implies

$$A_j = -\tfrac{1}{2}yA_{j-1} - \tfrac{1}{32}\pi^{-2}D^2y^{-1}A_{j-1} \qquad (j = 1, 2, 3, \ldots).$$

Since $A_0 = \exp(2\pi i y^2)$, this relation makes it possible to compute $A_1, A_2, A_3, \ldots$ in turn. Only the nonpositive powers of y in A_j are of interest, and to determine the coefficient of y^n in A_j it is only necessary to know the coefficients of y^{n+3} and y^{n-1} in A_{j-1}. Thus to determine all nonpositive terms in A_4, it is only necessary to begin with all terms through y^{12} in A_0 after which one easily computes

$$A_0 = 1 + 2\pi i y^2 - 2\pi^2y^4 - \frac{2^2\pi^3 i y^6}{3} + \frac{2\pi^4y^8}{3} + \frac{2^2\pi^5 i y^{10}}{3\cdot 5} - \frac{2^2\pi^6y^{12}}{3^2\cdot 5},$$

$$A_1 = -\frac{1}{2^4\pi^2y^3} - \frac{y}{2^3} - \frac{\pi i y^3}{2\cdot 3} + \frac{\pi^2y^5}{2^3} + \frac{\pi^3 i y^7}{3\cdot 5} - \frac{\pi^4y^9}{2^2\cdot 3^2},$$

$$A_2 = \frac{5}{2^7\pi^4y^6} + \frac{1}{2^5\pi^2y^2} + \frac{i}{2^5\cdot 3\pi} + \frac{y^2}{2^6} + \frac{\pi i y^4}{2^4\cdot 3} - \frac{\pi^2y^6}{2^3\cdot 3^2},$$

$$A_3 = \frac{-5\cdot 7}{2^9\pi^6y^9} - \frac{1}{2^5\pi^4y^5} - \frac{i}{2^9 3\pi^3y^3} - \frac{1}{2^6\pi^2y} - \frac{7iy}{2^8\cdot 3\pi} + \frac{y^3}{2^7\cdot 3^2},$$

$$A_4 = \frac{5^2\cdot 7\cdot 11}{2^{13}\pi^8y^{12}} + \frac{7\cdot 11}{2^{10}\pi^6y^8} + \frac{5i}{2^{12}\cdot 3\pi^5y^6} + \frac{19}{2^{10}\pi^4y^4} + \frac{i}{2^{10}\cdot 3\pi^3y^2} + \frac{11\cdot 13}{2^{11}3^2\pi^2}.$$

Multiplying by $\sum \Psi^{(m)}(p)y^m/m!$ and taking the constant term of the result then gives as the coefficients of $\omega^0, \omega^1, \omega^2, \omega^3, \omega^4$ in $b_0c_0 + \cdots + b_Kc_K$ the expressions

$$\Psi(p),$$

$$-\frac{\Psi^{(3)}(p)}{2^4\pi^2 3!},$$

$$\frac{5\Psi^{(6)}(p)}{2^7\pi^4 6!} + \frac{\Psi^{(2)}(p)}{2^5\pi^2 2!} + \frac{i\Psi(p)}{2^5\cdot 3\pi},$$

$$-\frac{5\cdot 7\Psi^{(9)}(p)}{2^9\pi^6 9!} - \frac{\Psi^{(5)}(p)}{2^5\pi^4 5!} - \frac{i\Psi^{(3)}(p)}{2^9 3\pi^3 3!} - \frac{\Psi^{(1)}(p)}{2^6\pi^2},$$

$$\frac{5^2\cdot 7\cdot 11\Psi^{(12)}(p)}{2^{13}\pi^8 12!} + \frac{7\cdot 11\cdot\Psi^{(8)}(p)}{2^{10}\pi^6 8!} + \frac{5i\Psi^{(6)}(p)}{2^{12}\cdot 3\pi^5 6!} + \frac{19\Psi^{(4)}(p)}{2^{10}\pi^4 4!} + \frac{i\Psi^{(2)}(p)}{2^{10}\cdot 3\pi^3 2!} + \frac{11\cdot 13\cdot\Psi(p)}{2^{11}3^2\pi^2},$$

respectively, provided, of course, that $K \geq 12$.

Since the actual remainder $R = Z(t) - 2\sum_1^N \cos[\vartheta(t) - t\log n]$ is a real number the imaginary terms in the above formulas must have no significance. In fact if the factor

$$\begin{aligned}\exp\left\{i\left[\frac{t}{2}\log\left(\frac{t}{2\pi}\right) - \frac{t}{2} - \frac{\pi}{8} - \vartheta(t)\right]\right\} &= \exp\left[-\frac{i}{48t} - \frac{7i}{5760t^3} + \cdots\right] \\ &= \exp\left[-\frac{i}{96\pi}\omega^2 + \cdots\right] \\ &= 1 - \frac{i}{96\pi}\omega^2 - \frac{1}{96^2\pi^2\cdot 2}\omega^4 + \cdots\end{aligned}$$

is taken into account, then the imaginary terms in front of ω^2 and ω^3 cancel and the coefficient of ω^4 becomes

$$\frac{1}{2^{23}\cdot 3^5\pi^8}\Psi^{(12)}(p) + \frac{11}{2^{17}\cdot 3^2\cdot 5\pi^6}\Psi^{(8)}(p) + \frac{19}{2^{13}3\pi^4}\Psi^{(4)}(p) + \frac{1}{2^7\pi^2}\Psi(p)$$

which not only eliminates the imaginary terms but also simplifies the coefficient of $\Psi(p)$.

In summary, *the remainder R in the formula*

$$Z(t) = 2\sum_{n^2<(t/2\pi)} n^{-1/2}\cos[\vartheta(t) - t\log n] + R$$

is approximately

$$R \sim (-1)^{N-1}\left(\frac{t}{2\pi}\right)^{-1/4} \times\left[C_0 + C_1\left(\frac{t}{2\pi}\right)^{-1/2} + C_2\left(\frac{t}{2\pi}\right)^{-2/2} + C_3\left(\frac{t}{2\pi}\right)^{-3/2} + C_4\left(\frac{t}{2\pi}\right)^{-4/2}\right],$$

where N is the integer part of $(t/2\pi)^{1/2}$, *p the fractional part, and*

$$C_0 = \Psi(p) = \frac{\cos 2\pi(p^2 - p - \frac{1}{16})}{\cos 2\pi p},$$

$$C_1 = -\frac{1}{2^5 3\pi^2}\Psi^{(3)}(p),$$

$$C_2 = \frac{1}{2^{11}\cdot 3^2\pi^4}\Psi^{(6)}(p) + \frac{1}{2^6\pi^2}\Psi^{(2)}(p),$$

$$C_3 = -\frac{1}{2^{16}\cdot 3^4\pi^6}\Psi^{(9)}(p) - \frac{1}{2^8\cdot 3\cdot 5\pi^4}\Psi^{(5)}(p) - \frac{1}{2^6\pi^2}\Psi^{(1)}(p),$$

$$C_4 = \frac{1}{2^{23}\cdot 3^5\pi^8}\Psi^{(12)}(p) + \frac{11}{2^{17}\cdot 3^2\cdot 5\pi^6}\Psi^{(8)}(p) + \frac{19}{2^{13}\cdot 3\pi^4}\Psi^{(4)}(p) + \frac{1}{2^7\pi^2}\Psi(p).$$

This† is the formula which Siegel found in Riemann's unpublished papers (see Fig. 2). The next section is devoted to some numerical applications of the formula and the following section to an analysis of the error.

7.6 SAMPLE COMPUTATIONS

To get some idea of the accuracy of the Riemann–Siegel formula, consider the computation of $Z(18)$, the value of which was found with three-place accuracy in Section 6.5. If $t = 18$, then $(t/2\pi)^{1/2} = 1.692\,569$, so in this case $N = 1$, $p = 0.692\,569$. The sum approximating $Z(18)$ consists of the single term $2\cos\vartheta(18) = 2\cos(0.080\,911) = 1.993\,457$. The denominator of $\Psi(p)$ is $\cos(2\pi \times 0.692\,569) = \cos(\pi + \frac{1}{3}\pi + 2\pi \times 0.025\,902) = (\sqrt{3}/2)\sin(0.162\,747) - \frac{1}{2}\cos(0.162\,747) = -0.353\,070$ and the numerator is $\cos 2\pi(p^2 - p - \frac{1}{16}) = \cos(-0.275\,417 \times 2\pi) = -\sin(0.159\,700) = -0.159\,022$; so

$$\begin{aligned} Z(18) &\sim 2\cos\vartheta(18) + (-1)^{1-1}(18/2\pi)^{-1/4}\Psi(0.692\,569) \\ &= 1.993\,457 + (0.768\,647)\frac{-0.159\,022}{-0.353\,070} \\ &= 1.993\,457 + 0.346\,197 = 2.339\,654 \end{aligned}$$

is the first approximation to $Z(18)$. Comparing this with the value 2.337 obtained in Section 6.5 shows that the Riemann–Siegel formula gives better than two-place accuracy even for this relatively small value of t and even when only the first approximation is used!

To use the higher order approximations, one must have some means of evaluating the more complicated functions $C_1, C_2, C_3, \ldots$ of Section 7.5. The simplest method of doing this is to compute the Taylor series coefficients of $\Psi(p)$, from which the Taylor series coefficients of the derivatives of Ψ and hence of the C's are easily computed. Since $0 < p < 1$, it is natural to ex-

†However, Siegel changed the coefficient of $\Psi(p)$ in C_4 from Riemann's value $11 \cdot 13(2^{11}3^2\pi^2)^{-1}$ to $(2^7\pi^2)^{-1}$ as on the preceding page. The above expression of the formula differs somewhat from both Riemann's expression of it and Siegel's (which differ from each other). Riemann expresses the series $\sum C_j\omega^j$ as a series $\sum B_j a^{-j}$, where $a = i(2\pi t)^{1/2}$ as before (α in Riemann's notation) and where, consequently, $B_j = (2\pi i)^j C_j$. Siegel expresses the series $\sum C_j\omega^j$ as a series $\sum A_j t^{-j/2}$, where, consequently, $A_j = C_j(2\pi)^{j/2}$. Moreover, Siegel expresses the A_j in terms of derivatives of the function

$$F(x) = \frac{\cos[x^2 + (3\pi/8)]}{\cos[(2\pi)^{1/2}x]} = \Psi\left[\frac{x}{(2\pi)^{1/2}} + \frac{1}{2}\right]$$

rather than in terms of derivatives of Ψ. His formulas can be deduced from those above using $\Psi^{(n)} = (2\pi)^{n/2}F^{(n)}$.

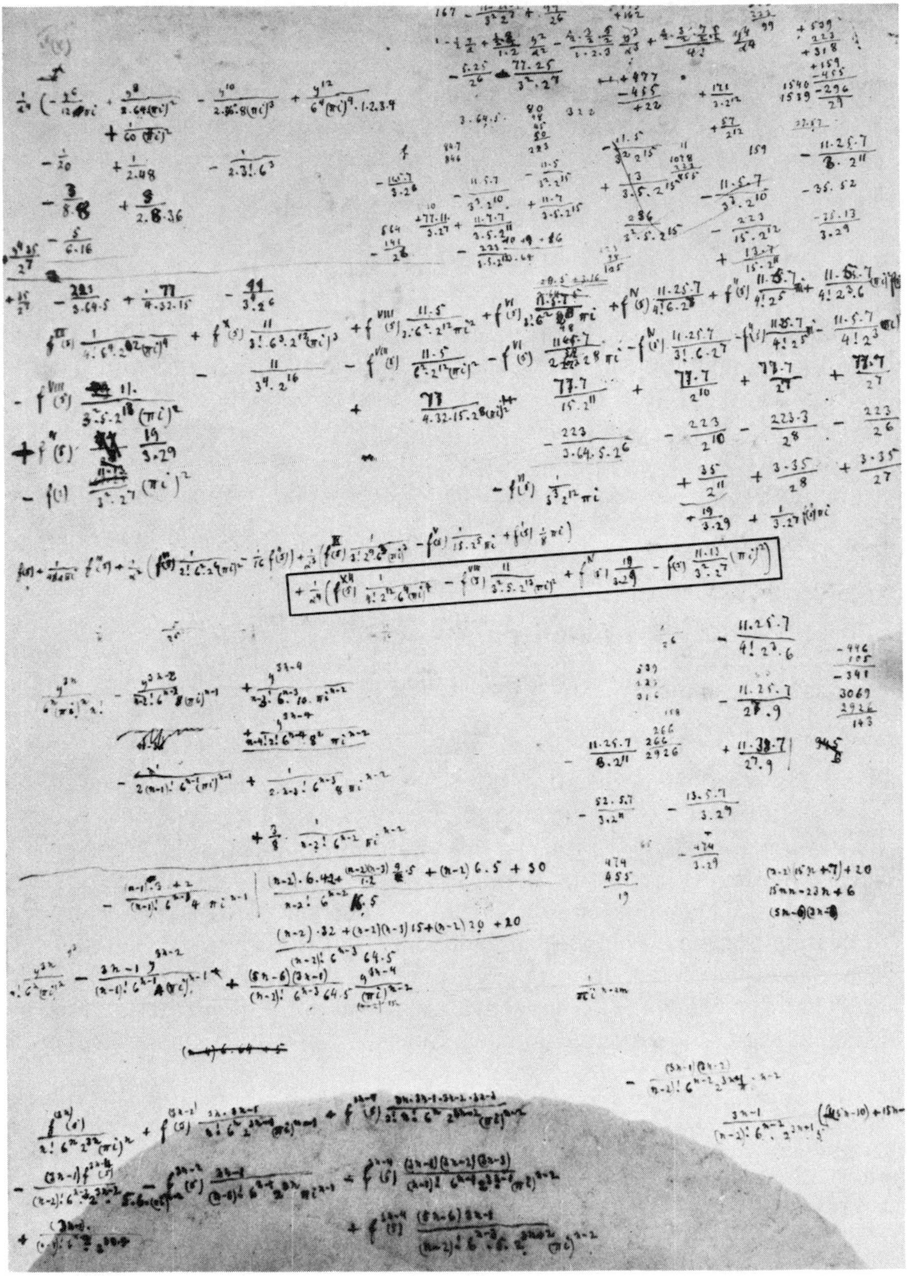

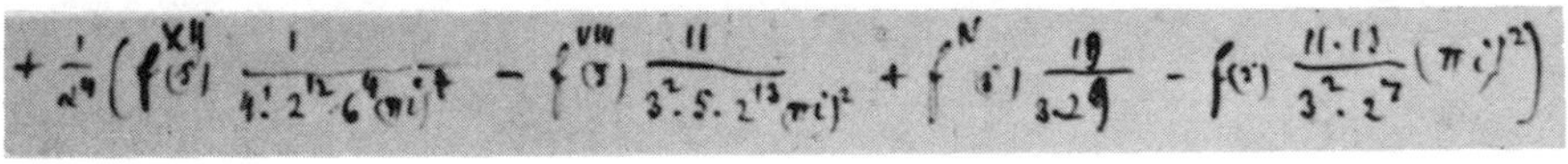

pand these functions in powers of $p - \frac{1}{2}$. Since $\Psi(\frac{1}{2} + a) = \cos[2\pi a^2 + (3\pi/8)]/\cos 2\pi a$, the expansion of Ψ in powers of $a = p - \frac{1}{2}$ is a quotient of known even power series and, as such, is an even power series whose coefficients can be explicitly computed. Then $C_0 = \Psi, C_2, C_4, \ldots$ will be even power series and $C_1, C_3, C_5, \ldots$ odd power series whose coefficients are easily found. Haselgrove [H8] gives a table of coefficients of C_0, C_1, C_2, C_3, C_4 in powers of $(1 - 2p)$ which is reproduced as Table V. Using these coefficients with the above value of p and therefore with $1 - 2p = -0.385\,138$, one finds easily $C_0 = 0.450\,401$, $C_1 = -0.009\,207$, $C_2 = 0.004\,996$, $C_3 = -0.000\,316$, $C_4 = 0.000\,323$ from which $2\cos\vartheta + (t/2\pi)^{-1/4}[C_0 + (t/2\pi)^{-1/2}C_1 + \cdots + (t/2\pi)^{-2}C_4]$ is $2.336\,796 \sim Z(18)$. Since the C_4 term is a 3 in the fifth place, one might hope for four-place accuracy and indeed this answer is extremely close to Haselgrove's six-place value $Z(18) = 2.336\,800$. Thus the error estimates of Section 7.3 are much too generous and Riemann was in fact in possession of the means to compute $\zeta(\frac{1}{2} + it)$ with amazing accuracy.

Using the Riemann–Siegel formula it is quite easy to locate the first few roots of $\zeta(\frac{1}{2} + it)$ by computation. The main term $2\cos\vartheta(t)$ is zero near $t = 14.5$ where $\vartheta(t)$ is near $-\pi/2$, as a simple computation using formula (2) of Section 6.5 for $\vartheta(t)$ shows. For $t = 14.5$ simple estimates give $t/2\pi \sim 2.30$, $(t/2\pi)^{1/2} \sim 1.5$, $N = 1$, $p \sim \frac{1}{2}$. Since $\Psi(\frac{1}{2}) = -\cos(5\pi/8) \sim 0.38$ and $(t/2\pi)^{-1/4} \sim (2/3)^{1/2} \sim 0.8$, the first correction term $(t/2\pi)^{-1/4}\Psi(p)$ is about $(0.8)(0.38) \sim 0.30$; so to move toward a zero of Z, the value of t should be reduced to where the main term $2\cos\vartheta(t)$ is -0.30. The derivative of this term is $-2\vartheta'(t)\sin\vartheta(t) \sim -2\cdot\frac{1}{2}\log(t/2\pi)\cdot(-1) \sim 0.83$, so t should be reduced from 14.5 to about $14.5 - [(0.30)/(0.83)] = 14.14$. Thus there might well be a root between 14.1 and 14.2. Now $Z(14.1)$ and $Z(14.2)$ can be computed by exactly the same method as was used for $Z(18)$ above. The results are shown in Table VI. The size of the C_4 term suggests an accuracy of about four places, and this is more than confirmed by Haselgrove's tables, which give $Z(14.1) = -0.027\,463$, $Z(14.2) = +0.052\,045$. Thus there is definitely a root in the interval and linear interpolation would place it at $14.1 + h$, where $h/0.1 = (0.027\,466)/(0.027\,466 + 0.052\,042) \sim 0.345$ so the root is at about 14.1345. Further computations with t in this range show that the

---

Fig. 2 This is the sheet on which the Riemann–Siegel formula appears in Riemann's unpublished papers in the Göttingen University Library. (Here it is somewhat reduced in size.) The enlargement shows the final terms of the formula, which include the coefficient that Siegel simplified. The lack of coherent organization and of any explanation are typical of these papers, which include, along with the unexplained formulas, various random jottings such as the Chebyshev note on p. 5 and a computation of $\sqrt{2}$ to 38 decimal places. (Reproduced with the permission of the Niedersächsische Staats- und Universitätsbibliothek, Handschriftenabteilung, Göttingen.)

TABLE V

TABLE OF COEFFICIENTS[a]

s	$c_{0,s}$	s	$c_{1,s}$
0	+0·38268 34323 65089 77173	1	+0·02682 51026 28375 35
2	+ ·43724 04680 77520 44936	3	− 1378 47734 26351 85
4	+ ·13237 65754 80343 52333	5	− 3849 12504 82235 08
6	− 1360 50260 47674 18865	7	− 987 10662 99062 08
8	− 1356 76219 70103 58088	9	+ 331 07597 60858 40
10	− 162 37253 23144 46528	11	+ 146 47808 57795 42
12	+ 29 70535 37333 79691	13	+ 1 32079 40624 88
14	+ 7 94330 08795 21469	15	− 5 92274 87018 47
16	+ 4655 61246 14504	17	− 59802 42585 37
18	− 14327 25163 09551	19	+ 9641 32245 62
20	− 1035 48471 12314	21	+ 1833 47337 22
22	+ 123 57927 08384	23	− 44 67087 57
24	+ 17 88108 38577	25	− 27 09635 09
26	− 33914 14393	27	− 77852 89
28	− 16326 63392	29	+ 23437 63
30	− 378 51094	31	+ 1583 02
32	+ 93 27423	33	− 121 20
34	+ 5 22184	35	− 14 58
36	− 33506	37	+ 29
38	− 3412	39	+ 9
40	+ 58		
42	+ 15		
	Sum $= A$		Sum $= \frac{1}{8}\pi B - \frac{1}{4}A$

s	$c_{2,s}$	s	$c_{3,s}$	s	$c_{4,s}$
	0·0		0·00		0·00
0	+0518 85428 30293	1	+133 97160 907	0	+046 48338 9
2	+ 30 94658 38807	3	−374 42151 364	2	−100 56607 4
4	−1133 59410 78229	5	+133 03178 920	4	+ 24 04485 6
6	+ 223 30457 41958	7	+226 54660 765	6	+102 83086 1
8	+ 519 66374 08862	9	− 95 48499 998	8	− 76 57860 9
10	+ 34 39914 40762	11	− 60 10038 459	10	− 20 36528 6
12	− 59 10648 42747	13	+ 10 12885 828	12	+ 23 21229 0
14	− 10 22997 25479	15	+ 6 86573 345	14	+ 3 26021 5
16	+ 2 08883 92217	17	− 5985 366	16	− 2 55790 5
18	+ 59276 65493	19	− 33316 599	18	− 41074 6
20	− 1642 38384	21	− 2191 929	20	+ 11781 2
22	− 1516 11998	23	+ 789 089	22	+ 2445 6
24	− 59 07803	25	+ 94 147	24	− 239 2
26	+ 20 91151	27	− 9 570	26	− 75 0
28	+ 1 78157	29	− 1 876	28	+ 1 3
30	− 16164	31	+ 45	30	+ 1 4
32	− 2380	33	+ 22		
34	+ 54				
36	+ 20				
	Sum $= B/96\pi$				Sum $= -A/18432\pi^2$

[a]From Haselgrove [H8].

Main term	1.993 457
C_0 term	0.346 199
C_1 term	−0.004 181
C_2 term	0.001 341
C_3 term	−0.000 050
C_4 term	0.000 030
$Z(18)$	2.336 796

Computation of the approximation to $Z(18)$.

value of Z given by the Riemann–Siegel formula with terms through C_4 changes sign between 14.134 727 and 14.134 729. Thus the Riemann–Siegel formula permits the computation of the first root $\frac{1}{2} + i\alpha$, $\alpha = 14.134\,725\ldots$ (see Section 6.1) with at least five-place accuracy. If the C_5 and C_6 terms were used, it is possible that even greater accuracy might be achieved. *Riemann computed this root*, finding its value to be 14.1386; the error in his value results, as the above shows, not from the inherent inaccuracy of the Riemann–Siegel formula, but merely from the fact that he must have carried out only rough computations.

Riemann also took some steps toward proving that this root 14.1... is the *first* root. By (4) of Section 3.8

$$\sum_{\operatorname{Im}\rho>0}\left(\frac{1}{\rho}+\frac{1}{1-\rho}\right)=1+\frac{1}{2}\gamma-\frac{1}{2}\log\pi-\log 2.$$

TABLE VI

t	14.1	14.2
$\vartheta(t)$	−1.742 722	−1.702 141
p	0.498 027	0.503 330
$1-2p$	0.003 946	−0.006 660
$(t/2\pi)^{-1/2}$	0.667 545	0.665 190
$(t/2\pi)^{-1/4}$	0.817 034	0.815 591
Main term	−0.342 160	−0.261 934
C_0 term	0.312 671	0.312 129
C_1 term	0.000 058	−0.000 097
C_2 term	0.001 889	0.001 872
C_3 term	0.000 001	−0.000 002
C_4 term	0.000 075	0.000 074
$Z(t)$	−0.027 466	+0.052 042

Hence†

(1) $$\sum_{\text{Im}\,\rho>0} [1/\rho(1-\rho)] = 0.02309\ldots.$$

Now the root $\rho = \frac{1}{2} + i14.1\ldots$ already found accounts for about 0.005 of the sum of the right. If there were a root in the upper halfplane with smaller imaginary part than this one, then there would have to be *two* such roots, either because it would not be on the line $\text{Re}\, s = \frac{1}{2}$, in which case there would be a symmetrical root on the opposite side of the line, or because it would be on the line, in which case $Z(t)$ would have to change sign a second time in order to be negative at 14.1 and at 0. Thus such a root would have to satisfy

$$1/\gamma^2 < \tfrac{1}{2}(0.018) = 0.009, \qquad \gamma > 10.$$

Using the Riemann–Siegel formula it is not difficult to see that such a root on the line is very improbable and therefore that the root just found is probably the first root on the line. If all 10 of the roots in the range $0 < \text{Im}\,\rho < 50$ are located, they account for about 0.0136 of the total in (1) and therefore suffice to prove that the above root is indeed the root in the upper halfplane with the least imaginary part.

The next root on the line would be expected in the vicinity of the next zero of $\cos\vartheta(t)$, which occurs when $\vartheta(t) \sim \pi/2$, $t \sim 20.7$. Assuming that the Riemann–Siegel formula is accurate, it is easy to prove that $Z(t)$ does indeed change from $+$ back to $-$ near this point and to locate the root 21.022... quite accurately. Of greater interest, however, is the *next* root, which occurs near $\vartheta(t) \sim 3\pi/2$, $t \sim 25.5$, because in this vicinity N increases from 1 to 2 and the approximation to $Z(t)$ passes through an apparent discontinuity. However, the discontinuity is illusory because if $(t/2\pi)^{1/2} = 2 - \epsilon$, then $N = 1$, p is nearly 1, and therefore $(-1)^{N-1}(t/2\pi)^{-1/4}\Psi(p)$ is $2^{-1/2}\cos(\pi/8)$. On the other hand, if $(t/2\pi)^{1/2} = 2 + \epsilon$, then the main sum contains a second term, namely $2\cdot 2^{-1/2}\cos[\vartheta(t) - t\log 2]$, and $(-1)^{N-1}(t/2\pi)^{-1/4}\Psi(p)$ changes sign to become $-2^{-1/2}\cos(\pi/8)$. But $\log 2 \sim \log(t/2\pi)^{1/2} = \frac{1}{2}\log(t/2\pi)$ so

$$\vartheta(t) - t\log 2 \sim \frac{t}{2}\log\left(\frac{t}{2\pi}\right) - \frac{t}{2} - \frac{\pi}{8} + \frac{1}{48t} - \frac{t}{2}\log\left(\frac{t}{2\pi}\right)$$

$$= -2^2\pi - \frac{\pi}{8} + \frac{1}{48\cdot 2^2\cdot 2\pi},$$

$$\cos[\vartheta(t) - t\log 2] \sim \cos(\pi/8),$$

†The numerical value of Euler's constant γ can be found by using logarithmic differentiation of $\Pi(s) = s\Pi(s-1)$ to find $\gamma = -\Pi'(0)/\Pi(0) = 1 + \frac{1}{2} + \frac{1}{3} + \cdots + (1/n) - \Pi'(n)/\Pi(n)$ and by then using formula (4) of Section 6.3. As Siegel reports, Riemann wrote down the above constant to 20 decimal places 0.02309 57089 66121 03381. Thus, although Riemann did not prove that the series (4) of Section 1.10 *converges*, he did compute its sum to 20 decimal places. (I have not checked the accuracy of his answer.)

and the two new terms add up to $2^{-1/2}\cos(\pi/8)$ which is the old term. Thus to a first approximation the Riemann–Siegel formula is continuous at $t = 2^2 \cdot 2\pi$ and, more generally, at all points $t = k^2 2\pi$, where N changes. To a second approximation the C_2 term changes sign [the C_1 term does not change sign because C_1 is an odd function of $(1-2p)$ and the two sign changes cancel] from $\pm k^{-1/2}k^{-2}C_2(0)$ to $\mp k^{-1/2}k^{-2}C_2(0)$, whereas the new term of the sum is $\pm 2k^{-1/2}\cos[-(\pi/8) + (1/k^2 96\pi)]$ which consists of the term already accounted for plus $\pm 2k^{-1/2}\sin(\pi/8)\sin(1/k^2 96\pi) \sim \pm 2k^{-1/2}k^{-2}\sin(\pi/8)/96\pi$; hence there is no discontinuity at this level of approximation provided $C_2(0) = \sin(\pi/8)/96$, which is in fact the case as can be shown by straightforward evaluation of $C_2(0)$.

Although there is no serious discontinuity near $t = 2^2 \cdot 2\pi \sim 25.133$, it is still quite conceivable that the Riemann–Siegel formula would not be as accurate in this region as it proved to be in the examples above. To find a root of Z corresponding to the root $t \sim 25.5$ of $\cos\vartheta(t) = 0$, one would observe first that the terms after $2\cos\vartheta(t)$ are about $2^{-1/2}\cos(\pi/8) \sim 0.65$ in this range of t, as was just seen; hence t should be changed to make $2\cos\vartheta(t) \sim -0.65$. Since the derivative of $2\cos\vartheta(t)$ is about $-2\cdot\frac{1}{2}\log(t/2\pi)\sin\vartheta(t) \sim \log(t/2\pi) \sim 1.4$, this suggests $t = 25.5 - (0.65)/(1.4) \sim 25.0$. Now for $t = 25$, N is 1 and p is 0.994 711 4, so $1 - 2p$ is $-0.989\,422\,8$. Careful computations for $t = 25$ (which are impeded by the fact that p is nearly 1 and that the series for C_i consequently converge slowly) give $Z(25) \sim -0.014\,873\,455$ which agrees with Haselgrove's six-place value $-0.014\,872$ except in the last place. Thus even here, in the neighborhood of an apparent discontinuity, the Riemann–Siegel formula is astonishingly accurate and there is no conclusive empirical evidence of *any* inherent error in the formula, much less an inherent error of the order of magnitude allowed by the crude error estimates above.

The next approximation to the root near $t = 25$ would be obtained by increasing t enough to increase $2\cos\vartheta(t)$ by 0.0149. The derivative at $t = 25$ is about $-\log(25/2\pi)\sin[(3\pi/2) - 0.34)] = \cos(0.34)\log(25/2\pi) \sim (0.94) \times (1.38) \sim 1.30$, so t should be increased by $(0.0149)/(1.30) \sim 0.011$ to about

main term	−0.670 310 810
C_0 term	0.645 191 368
C_1 term	0.010 011 009
C_2 term	0.000 216 855
C_3 term	0.000 017 159
C_4 term	0.000 000 964
$Z(25)$	−0.014 873 455

Computation of the approximation to $Z(25)$.

25.011. The actual root is, of course, at 25.01085. . . . *Riemann also computed this root*, but the value 25.31 which he obtained is very far off—so far off, in fact, that it must surely indicate a computational error because even the rough calculation at the beginning of the preceding paragraph yielded the better value 25.0. I have not been able to follow the details of Riemann's computation, but I have followed enough of it to see that after starting with $t = 4 \cdot 2\pi$ he goes to the *larger* value $(4.030. . .) \cdot 2\pi$, which means he has gone in the wrong direction. This error should have revealed itself when he computed the new value of Z, but presumably the error in the first computation, whatever it was, was repeated in the subsequent computations.

Although the above computations of $Z(t)$ [from which $\zeta(\frac{1}{2} + it)$ can be computed immediately] are considerably shorter than the corresponding computations using Euler–Maclaurin summation would be, the real superiority of the Riemann–Siegel formula is for larger values of t, both because the number of terms it requires increases slowly and because the inherent error decreases. For example, to compute $Z(1000)$ using the Riemann–Siegel formula requires the evaluation of $[(t/2\pi)^{1/2}] = 12$ terms in the main sum. The inherent error is, judging by the above calculations, much smaller than the C_4 term which is of the order of magnitude of $(t/2\pi)^{-9/4}(0.001) \sim 12^{-9/2} \times 10^{-3} < 2 \times 10^{-8}$. To achieve comparable accuracy with the Euler–Maclaurin formula would require hundreds of terms as opposed to just 12, and would require so much arithmetic that computations would have to be carried out with great accuracy to counteract the accumulation of roundoff error.

7.7 ERROR ESTIMATES

The computational examples of the preceding section suggest that the usual rule of thumb for asymptotic series (see Section 6.2) applies to the Riemann–Siegel formula; that is, *as long as the terms are decreasing rapidly the bulk of the error is in the first term omitted.* Moreover, the first four remainder terms are rapidly decreasing in size even when t is only 14, and the C_4 term is already less than 10^{-4} when t is in this range and is very much smaller for larger t. This suggests that even though the Riemann–Siegel formula has an inherent error, this error is extremely small and the formula in fact makes possible the computation of $Z(t)$ with an accuracy of several decimal places.

Unfortunately none of the estimates of the error in the Riemann–Siegel formula come anywhere near to justifying these conjectures about its accuracy. At the present time the only published error estimate seems to be Titchmarsh's [T5], which shows that *if* $t > 125 \cdot 2\pi \sim 786$ *and if the* C_1 *term is the first term omitted, then the error is less than* $(3/2)(t/2\pi)^{-3/4}$ *in magnitude.* (Actually the estimates of Titchmarsh [T5] are a good deal more complicated and cover a wider range of t. The simplified version given here is taken from Titchmarsh's book [T8, p. 331].) In proving the existence of zeros of $Z(t)$, and hence of roots ρ on $\operatorname{Re} s = \frac{1}{2}$, the principal requirement is to be able to determine the *sign* of $Z(t)$ with certainty, and in most cases it has been possible to do this by finding values of t where $Z(t)$ is far enough from zero that its sign is rigorously determined by Titchmarsh's estimate of the error. However, cases do arise in which $Z(t)$ changes sign but remains very small in absolute value (Lehmer's phenomenon—see Section 8.3) and in these cases the presence of zeros of Z cannot be rigorously established without a stronger error estimate and, in particular, without one which uses more than one term of the Riemann–Siegel formula.

Rosser, Yohe, and Shoenfeld [R3] have announced that *if the* C_3 *term is the first term omitted, then the error is less than* $(2.88)(t/2\pi)^{-7/4}$ *provided* $t > 2000 \cdot 2\pi \sim 12{,}567$. Their proof of this result has not yet appeared. They have also established rigorous estimates of the error in their procedures for computing the C_0, C_1, and C_2 terms and in this way have been able to determine with certainty the sign of $Z(t)$ in the range covered by their calculations (see Section 8.4). In fact, they report that in five million evaluations of $Z(t)$ they found only four values of t where $|Z(t)|$ was so small that their program was unable to determine its sign with certainty.

Siegel himself proved that *all terms of the Riemann–Siegel formula are significant* in the sense that for every j there exist constants t_0, K such that if the C_j term is the first term omitted, then the error is less than $K(t/2\pi)^{-(2j+1)/4}$ provided $t > t_0$. Thus Titchmarsh's estimate gives specific values for K, t_0 in the case $j = 1$, and the estimate of Rosser *et al.* does the same for $j = 3$. Siegel's theorem can also be stated in the form: *The Riemann–Siegel formula is an asymptotic expansion of* $Z(t)$ in the sense of "asymptotic" defined in Section 5.4 provided the "order of magnitude" of the C_{j-1} term is interpreted† as meaning $(t/2\pi)^{-(2j-1)/4}$. The actual values of t_0 which Siegel's proof provides for various values of j are extremely large and are of no use in actual computation.

Since the location of the roots ρ has been reliably carried out by several computer programs up to the level where the estimate of Rosser *et al.* applies, and since it would seem that this estimate is sufficient for locating the roots beyond this level (it is conceivable, however, that there might be occurrences of Lehmer's phenomenon so extreme that the C_3 term is needed to determine

†The hitch here is that the C_{j-1} term for some values of t might be *zero*. However, it is of the order of magnitude of $(t/2\pi)^{-(2j-1)/4}$ in the sense that it is always less than a constant times this amount but not always less than a constant times any higher power of t^{-1}.

the sign of Z), the known estimates appear to be sufficient for the location of the roots ρ. Nonetheless, it would be of interest to have an estimate which comes closer to the "rule of thumb" estimate which seems justified by the computations, even for $t \sim 15$. Also of interest would be an answer to the question raised by Siegel as to whether the Riemann–Siegel formula *converges* for fixed t as more and more terms are used. Presumably it does not—by the analogy with Stirling's formula—but this has never been proved.

7.8 SPECULATIONS ON THE GENESIS OF THE RIEMANN HYPOTHESIS

We can never know what led Riemann to say it was "probable" that the roots ρ all lie on the line $\text{Re } s = \frac{1}{2}$, and the contents of this chapter show very clearly how foolhardy it would be to try to say what mathematical ideas may have lain behind this statement. Nonetheless it is natural to *try* to guess what might have led him to it, and I believe that the Riemann–Siegel formula gives some grounds for a plausible guess.

Even today, more than a hundred years later, one cannot really give any solid reasons for saying that the truth of the Riemann hypothesis is "probable." The theorem of Bohr and Landau (Section 9.6) stating that *for any* $\delta > 0$ *all but an infinitesimal proportion of the roots* ρ *lie within* δ *of* $\text{Re } s = \frac{1}{2}$ is the only positive result which lends real credence to the hypothesis. Also the verification of the hypothesis for the first three and a half million roots above the real axis (Section 8.4) perhaps makes it more "probable." However, any real *reason*, any plausibility argument or heuristic basis for the statement, seems entirely lacking. Siegel states quite positively that the Riemann papers contain† no steps toward a proof of the Riemann hypothesis, and therefore one is safe in assuming that they do not contain any plausibility arguments for the Riemann hypothesis either. Thus the question remains: Why did Riemann think it was "probable"?

My guess is simply that Riemann used the method followed in Section 7.6 to locate roots and that he observed that normally—as long as t is not so large that the Riemann–Siegel formula contains too many terms and as long as the terms do not exhibit too much reinforcement—this method allows one to go from a zero of the first term $2 \cos \vartheta(t)$ to a nearby zero of $Z(t)$. This heuristic argument implies there are "about" as many zeros of $Z(t)$ as there are of $2 \cos \vartheta(t)$, that is, about $\pi^{-1}\vartheta(t)$ zeros. But Riemann already knew (we do not know how) that this was the approximate formula for the total num-

†Of course there is no indication that these one hundred or so pages contain all of Riemann's studies of the zeta function, and they almost certainly do not.

ber of roots, on the line or off. Would it not be natural to explain this approximate equality by the hypothesis that, for some reason which might become clear on further investigation, the roots are all on the line? And would it not be natural to express this hypothesis in exactly the words which Riemann uses?

This guess, if it is correct, has two important consequences. In the first place it implies that when Riemann says that the number of roots on the line is "about" equal to the total number of roots, he does *not* imply asymptotic equality, as has often been assumed, but simply that when t is not too large and the terms do not reinforce too much, there is a one-to-one correspondence between zeros of $Z(t)$ and zeros of $\cos \vartheta(t)$. To go from a zero $\hat{t}$ of $\cos \vartheta(t)$ to one of $Z(t)$, one would evaluate the terms of the Riemann–Siegel formula at $\hat{t}$. Since the terms other than the first are individually small, since they would normally cancel each other to some extent, and since they change more slowly than the first term, it would seem likely that in most cases one could move to a point near $\hat{t}$ where the first term $2 \cos \vartheta(t)$ had a value equal to the negative of the value of the remaining terms at $\hat{t}$ and that the value of the remaining terms would not change too much in the process. If so, then the value of Z at the new point is small and can be made still smaller by successive approximations based on the assumption that for a small change in t the bulk of the change in $Z(t)$ occurs in the first term, hence converging to a root of Z. The method can certainly fail. To see how completely it can fail, it suffices to consider an extreme failure of Gram's law, for example the failure between g_{6708} and g_{6709} in Lehmer's graph (Fig. 3, Section 8.3). If $\hat{t}$ is the zero of $\cos \vartheta(t)$ in this interval, then $Z(\hat{t})$ is about -2, perhaps even less; so it is not possible to increase the first term $2 \cos \vartheta(t)$ enough to make up for the deficit in the other terms. If one increases $2 \cos \vartheta(t)$ the full amount by moving to g_{6708} where it is $+2$, the total value of Z is still negative, approximately $-\frac{1}{2}$. To reach the zero of Z corresponding to $\hat{t}$, one must move even further to the left and hence one must *decrease* $2 \cos \vartheta(t)$ in the hope that the other terms will increase and increase enough to make up for the decrease in $2 \cos \vartheta(t)$ and the remaining deficit of $-\frac{1}{2}$. Thus one must abandon the proposed method entirely and hope that the desired zero is there anyway. Viewed in this way, the heuristic argument which I impute to Riemann is virtually identical with Gram's law but with the very important difference that its rationale is based on the Riemann–Siegel formula rather than the Euler–Maclaurin formula, so that, unlike Gram's rationale, it is not at all absurd to expect the main term to dominate the sign of the series for t into the hundreds or even the thousands because the formula for $Z(t)$ has only ten or twenty terms in this range, not hundreds of terms. It seems entirely possible that Riemann would have been able to judge that the failures in this range would be relatively rare, as has now been verified by computation, and to conclude that the number of zeros of $Z(t)$ between 0 and T *for T in this medium range* is about $\pi^{-1}\vartheta(T)$, that is, about† $(T/2\pi)\log(T/2\pi) - (T/2\pi)$. In support of this interpretation of Riemann's use of the word "about" (*etwa*) in this place, one might observe that he prefaces it with the phrase "one finds in fact" (meaning computationally?) and that, unlike his use of "about" in the previous sentence, he gives no estimate of the error in the approximation.

The second consequence of my guess is that it implies that Riemann based his hypothesis on no insights about the function ξ which are not available to us today (now that we have the Riemann–Siegel formula) and that, on the contrary, had he known some of the facts which have since been discovered, he might well have been led to reconsider. After all, he had just discovered the extension of the zeta function to the entire complex plane, the functional equation, and an effective numerical technique for locating many roots on the line, so it would be perfectly natural for him to be looking for regularities and perfectly natural for him to expect that an observed regularity of this sort would hold and would yield to the power of his function-theoretic concepts and techniques. However, it did not yield, and Riemann lived for several years after he made the hypothesis. Moreover, the discoveries of Lehmer's phenomenon (Section 8.3) and of the fact that $Z(t)$ is unbounded (Section 9.2) completely vitiate any argument based on the Riemann–Siegel formula and suggest that, unless some basic cause is operating which has eluded mathematicians for 110 years, occasional roots ρ off the line are altogether possible. In short, although Riemann's insight was stupendous it was not supernatural, and what seemed "probable" to him in 1859 might seem less so today.

7.9 THE RIEMANN-SIEGEL INTEGRAL FORMULA

In 1926 Bessel–Hagen found (according to Siegel [S4]) in the Riemann papers a new representation of the zeta function in terms of definite integrals. Naturally Siegel included an exposition of this formula in his 1932 account of the portions of Riemann's *Nachlass* relating to analytic number theory. As stated by Siegel, the formula is essentially

$$\frac{2\xi(s)}{s(s-1)} = F(s) + \overline{F(1-\bar{s})} \tag{1}$$

where F is defined by the formula

$$F(s) = \Pi\left(\frac{s}{2} - 1\right)\pi^{-s/2}\int_{0\searrow 1}\frac{e^{-i\pi x^2}x^{-s}\,dx}{e^{i\pi x} - e^{-i\pi x}}$$

†Of course, if the roots are all on the line, which Riemann thought was the probable explanation of the near equality of the two estimates, then the same estimate applies for all T and the slight ambiguity in the range of T does no harm.

in which the symbol $0 \searrow 1$ means that the path of integration is a line of slope -1 crossing the real axis between 0 and 1 and directed from upper left to lower right, and in which x^{-s} is defined on the slit plane (excluding 0 and negative real numbers) in the usual way by taking $\log x$ to be real on the positive real axis and setting $x^{-s} = e^{-s \log x}$. Because $\exp(-i\pi x^2)$ approaches zero very rapidly as $|x| \to \infty$ along any line of the form $0 \searrow 1$ and because the integrand is nonsingular on the slit plane except for simple poles at the positive integers, it is easily seen that $F(s)$ is an analytic function of s defined for all s except possibly for $s = 0, -2, -4, \ldots$, where the factor in front has simple poles. [Formula (1), once it is proved, implies that $F(s)$ is analytic at $-2, -4, \ldots$ and has a simple pole at 0.]

Siegel deduces formula (1) from an alternative form of the identity

(2) $$e^{i\pi/8} e^{-2\pi i p^2} \frac{1}{2\pi i} \int_\Gamma \frac{e^{iu^2/4\pi} e^{2pu}\, du}{e^u - 1} = \frac{\cos 2\pi(p^2 - p - \frac{1}{16})}{\cos 2\pi p}$$

[formula (5) of Section 7.4]. The change of variable $u = 2\pi i w$ puts the path of integration Γ in the form $0 \searrow 1$ but with the orientation reversed, and puts the identity itself in the form

$$\int_{0 \searrow 1} \frac{e^{-i\pi w^2} e^{4\pi i p w}\, dw}{e^{2\pi i w} - 1} = -e^{-i\pi/8} e^{2\pi i p^2} \frac{\cos 2\pi(p^2 - p - \frac{1}{16})}{\cos 2\pi p}$$

which with $p = \frac{1}{2}(v + \frac{1}{2})$ can be simplified using $2p = v + \frac{1}{2}$, $4p^2 - 4p - \frac{1}{4} = (v + \frac{1}{2})^2 - 2(v + \frac{1}{2}) - \frac{1}{4} = v^2 - v - 1$ to be

$$\begin{aligned}
\int_{0 \searrow 1} \frac{e^{-i\pi w^2} e^{2\pi i[v + (1/2)]w}\, dw}{e^{i\pi w}(e^{i\pi w} - e^{-i\pi w})} &= -e^{-i\pi/8} e^{i\pi[v + (1/2)]^2/2} \frac{\cos[\pi(v^2 - v - 1)/2]}{\cos \pi(v + \frac{1}{2})} \\
&= -e^{i\pi v^2/2} e^{i\pi v/2} \frac{e^{i\pi(v^2 - v - 1)/2} + e^{-i\pi(v^2 - v - 1)/2}}{e^{i\pi[v + (1/2)]} + e^{-i\pi[v + (1/2)]}} \\
&= -\frac{e^{i\pi v^2} e^{-i\pi/2} + e^{i\pi v} e^{i\pi/2}}{e^{i\pi v} e^{i\pi/2} + e^{-i\pi v} e^{-i\pi/2}}
\end{aligned}$$

and finally

(3) $$\int_{0 \searrow 1} \frac{e^{-i\pi w^2} e^{2\pi i v w}\, dw}{e^{i\pi w} - e^{-i\pi w}} = \frac{e^{i\pi v^2}}{e^{i\pi v} - e^{-i\pi v}} - \frac{1}{1 - e^{-2\pi i v}}$$

which is the alternative form of (2). Let s be a negative real number, multiply both sides of this equation by $v^{-s}\, dv$, and integrate along the ray from $v = 0$ to $v = \infty\, i^{1/2}$. The double integral on the left converges absolutely, so the order of integration can be interchanged. Since by elementary manipulation of definite integrals

$$\begin{aligned}
\int_0^{\infty\, i^{1/2}} v^{-s} e^{2\pi i v w}\, dv &= \int_0^{\infty w i^{-1/2}} \left(\frac{ix}{2\pi w}\right)^{1-s} e^{-x}\, d \log x \\
&= \left(\frac{i}{2\pi w}\right)^{1-s} \int_0^\infty x^{-s} e^{-x}\, dx \\
&= i e^{-i\pi s/2} (2\pi)^{s-1} w^{s-1} \Pi(-s)
\end{aligned}$$

(for w on $0 \searrow 1$), it follows that the left side becomes

$$\int_0^{\infty i^{1/2}} \int_{0\searrow 1} \frac{e^{-i\pi w^2} e^{2\pi i v w} v^{-s}\, dw\, dv}{e^{i\pi w} - e^{-i\pi w}}$$
$$= ie^{-i\pi s/2}(2\pi)^{s-1}\Pi(-s) \int_{0\searrow 1} \frac{e^{-i\pi w^2} w^{s-1}\, dw}{e^{i\pi w} - e^{-i\pi w}}.$$

The second term on the right becomes†

$$\int_0^{\infty i^{1/2}} \left(\frac{-1}{1 - e^{-2\pi i v}}\right) v^{-s}\, dv = \int_0^{\infty i^{1/2}} \sum_{n=1}^{\infty} e^{2\pi i n v} v^{-s}\, dv$$
$$= \sum_{n=1}^{\infty} \int_0^{\infty i^{1/2}} e^{2\pi i n v} v^{-s}\, dv$$
$$= \sum_{n=1}^{\infty} \int_0^{\infty i^{1/2}} e^{2\pi i w} \left(\frac{w}{n}\right)^{1-s} d\log w$$
$$= \zeta(1 - s) \int_0^{\infty i^{1/2}} w^{-s} e^{2\pi i w}\, dw$$
$$= \zeta(1 - s) ie^{-i\pi s/2}(2\pi)^{s-1} 1^{s-1}\Pi(-s),$$

by the same calculation. The first term on the right can be expressed in terms of the definite integral

$$(4) \qquad \int_{0\nearrow 1} \frac{e^{i\pi u^2} u^{-s}\, du}{e^{i\pi u} - e^{-i\pi u}}$$

(where $0 \nearrow 1$ denotes the complex conjugate of a path $0 \searrow 1$) because for negative real s the path $0 \nearrow 1$ can be moved over to the line of slope 1 through the origin so that (4) can be expressed as

$$\int_{-\infty i^{1/2}}^{0} \frac{e^{i\pi u^2} u^{-s}\, du}{e^{i\pi u} - e^{-i\pi u}} + \int_0^{\infty i^{1/2}} \frac{e^{i\pi u^2} u^{-s}\, du}{e^{i\pi u} - e^{-i\pi u}}$$
$$= \int_{\infty i^{1/2}}^{0} \frac{e^{i\pi(-u)^2}(-u)^{-s}\, d(-u)}{e^{-i\pi u} - e^{i\pi u}} + \int_0^{\infty i^{1/2}} \frac{e^{i\pi u^2} u^{-s}\, du}{e^{i\pi u} - e^{-i\pi u}}$$
$$= \int_0^{\infty i^{1/2}} \frac{e^{i\pi u^2}[u^{-s} - (-u)^{-s}]\, du}{e^{i\pi u} - e^{-i\pi u}}$$
$$= (1 - e^{i\pi s}) \int_0^{\infty i^{1/2}} \frac{e^{i\pi v^2} v^{-s}\, dv}{e^{i\pi v} - e^{-i\pi v}}$$

because $\log(-u) = \log u - i\pi$ for u on the ray $\operatorname{Im} \log u = \pi/4$, and hence $(-u)^{-s} = u^{-s} e^{-s(-i\pi)}$. Thus (3) becomes

$$ie^{-i\pi s/2}(2\pi)^{s-1}\Pi(-s) \int_{0\searrow 1} \frac{e^{-i\pi w^2} w^{s-1}\, dw}{e^{i\pi w} - e^{-i\pi w}}$$
$$= \frac{1}{1 - e^{i\pi s}} \int_{0\nearrow 1} \frac{e^{i\pi u^2} u^{-s}\, du}{e^{i\pi u} - e^{-i\pi u}} + ie^{-i\pi s/2}(2\pi)^{s-1}\Pi(-s)\zeta(1 - s).$$

†Justification of the interchange of summation and integration is not altogether elementary. One method is to observe that $\lim_{N\to\infty} \int_0^{\infty i^{1/2}} (e^{2\pi i N v}/e^{-2\pi i v} - 1) v^{-s}\, dv = 0$ by the Riemann–Lebesgue lemma.

Now

$$(1 - e^{i\pi s})ie^{-i\pi s/2}(2\pi)^{s-1}\Pi(-s) = 2[\sin(s\pi/2)](2\pi)^{s-1}\Pi(-s)$$

is the factor which appears in the functional of ζ [formula (4) of Section 1.6], and therefore, as in Section 1.6, it can be written as

$$\frac{\Pi[\frac{1}{2}(1-s)-1]\pi^{-(1-s)/2}}{\Pi(\frac{1}{2}s-1)\pi^{-s/2}}.$$

Therefore, when the above formula is multiplied first by $(1 - e^{i\pi s})$ and then by $\Pi\left(\frac{s}{2}-1\right)\pi^{-s/2}$, it becomes

$$\Pi\left(\frac{1-s}{2}-1\right)\pi^{-(1-s)/2}\int_{0\searrow 1}\frac{e^{-i\pi w^2}w^{s-1}\,dw}{e^{i\pi w}-e^{-i\pi w}}$$
$$= \Pi\left(\frac{s}{2}-1\right)\pi^{-s/2}\int_{0\nearrow 1}\frac{e^{i\pi u^2}u^{-s}\,du}{e^{i\pi u}-e^{-i\pi u}} + \frac{2\xi(1-s)}{(1-s)(-s)}$$

because by definition

$$(5) \qquad \frac{2\xi(s)}{s(s-1)} = \Pi\left(\frac{s}{2}-1\right)\pi^{-s/2}\zeta(s).$$

The left side of this equation is $F(1-s)$ and the first term on the right is $-\overline{F(\bar{s})}$; hence

$$\frac{2\xi(1-s)}{(1-s)(-s)} = F(1-s) + \overline{F(\bar{s})}$$

and the desired formula (1) is proved by substituting $1-s$ for s.

The Riemann–Siegel formula puts in evidence the fact that ξ satisfies the functional equation $\xi(s) = \overline{\xi(1-\bar{s})}$ because it shows that

$$\frac{2\xi(s)}{s(s-1)} = F(s) + \overline{F(1-\bar{s})} = \overline{F(1-\bar{s}) + \overline{F(s)}}$$
$$= \text{complex conjugate of } \frac{2\xi(1-\bar{s})}{(1-\bar{s})(-\bar{s})} = \frac{\overline{2\xi(1-\bar{s})}}{s(s-1)}.$$

On the other hand, ξ is real on the real axis by (5), and therefore, by the reflection principle, $\overline{\xi(s)} = \xi(\bar{s})$. The Riemann–Siegel integral formula therefore gives a new proof of the functional equation $\xi(s) = \xi(1-s)$. This proof differs from Riemann's first proof in that it uses s and $1-s$ more symmetrically, and it differs from his second proof in that it does not depend on the identity $1 + 2\psi(x) = x^{-1/2}[1 + 2\psi(x^{-1})]$ from the theory of theta functions.

Since the theta function identity $1 + 2\psi(x) = x^{-1/2}[1 + 2\psi(x^{-1})]$ can be deduced from $\xi(s) = \xi(1-s)$ fairly easily (by Fourier inversion—see Chapter 10), the proof of this section gives an alternative proof of the theta function identity based on the evaluation of the definite integral (2). More generally, Siegel states that Riemann in his unpublished lectures derived the transformation theory of theta functions from a study of the integral

$$\Phi(\tau, u) = \int_{0\searrow 1} \frac{e^{i\pi\tau x^2} e^{2\pi i u x}\, dx}{e^{-i\pi x} - e^{i\pi x}}$$

of which the special case $\tau = -1$ was considered above.

Chapter 8

Large-Scale Computations

8.1 INTRODUCTION

The discovery of the Riemann–Siegel formula made it quite feasible to extend the program begun by Gram and Hutchinson (see Chapter 6) well beyond the Gram point g_{137} reached by Hutchinson in 1925. Since the Gram points g_n are easily computed, this extension is simply a matter of using the Riemann–Siegel formula to evaluate $Z(g_n)$ and of finding points g_n' near g_n for which $(-1)^n Z(g_n') > 0$ in those presumably rare cases when Gram's law† $\operatorname{Re}\zeta(\frac{1}{2} + ig_n) = Z(g_n)\cos\vartheta(g_n) = (-1)^n Z(g_n) > 0$ fails. In this way it should be possible, unless the failures of Gram's law become too frequent, to locate many more roots ρ on the line $\operatorname{Re} s = \frac{1}{2}$. Such computations were carried out by Titchmarsh and Comrie [*T*5, *T*6] in 1935–1936 extending up to the Gram point g_{1040} and thus locating 1041 roots on the line. Moreover, by a suitable generalization of the techniques of Backlund and Hutchinson using the Riemann–Siegel formula (in a more general form which is applicable for $\operatorname{Re} s \neq \frac{1}{2}$) in place of the Euler–Maclaurin formula, Titchmarsh and Comrie were able to show that $N(g_{1040}) = 1041$ and to conclude therefore that *the roots ρ in the range $\{0 \leq \operatorname{Im} s \leq g_{1040}\}$ are all simple zeros on the line $\operatorname{Re} s = \frac{1}{2}$.*

No doubt this program of computation would have been carried further if World War II had not intervened. By the time the war was over, the computer revolution was well under way and automatic electronic digital com-

†Strictly speaking the statement $(-1)^n Z(g_n) > 0$ should perhaps be called the *weak* Gram's law, since, as defined by Hutchinson, "Gram's law" is the stronger statement that there are precisely $n + 1$ zeros of $Z(t)$ between 0 and g_n. This distinction is not very important and is ignored in what follows.

puters were rapidly being developed. These new tools made it feasible to extend the computations to cover tens of thousands and even hundreds of thousands of Gram points, but, apparently because computer technology was changing so rapidly that no single computer remained in operation long enough for such a low-priority project to be programmed and run on it before a new computer requiring a new program would replace it, it was not until 1955–1956 that the computations were carried significantly past the level reached by Titchmarsh and Comrie 20 years before.

These computations, carried out by D.H. Lehmer [L7, L8], showed that for the first 25,000 Gram points g_n the exceptions to Gram's law $(-1)^n Z(g_n) > 0$ are not great and *all roots* ρ *in the range* $\{0 \le \operatorname{Im} s \le g_{25000}\}$ *are simple zeros on the line* $\operatorname{Re} s = \frac{1}{2}$. Lehmer had at his disposal, in addition to the Riemann–Siegel formula and the new electronic computers, a new method introduced by Turing [T9] in 1953 for determining the number of roots in a given range. Turing's method is much easier to apply in practice than is the method of Backlund (see Section 6.6) which it supplants. Turing's method is described in the next section.

Although Lehmer's computations confirmed the Riemann hypothesis as far as they went, they disclosed certain irregularities in the behavior of $Z(t)$ which made it seem altogether possible that further computations might actually produce a counterexample to the Riemann hypothesis (see Section 8.3). Further computations were carried out—by Lehman [L5] to the two hundred and fifty thousandth zero and by Rosser *et al.* [R3] to the *three and a half millionth zero*—without, however, producing a counterexample and in fact proving that all roots ρ in the range $\{0 \le \operatorname{Im} s \le g_{3500000}\}$ are simple zeros on the line $\operatorname{Re} s = \frac{1}{2}$ (see Section 8.4).

8.2 TURING'S METHOD

As before, let $N(T)$ denote the number of roots ρ in the range $\{0 \le \operatorname{Im} s \le T\}$ (counted with multiplicities). Recall that Backlund's method of evaluating $N(T)$ is to prove if possible that $\operatorname{Re} \zeta(\sigma + iT)$ is never zero for $\frac{1}{2} \le \sigma \le 1\frac{1}{2}$, from which it follows that $N(T)$ is the integer nearest $\pi^{-1}\vartheta(T) + 1$ (see Section 6.6). The disadvantage of this method is that it requires the evaluation of $\zeta(s)$ at points not on the line $\operatorname{Re} s = \frac{1}{2}$. By contrast, Turing's method not only does not require the evaluation of $\zeta(s)$ at points not on $\operatorname{Re} s = \frac{1}{2}$, but in fact it requires only the information which is naturally acquired in looking for changes of sign in $Z(t)$, namely, a list of those Gram points g_n for which Gram's law $(-1)^n Z(g_n) > 0$ fails and a determination of how far each of them must be moved in order to give Z the desired sign $(-1)^n$.

More precisely, assume that for all integers n in a certain range numbers h_n have been found such that $(-1)^n Z(g_n + h_n) > 0$, such that the sequence $\ldots, g_{n-1} + h_{n-1}, g_n + h_n, g_{n+1} + h_{n+1}, \ldots$ is strictly increasing, and such that h_n is small and is zero whenever possible. Such a list of numbers h_n would naturally be generated in using Gram's law to locate changes of sign of Z. Turing showed that *if $h_m = 0$ and if the values of h_n for n near m are not too large, then $N(g_m)$ must have the value predicted by Gram's law*, namely, $N(g_m) = m + 1$.

Turing's method is based on the following theorem of Littlewood (see Section 9.5). Let $S(T)$ denote† the error in the approximation $N(T) \sim \pi^{-1}\vartheta(T) + 1$, that is, let $S(T) = N(T) - \pi^{-1}\vartheta(T) - 1$. Von Mangoldt proved (see Section 6.7) that the absolute value of $S(T)$ grows no faster than a constant times $\log T$ as $T \to \infty$. Littlewood in 1924 proved a different kind of estimate of $S(T)$, namely, that $\int_0^T S(t)\,dt$ grows no faster than a constant times $\log T$ as $T \to \infty$. Although this does not imply von Mangoldt's result—it is possible for $S(t)$ to have arbitrarily large absolute values without its integral necessarily being large—Littlewood's theorem is in a sense much stronger in that it shows that on the average $S(T)$ approaches zero,‡ $\lim_{T\to\infty} 1/T \int_0^T S(t)\,dt = 0$, whereas von Mangoldt's result only shows that $S(T)$ does not become large too fast.

Now suppose that $h_m = 0$ and that the h_n for n near m are small. Then $S(g_m)$ must be an integer because $S(g_m) = N(g_m) - m - 1$ and this integer must be even because the parity of the number of roots *on* the line segment from $\frac{1}{2}$ to $\frac{1}{2} + ig_m$ (counted with multiplicities) is determined by the sign of $Z(g_m)$ which is $(-1)^m$ by assumption, and because the roots *off* the line $\operatorname{Re} s = \frac{1}{2}$, if any, occur in pairs. Thus to prove $S(g_m) = 0$ it suffices to prove $S(g_m) < 2$ and $S(g_m) > -2$. Assume first that $S(g_m) \geq 2$. If $h_{m+1} = 0$, then $(-1)^{m+1} Z(g_{m+1}) > 0$, so there must be a zero of Z between g_m and g_{m+1}, and N must increase by at least one as t goes from g_m to g_{m+1}; on the other hand $\pi^{-1}\vartheta(t) + 1$ increases by exactly one as t goes from g_m to g_{m+1}, so $S(g_{m+1}) \geq 2$ and $S(t)$ never falls below one for $g_m \leq t \leq g_{m+1}$. If $h_{m+1} < 0$, then it is true *a fortiori* that $S(g_{m+1}) \geq 2$ and that $S(t)$ never falls below one for $g_m \leq t \leq g_{m+1}$, whereas if h_{m+1} is a small positive number, then t must pass g_{m+1} and go to $g_{m+1} + h_{m+1}$ in order to bring $S(t)$ back up to a value near two and in the process $S(t)$ falls only slightly below one. If h_{m+2} and the succeeding h's are zero or are small this argument can be continued to show that $S(g_n + h_n)$ is nearly two for $n = m + 1, m + 2, m + 3, \ldots$ and that $S(t)$

†The formula $N(T) = \pi^{-1}\vartheta(T) + 1 + \pi^{-1} \operatorname{Im} \int_C [\zeta'(s)/\zeta(s)]\,ds$ of Section 6.6 shows that $S(T)$ can also be defined as $\pi^{-1} \operatorname{Im} \log \zeta(\frac{1}{2} + iT)$ when $\log \zeta$ is defined by analytic continuation from the positive real axis as in Section 6.6.

‡In particular, Littlewood's theorem shows that the constant 1 in the definition of $S(T)$ has significance.

falls only slightly below one to the right of g_m. Since by Littlewood's theorem this can not continue indefinitely [because then the average value of $S(t)$ would not approach zero] either some of the values $h_{m+1}, h_{m+2}, h_{m+3}, \ldots$ must be large and positive or the original assumption $S(g_m) \geq 2$ must be false. Turing's idea was to prove a *quantitative* version of Littlewood's theorem in order to obtain a quantitative description of the sizes of the h_n implied by $S(g_m) \geq 2$; then to prove $S(g_m) \leq 0$ it suffices to show that the h_n are in fact less than this amount.

Specifically, Turing showed that Littlewood's proof (see Section 9.5) can be made to yield the inequality

$$\left| \int_{t_1}^{t_2} S(t)\, dt \right| \leq 2.30 + 0.128 \log\left(\frac{t_2}{2\pi}\right) \tag{1}$$

for all $t_2 > t_1 > 168\pi$. This can be used to derive a relationship between $S(g_m)$ and the sizes of $h_{m+1}, h_{m+2}, h_{m+3}, \ldots$ as follows. Assume as before that $h_m = 0$ and let $L(t)$ denote the step function which is zero at g_m and which has jumps of one at $g_{m+1} + h_{m+1}, g_{m+2} + h_{m+2}, \ldots$. Then since Z changes sign between successive points $g_n + h_n$, the number of roots N must go up by at least one and $N(t) \geq N(g_m) + L(t)$ for $t \geq g_m$. On the other hand let $L_1(t)$ denote the step function which is zero at g_m and which has jumps of one at $g_{m+1}, g_{m+2}, \ldots$. Then $L_1(t)$ increases by one when $\pi^{-1}\vartheta(t)$ increases by one, from which it follows that $\pi^{-1}\vartheta(t) + 1 \leq \pi^{-1}\vartheta(g_m) + 1 + L_1(t) + 1$ for $t \geq g_m$ and hence $S(t) \geq S(g_m) + L(t) - L_1(t) - 1$. Now $L(t) - L_1(t)$ is normally zero, but if h_n is positive, then $L_1(t)$ jumps before $L(t)$ does and $L(t) - L_1(t)$ is -1 on the interval from g_n to $g_n + h_n$. Similarly if $h_n < 0$, then $L(t) - L_1(t)$ is $+1$ on the interval from $g_n + h_n$ to g_n. Assume for the sake of convenience that $h_{m+k} = 0$. Then

$$\begin{aligned} \int_{g_m}^{g_{m+k}} S(t)\, dt &\geq \int_{g_m}^{g_{m+k}} [S(g_m) - 1]\, dt + \int_{g_m}^{g_{m+k}} [L(t) - L_1(t)]\, dt \\ &= (g_{m+k} - g_m)[S(g_m) - 1] - \sum_{j=m+1}^{m+k-1} h_j, \end{aligned}$$

$$S(g_m) - 1 \leq \frac{\int_{g_m}^{g_{m+k}} S(t)\, dt + \sum_{j=m+1}^{m+k-1} h_j}{g_{m+k} - g_m},$$

$$S(g_m) \leq 1 + \frac{2.30 + 0.128 \log(g_{m+k}/2\pi) + \sum_{j=m+1}^{m+k-1} h_j}{g_{m+k} - g_m}. \tag{2}$$

Since $k\pi = \vartheta(g_{m+k}) - \vartheta(g_m) = \int_{g_m}^{g_{m+k}} \vartheta'(t)\, dt$ is approximately $(g_{m+k} - g_m) \cdot \frac{1}{2} \log(g_m/2\pi)$ (see Section 6.5), the second term on the right is about

$$\frac{1}{2k\pi}\left[2.30 \log \frac{g_m}{2\pi} + 0.128\left(\log \frac{g_m}{2\pi}\right)^2 + \left(\sum h_j\right) \log \frac{g_m}{2\pi}\right].$$

As k increases, this term rapidly becomes less than one unless $\sum h_j$ is rather large and positive. In actual fact, in the range of the calculations of Rosser *et al.* (in which g_m is comfortably less than two million so $\log(g_m/2\pi)$ is comfortably less than 14), it was always possible to prove in this way that $S(g_m) < 2$ for all values of m such that $h_m = 0$, and for this purpose it was never necessary to use any value of k larger than 15 (see Section 8.4). In this way it was proved that $N(g_m) \leq m + 1$ for Gram points g_m in the vicinity of $m =$ 3,500,000. Since h's had been found all the way up to this level, no lower bound of N was necessary and it followed that all $m + 1$ of these roots are simple zeros on $\operatorname{Re} s = \frac{1}{2}$.

In the same way one can obtain a lower bound for $S(g_m)$, namely,

$$S(g_m) \geq -1 - \frac{2.30 + 0.128 \log(g_m/2\pi) - \sum_{j=1}^{k-1} h_{m-j}}{g_m - g_{m-k}}$$

where m, k are such that $h_m = 0$, $h_{m-k} = 0$. Using a bound similar to this one, Rosser *et al.* were able to prove that $N(g_m) = m + 1$ for certain Gram points g_m in the vicinity of $m = 13{,}400{,}000$ and then to prove that 41,000 consecutive roots in this range are all simple zeros on the line $\operatorname{Re} s = \frac{1}{2}$.

8.3 LEHMER'S PHENOMENON

Lehmer's 1955–1956 computations followed Hutchinson's scheme of determining the sign of $Z(g_n)$ for the Gram points g_n and, in the cases where $(-1)^n Z(g_n) > 0$ fails to hold, of finding small numbers h_n such that $(-1)^n Z(g_n + h_n) > 0$ holds. As long as this is possible it shows that there are at least $m + 1$ zeros of $Z(t)$ (not counting multiplicities) in the range $0 < t < g_m$ and, using Turing's method, it should be possible to show that there are no more than $m + 1$ roots ρ in the range $\{0 \leq \operatorname{Im} s \leq g_m\}$ (counting multiplicities) and hence that all roots ρ in this range are simple zeros on the line $\operatorname{Re} s = \frac{1}{2}$.

The first step in carrying out this scheme is naturally to write a program for computing $Z(t)$ economically and not necessarily very accurately, with a view simply to determining the sign of $Z(t)$ for a given t. Lehmer's program computed the main sum in the Riemann–Siegel formula with an accuracy of several decimal places and computed the C_0 term very roughly. If the absolute value of the resulting estimate of $Z(t)$ was safely larger than Titchmarsh's estimate $\frac{3}{2}(t/2\pi)^{-3/4}$ of the error in the Riemann–Siegel formula, then its sign was considered to be the sign of $Z(t)$. Lehmer then began running through the list of Gram points (previously computed by a method similar to the method outlined in Section 6.5) and in those few cases† where $Z(g_n)$ could not be shown to have the sign $(-1)^n$, found small h_n for which $Z(g_n + h_n)$ could. No difficulty was encountered until the Gram point g_{4763} was reached, but here, despite the fact that the Riemann–Siegel formula is very accurate for such large t, the values of Z were too small to find an h, although there was a range in which the sign of $Z(g + h)$ was not conclusively determined. More exact computations‡ did show that Z has the required number of sign changes in this region, but the appearance of this phenomenon of a region in which Z just *barely* changes sign was of very great interest.

If there were a point at which the graph of $Z(t)$ *came near to* $Z = 0$ *but did not actually cross it*—that is, if Z had a small positive local minimum or a small negative local maximum—*then the Riemann hypothesis would be contradicted.* This theorem can be proved as follows. It will suffice to show that the Riemann hypothesis implies Z'/Z is monotone because this, in turn, implies that Z'/Z has at most one zero between successive zeros of Z and therefore that Z' has at most one zero between successive zeros of Z. Now $\xi(\frac{1}{2} + it) = -f(t)Z(t)$, where

$$f(t) = |\Pi[(s/2) - 1]|\pi^{-1/4}\tfrac{1}{2}(t^2 + \tfrac{1}{4})$$

(see Section 6.5). Thus

$$\frac{(d/dt)\xi(\frac{1}{2} + it)}{\xi(\frac{1}{2} + it)} = \frac{f'(t)}{f(t)} + \frac{Z'(t)}{Z(t)}$$

$$-\frac{Z'(t)}{Z(t)} = -i\frac{\xi'(\frac{1}{2} + it)}{\xi(\frac{1}{2} + it)} + \frac{f'(t)}{f(t)}.$$

Logarithmic differentiation of the product formula for ξ (justified in Chapter 3) puts the first term on the right in the form

$$-i\sum_{\rho}\frac{1}{(\frac{1}{2} + it) - \rho} = \sum_{\alpha}\frac{-i}{it - i\alpha} = \sum_{\alpha}\frac{1}{\alpha - t},$$

where, as before, $\rho = \frac{1}{2} + i\alpha$ runs over all roots ρ. Thus the derivative of the first term on the right is $\sum(\alpha - t)^{-2}$ and, if the Riemann hypothesis is true, this derivative is not only positive (because all terms are positive) but also very large [because by von Mangoldt's estimate of $N(t)$ the α's must be quite dense] between successive zeros of Z. The second term on the right can be rewritten

$$\frac{f'(t)}{f(t)} = \frac{d}{dt}\,\mathrm{Re}\log\Pi\left(\frac{\frac{1}{2} + it}{2} - 1\right) + \frac{2t}{t^2 + \frac{1}{4}}.$$

†There were about 360 such cases altogether in the first 5000 Gram points, and about 100 such in each succeeding 1000 up to 25,000.

‡Lehmer used the Euler–Maclaurin formula for these in order to have a rigorous error estimate.

The second term in this formula does decrease as t increases, but its decrease is obviously insignificant compared to the increase of the term already considered. The derivative of the first term can be estimated using Stirling's series

$$\frac{d^2}{dt^2}\operatorname{Re}\log\Pi\left(-\frac{3}{4}+\frac{it}{2}\right)=\operatorname{Re}\frac{d^2}{dt^2}\log\Pi\left(-\frac{3}{4}+\frac{it}{2}\right)$$
$$=\left(\frac{i}{2}\right)^2\operatorname{Re}\frac{d^2}{ds^2}\bigg|_{s=-(3/4)+(it/2)}\log\Pi(s)$$
$$=-\frac{1}{4}\operatorname{Re}\left\{\frac{1}{s}-\frac{1}{2s^2}+\frac{B_2}{s^3}+\frac{B_4}{s^5}+\cdots\right\}\bigg|_{s=-(3/4)+(it/2)}$$

The first two terms give

$$\operatorname{Re}\left\{\frac{1-2s}{8s^2}\right\}=\operatorname{Re}\left\{\frac{(1-2s)\bar{s}^2}{8|s|^4}\right\}$$
$$=\frac{\operatorname{Re}\left\{\left(\frac{5}{2}-it\right)\left(\frac{9}{16}+\frac{3it}{4}-\frac{t^2}{4}\right)\right\}}{8|s|^4}$$
$$=\frac{\frac{45}{32}-\frac{5}{8}t^2+\frac{3}{4}t^2}{8|s|^4}$$

which is positive and of the order of magnitude of t^{-2} for large t. The other terms of Stirling's series are insignificant in comparison with this one, and it follows that the first term in the formula for f'/f increases with t. Therefore $-Z'/Z$ is an increasing function of t (for t at all large) and the theorem is proved. This shows that the place near g_{4763} where Z just barely crosses the axis is "almost" a counterexample to the Riemann hypothesis.

Pursuing the calculations on up to $g_{25,000}$, Lehmer found more of these "near counterexamples" to the Riemann hypothesis. One of these on which he gives very complete information occurs for g_{6708} (see Fig. 3†). There are three zeros between g_{6704} and g_{6705}, and to achieve the correct sign for $Z(g_n + h_n)$ in these cases, a positive value of h_{6704} and a negative value of h_{6705} are necessary. Then $(-1)^n Z(g_n) > 0$ for $n = 6706, 6707$, but to obtain the correct sign for $Z(g_n + h_n)$ when $n = 6708$, h_n must be negative and must be chosen so that $g_n + h_n$ lies in the very short range of the "near counterexample" where the graph of Z just barely crosses over to positive values between g_{6707} and g_{6708}. This range where Z is positive has length 0.0377 (difference of t values of the two zeros), and the largest value of Z which

†One question which naturally presents itself when one examines the diagram is the question of the size of $Z(t)$. Will it always remain this small? A proof that $|Z(t)| = |\zeta(\frac{1}{2} + it)|$ is in fact *unbounded* as $t \longrightarrow \infty$ is outlined in Section 9.2. The *Lindelöf hypothesis* is that $|Z(t)|$ grows more slowly than any positive power of t—that is, for every $\epsilon > 0$ it is true that $Z(t)/t^\epsilon \longrightarrow 0$ as $t \longrightarrow \infty$—but this has never been proved or disproved.

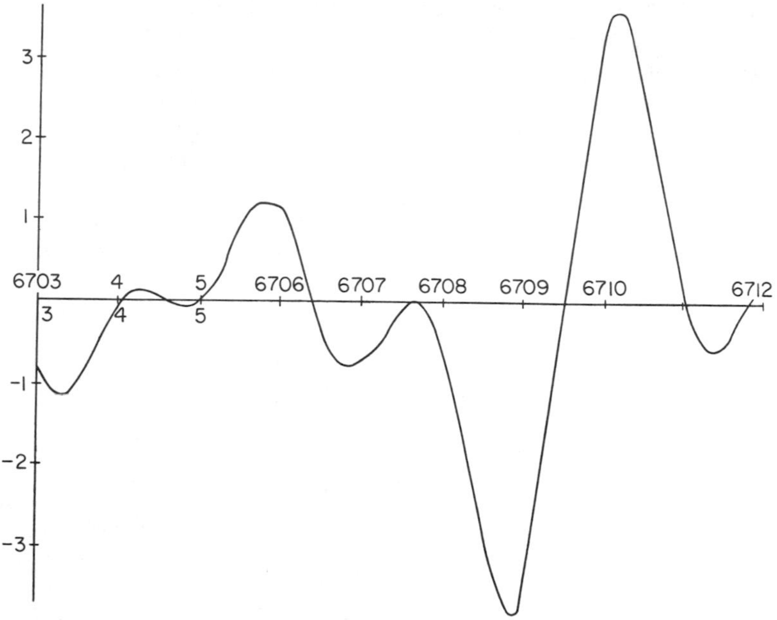

Fig. 3 The scale on the vertical axis is the value of $Z(t)$. The scale on the horizontal axis shows the location of the Gram points g_{6703}, g_{6704}, etc. (From D. H. Lehmer, On the roots of the Riemann zeta-function. *Acta Mathematica* **95**, 291–298, 1956, with the permission of the author).

occurs in it is only about 0.00397, so the choice of h_{6708} is a delicate task requiring quite accurate values of Z. Note that this "near counterexample" is followed by a very strong oscillation of Z as the terms go from a high degree of cancellation to a high degree of reinforcement. The degree of irregularity of Z shown by this graph of Lehmer, and especially the low maximum value between g_{6707} and g_{6708}, must give pause to even the most convinced believer in the Riemann hypothesis.

The extremum of lowest absolute value in the range of the first 25,000 Gram points is reported by Lehmer to be the value $+0.002$ at $t = 17143.803905$. Here the Riemann–Siegel formula gives $Z(t) \sim 0.002\,153\,336$. This low local maximum occurs between the two most nearly coincident zeros of Z found by Lehmer, namely, the zeros at† $t = 17143.786\,536$ and $t = 17143.821\,844$. To see how very close this low maximum comes to being a counterexample to the Riemann hypothesis, note that it completely de-

†The values given by Lehmer [L8] are incorrect. The values above are taken from Haselgrove's tables [H8].

main sum	0.073 478 610
C_0 term	−0.071 297 360
C_1 term	−0.000 027 686
C_2 term	−0.000 000 227
C_3 term	+0.000 000 001
$Z(17143.803905)$	+0.002 153 336.

Computation of the approximation to $Z(17143.803905)$.

stroys the rationale of Gram's law, according to which $Z(g_n)$ should have a tendency to have the sign $(-1)^n$ because the first term of the Riemann–Siegel formula is $2 \cdot (-1)^n$ and because the terms decrease in absolute value. In the example above, whether because the terms of the main sum tend to be zero or whether because there is large-scale cancellation of terms, the *entire* main sum amounts to only 0.073; this is of the same order of magnitude as the C_0 term which, as was seen in Section 7.6, is of the same order of magnitude as the last terms of the main sum. In short, the determination of the sign of $Z(t)$ can be a very delicate matter involving even the smallest terms in the main sum of the Riemann–Siegel formula, and although *on the average* one can expect the sign to be determined by the largest terms, there is no obvious reason why the exceptions to this statement could not include a counterexample to the Riemann hypothesis.

Subsequent calculations have so far fulfilled Lehmer's prediction that this phenomenon would recur infinitely often. For example, Rosser *et al.* found a pair of zeros in the vicinity of the 13,400,000th zero which were separated by just 4.4×10^{-4} (whereas the distance between successive Gram points in this region is about $0.07 = 700 \times 10^{-4}$) and between which $|Z|$ is less than 7.1×10^{-5}. It would be interesting to have a graph, such as Lehmer's above, of the behavior of Z in this region.

8.4 COMPUTATIONS OF ROSSER, YOHE, AND SCHOENFELD

At the Mathematics Research Center in Madison, Wisconsin, there are three reels of magnetic tape containing three and a half million triples of numbers (G_n, Z_n, ϵ_n) such that $G_{n+1} > G_n$, $(-1)^n Z_n > 0$, $|Z_n| > \epsilon_n > 0$, and such that, according to a rigorous analysis of the error in the Riemann–Siegel formula,† the value of $Z(G_n)$ differs from Z_n by less than ϵ_n. These tapes, unless they contain an error‡ prove the existence of three and a half million

†In the computations the C_3 term was the first term omitted.
‡This is a proviso which applies, after all, to any "proof" (see Lakatos [L1]).

roots ρ on the line $\operatorname{Re} s = \frac{1}{2}$ and locate them between $\frac{1}{2} + iG_n$ and $\frac{1}{2} + iG_{n+1}$. Moreover, by applying Turing's method to the last 15 or so of the G_n $(= g_n + h_n$ in the notation of Section 8.2) these data prove that there are *only* three and a half million roots with imaginary parts in this range (counting multiplicities), hence that for each n there is precisely one root ρ in the range $\{G_n \leq \operatorname{Im} s \leq G_{n+1}\}$ and it is a simple zero on the line $\operatorname{Re} s = \frac{1}{2}$. There is a fourth reel of tape which proves in similar fashion that 41,600 consecutive roots ρ beginning with the 13,400,000th are simple zeros on the line.

In the course of the computations of which these tapes are the record, Rosser, Yohe, and Schoenfeld discovered the following interesting phenomenon. Let a Gram point be called "good" if $(-1)^n Z(g_n) > 0$ and "bad" otherwise. Rosser *et al.* called the interval between two consecutive good Gram points a "Gram block"—that is, a Gram block is an interval $\{g_n \leq t \leq g_{n+k}\}$, where g_n, g_{n+k} are good Gram points but $g_{n+1}, g_{n+2}, \ldots, g_{n+k-1}$ are bad—and they found somewhat to their surprise that in the range of their computations *every Gram block contains the expected number of roots.* Let this be called "Rosser's rule." This phenomenon, as long as it continues, is of obvious usefulness in locating roots. However, Rosser *et al.* express a belief that it will not continue forever and this belief can be proved† to be correct as follows.

To be specific, let Rosser's rule be the statement that in any Gram block $\{g_n \leq t \leq g_{n+k}\}$ there are at least k changes of sign of Z. This implies, since $\pi^{-1}\vartheta(t) + 1$ increases by exactly k on the block, that $S(g_{n+k}) \geq S(g_n)$. Therefore, by induction it implies $S(g_n) \geq 0$ for all good Gram points g_n. On the other hand, if g_m, g_{m+1} are both bad, then Z changes sign on $\{g_m \leq t \leq g_{m+1}\}$, from which $S(g_{m+1}) \geq S(g_m)$. Thus after a good Gram point the value of S could drop by one at the next Gram point (assuming it is bad), but thereafter it could drop no further until a good Gram point were reached, at which time it would have to return at least to its former value by Rosser's rule. Thus, in particular, Rosser's rule implies $S(g_n) \geq -1$ for all Gram points g_n. Now this implies that the Riemann hypothesis is true because, if it were false, then there would be an increase of 2 in $N(T)$ not counted in the above estimates (which counted only sign changes of Z, hence roots on the line) which together with the above estimates gives $S(g_n) \geq 1$ for all Gram points g_n past the supposed counterexample to the Riemann hypothesis and hence $S(t) \geq 0$ beyond this point. This pretty clearly contradicts Littlewood's theorem that the average of S is zero and a rigorous proof is not hard to give. However, the actual estimates can be avoided by the method given below. For the moment assume it has been shown that Rosser's rule implies there

†This proof is in essence the one given by Titchmarsh [T5] to prove that Gram's law $\operatorname{Re} \zeta(\frac{1}{2} + ig_n) > 0$ fails infinitely often.

are no counterexamples to the Riemann hypothesis. Then by a 1913 theorem of Bohr and Landau (see Section 9.8) it follows that $S(t)$ is neither bounded above nor bounded below. In particular $S(t) \geq -2$ is false and this contradiction proves that Rosser's rule cannot hold.

More generally this argument can be used to show that there are *infinitely many* exceptions to Rosser's rule, that is, infinitely many Gram blocks with fewer than the expected number of sign changes. One need only observe that otherwise the argument above shows that once one is past all exceptions to Rosser's rule, the value of $S(g_n)$ and hence of $S(t)$ is bounded below. Then there are at most finitely many exceptions to the Riemann hypothesis, since otherwise $S(t)$ would eventually be large and positive, contradicting Littlewood's theorem. However, as Titchmarsh observes, the theorem of Bohr and Landau remains true if there are only finitely many exceptions to the Riemann hypothesis and it implies that $S(t)$ cannot then be bounded below. (Thus the proof above that Rosser's rule implies *no* exceptions to the Riemann hypothesis can be omitted.) This contradiction proves that there must be infinitely many exceptions to Rosser's rule.

From the fact that not a single exception to Rosser's rule has yet been found it is tempting to conclude that the computations have not yet reached the real irregularities of $Z(t)$. But actually Rosser's rule is not in any way a measure of the "regularity" of $Z(t)$. On the contrary, it measures only the success of a rather crude attempt to predict the oscillations in sign of $Z(t)$ [and hence of $\xi(\frac{1}{2} + it)$], an attempt which in fact has proved far more successful than in all likelihood Gram imagined it would be when he first proposed it.

Note added in second printing: This fact, that Rosser's rule fails infinitely often, was proved by R. Sherman Lehman (On the distribution of zeros of the Riemann zeta function, *Proc. Lon. Math. Soc.* (3) **XX** (1970) 303–320). In this same paper Lehman points out errors in Turing's proof of the main estimate (1) of Section 8.2. However, he replaces Turing's proof with his own proof of a slightly stronger inequality.

Chapter 9

The Growth of Zeta as $t \to \infty$ and the Location of Its Zeros

9.1 INTRODUCTION

The problem of locating the roots ρ of ζ, and consequently the problem of estimating the error in the prime number theorem, is closely related to the problem of estimating the growth of ζ in the critical strip $\{0 \leq \text{Re}\, s \leq 1\}$ as $\text{Im}\, s \to \infty$. Evidence of the relation between these two problems can be seen in Section 4.2 where the main step in the proof of the prime number theorem depended on estimates of $\text{Re} \log \zeta(\sigma + it) = \log |\zeta(\sigma + it)|$ for σ near 1 and for all t, in Section 5.2 where the main step in de la Vallée Poussin's estimate of the error in the prime number theorem depended on estimates of $\zeta'(\sigma + it)/\zeta(\sigma + it)$ for large t, and in Section 6.7 where Backlund's proof of Riemann's estimate of $N(T)$ depended on estimates of the growth of $|\xi(s)|$ in the strip $0 \leq \text{Re}\, s \leq 4$.

A major landmark in the study of ζ in the critical strip is Lindelöf's 1908 paper [L11] in which he not only proved some estimates which were far stronger than those that had been established previously and introduced new techniques and theorems basic to subsequent studies, but in which he also enunciated the famous "Lindelöf hypothesis." This paper is the subject of Section 9.2. In Section 9.3 a brief but important note [L12] written by Littlewood in 1912 is discussed; this note is important because it introduced the use of the three circles theorem and showed that the Riemann hypothesis implies the Lindelöf hypothesis. In 1918 Backlund proved a more exact result to the effect that the Lindelöf hypothesis implies and is implied by a certain statement about the location of the roots ρ which is much weaker than the Riemann hypothesis; this result is proved in Section 9.4. The following section, 9.5, is devoted to Littlewood's theorem, mentioned in Section 8.2, that

$\int_0^T S(t)\,dt$ grows no faster than a constant times $\log T$ as $T \to \infty$ [where $S(T)$ is the error in the approximation $N(T) \sim \pi^{-1}\vartheta(t) + 1$ or, what was seen in Section 6.7 to be the same, is $\pi^{-1}\operatorname{Im}\log\zeta(\frac{1}{2} + iT)$]. The difficult step in this proof is the estimation of $\operatorname{Re}\log\zeta(s)$ for s in the critical strip with large imaginary part. In Section 9.6 the theorem of Bohr and Landau is proved which states (see Section 1.9) that the relative error in the approximation "the number of roots ρ with imaginary parts between 0 and T which lie within δ of $\operatorname{Re} s = \frac{1}{2}$ is, for every $\delta > 0$, approximately equal to the total number of roots ρ with imaginary parts in this range" approaches zero as $T \to \infty$ for fixed δ. Here the estimate of $|\zeta(s)|$ which is needed is an estimate of the *average* value of $|\zeta(s)|^2$ on lines $\operatorname{Re} s =$ const. These averages are evaluated in Section 9.7. Finally, Section 9.8 is devoted to the enunciation, without proof, of various other theorems of this sort on the growth of ζ in the critical strip and the location of the roots ρ.

9.2 LINDELÖF'S ESTIMATES AND HIS HYPOTHESIS

The estimate $|\zeta(\sigma + it)| = |\Sigma n^{-\sigma-it}| \leq \Sigma n^{-\sigma} = \zeta(\sigma)$ shows that for $\sigma > 1$ the modulus of $\zeta(\sigma + it)$ is *bounded* as $t \to \infty$. On the other hand, it is not difficult to show† that $\zeta(\sigma)$ is the *least* upper bound of $|\zeta(\sigma + it)|$ as $t \to \infty$ because values of t can be chosen to make $n^{-\sigma-it} = n^{-\sigma}[\cos(t\log n) - i\sin(t\log n)]$ nearly equal to $n^{-\sigma}$ for as many values of n as desired so that for any $\epsilon > 0$ arbitrarily large t can be chosen to make $|\zeta(\sigma + it)| > \zeta(\sigma) - \epsilon$. Since $\zeta(\sigma) \to \infty$ as $\sigma \downarrow 1$, this shows that $|\zeta(s)|$ is *not* bounded on the quarterplane $\{\operatorname{Re} s > 1, \operatorname{Im} s > 1\}$. As for the line $\{\operatorname{Re} s = 1\}$, Mellin [M4] showed in 1900 that on it the growth of $|\zeta(s)|$ is no more rapid than the growth of $\log t$ as $t \to \infty$, an estimate which Lindelöf proves very simply by using Euler–Maclaurin summation to write

$$\zeta(s) = \sum_{n-1}^{N-1} n^{-s} + \frac{N^{1-s}}{s-1} + \frac{1}{2}N^{-s} - s\int_N^\infty \bar{B}_1(x)x^{-s-1}\,dx,$$

$$|\zeta(1 + it)| \leq 1 + \frac{1}{2} + \frac{1}{3} + \cdots + \frac{1}{N-1} + \frac{1}{t} + \frac{1}{2N} + |1 + it|\int_N^\infty \frac{1}{2}x^{-2}\,dx$$

which with N equal to the greatest integer less than or equal to t gives easily $|\zeta(1 + it)| < \log t +$ const as desired. This makes it reasonable to expect

†See, for example, Titchmarsh [T3, pp. 6–7].

that although $|\zeta(s)|$ is unbounded on $\{\text{Re}\, s \geq 1\}$, its growth is less rapid than $\log t$. This is an immediate consequence of the following important generalization of the maximum modulus theorem to a particular type of noncompact domain.

Lindelöf's Theorem Let $f(s)$ be defined and analytic in a halfstrip $D = \{s\colon \sigma_1 \leq \text{Re}\, s \leq \sigma_2,\ \text{Im}\, s \geq t_0 > 0\}$. If the modulus of f is less than or equal to M on the boundary ∂D of D *and* if there is a constant A such that $|f(\sigma + it)|t^{-A}$ is bounded on D, then the modulus of f is less than or equal to M throughout D.

Proof Consider the function $\log|f(s)|$ which is a real-valued harmonic function defined throughout D except for singularities at the zeros of $f(s)$ where it approaches $-\infty$. The additional growth condition on f states that $\log|f(\sigma + it)| < A \log t + \text{const}$ on D for A sufficiently large. This implies that for any $\epsilon > 0$ the harmonic function $\log|f(s)| - \epsilon t$ is less than any given constant on the line segment $\{\text{Im}\, s = T, \sigma_1 \leq \text{Re}\, s \leq \sigma_2\}$ provided T is large enough to make ϵT much larger than $A \log T$. In particular, for T sufficiently large $\log|f(s)| - \epsilon t$ is less than $\log M$ on the boundary of the rectangle $\{\sigma_1 \leq \text{Re}\, s \leq \sigma_2, t_0 \leq \text{Im}\, s \leq T\}$; hence the same inequality holds throughout the rectangle and therefore throughout the half-strip, and it follows that $|f(s)| \leq e^{\epsilon t}M$ throughout the half-strip. Since ϵ was arbitrary, this implies Lindelöf's theorem.

Corollary 1 $|\zeta(\sigma + it)|/\log t$ is bounded for $\sigma \geq 1$, $t \geq 2$.

Proof Since $|\zeta(\sigma + it)|$ is bounded for $\sigma \geq 1 + \delta$ by $|\zeta(\sigma + it)| \leq \zeta(\sigma) \leq \zeta(1 + \delta)$, it suffices to consider the half-strip $\{1 \leq \text{Re}\, s \leq 1 + \delta, \text{Im}\, s \geq 2\}$. Within this half-strip $\log s$ differs by a bounded amount from $\log t$. Moreover, $|\zeta(\sigma + it)| \leq \text{const}\, t^2$ in the half-strip, as was proved in Section 6.7 using Euler–Maclaurin summation. Combining these observations with Mellin's estimate of $|\zeta(1 + it)|$ shows that Lindelöf's theorem applies to $\zeta(s)/\log s$ on the half-strip and the corollary follows.

Corollary 2 $|\zeta(s)|$ is not bounded on any line $\text{Re}\, s = \sigma$ for $\sigma \leq 1$.

Proof If it were bounded, then Lindelof's theorem would show that $|\zeta(s)|$ was bounded on a half-strip which included $\{1 \leq \sigma \leq 2, t \geq 1\}$, contrary to the fact that $|\zeta(s)|$ is unbounded on $\{\sigma > 1, t \geq 1\}$.

Thus the general pattern is for the values of $|\zeta(s)|$ on $\text{Re}\, s = \sigma$ to be greater as σ decreases, at least in the range considered above. In the range

$\operatorname{Re} s \leq 0$ the functional equation combined with Stirling's formula can be used to estimate the growth of $|\zeta(s)|$ as $\operatorname{Im} s \to \infty$ as follows:

$$\begin{aligned}
\log|\xi(\sigma + it)| &= \operatorname{Re}\log\xi(\sigma + it) \\
&= \operatorname{Re}\log\Pi\left(\frac{\sigma + it}{2}\right) - \frac{\sigma}{2}\log\pi + \log|\sigma - 1 + it| \\
&\quad + \log|\zeta(\sigma + it)| \\
&\sim \frac{\sigma + 1}{2}\log\left|\frac{s}{2}\right| - \frac{t}{2}\operatorname{Im}\log\frac{s}{2} - \frac{\sigma}{2} + \frac{1}{2}\log 2\pi \\
&\quad - \frac{\sigma}{2}\log\pi + \log|t| + \log|\zeta(\sigma + it)| \\
&\sim \frac{\sigma}{2}\left(\log\frac{t}{2} - 1 - \log\pi\right) - \frac{t}{2}\cdot\frac{\pi}{2} + \frac{3}{2}\log t \\
&\quad - \frac{1}{2}\log 2 + \frac{1}{2}\log 2\pi + \log|\zeta(\sigma + it)| \\
&\sim \frac{\sigma}{2}\log\frac{t}{2\pi e} - \frac{t\pi}{4} + \frac{3}{2}\log t + \frac{1}{2}\log\pi \\
&\quad + \log|\zeta(\sigma + it)|
\end{aligned}$$

where the error in the approximation approaches zero as $t \to \infty$ for fixed σ. Thus

$$\begin{aligned}
0 &= \log|\xi(\sigma + it)| - \log|\xi(1 - \sigma + it)| \\
&\sim \frac{\sigma}{2}\log\frac{t}{2\pi e} - \frac{1 - \sigma}{2}\log\frac{t}{2\pi e} + \log|\zeta(\sigma + it)| \\
&\quad - \log|\zeta(1 - \sigma + it)|
\end{aligned}$$

and finally

$$1 \sim \left(\frac{t}{2\pi e}\right)^{\sigma - (1/2)}\left|\frac{\zeta(\sigma + it)}{\zeta(1 - \sigma + it)}\right|, \tag{1}$$

where the error in the approximation approaches zero as $t \to \infty$ for fixed σ. Thus for $\operatorname{Re} s = 0$ the modulus of $\zeta(s)$ is less than a constant times $t^{1/2}\log t$ and on any line $\operatorname{Re} s = \sigma < 0$ it is less than a constant times $t^{(1/2)-\sigma}$. Moreover, the last estimate is a least upper bound. This gives a satisfactory description of the growth of $|\zeta(s)|$ on lines $\operatorname{Re} s = \sigma \leq 0$ and shows that this growth becomes more rapid as σ decreases.

For lines $\operatorname{Re} s = \sigma$ inside the critical strip $0 < \sigma < 1$ the above estimates do not apply. However, Lindelöf observed that an upper bound for the growth of $|\zeta(s)|$ on such lines can be obtained by linear interpolation of the estimates for $\sigma = 0$ and $\sigma = 1$. More precisely, he observed that there is a constant K such that $|\zeta(\sigma + it)| < Kt^{(1/2)-(1/2)\sigma}\log t$ throughout the half-strip

$\{0 \leq \text{Re}\, s \leq 1, \text{Im}\, t \geq 1\}$. (The exponent $\frac{1}{2} - \frac{1}{2}\sigma$ is the affine function which is $\frac{1}{2}$ for $\sigma = 0$ and 0 for $\sigma = 1$.) This is a consequence of the following general theorem.

Modified Lindelöf's Theorem Let $f(s)$ be defined and analytic in a half-strip $D = \{s: \sigma_1 \leq \text{Re}\, s \leq \sigma_2, \text{Im}\, s \geq t_0 > 0\}$. If p, q are such that the modulus of f is less than a constant times t^p on $\text{Re}\, s = \sigma_1$ and less than a constant times t^q on $\text{Re}\, s = \sigma_2$ *and* if there is a constant A such that $|f(\sigma + it)|t^{-A}$ is bounded on D, then there is a constant K such that $|f(\sigma + it)| \leq Kt^{k(\sigma)}$ throughout D, where $k(\sigma) = [(q - p)/(\sigma_2 - \sigma_1)](\sigma - \sigma_1) + p$ is the affine function which is p at σ_1 and q at σ_2.

Proof Apply the previous argument using the harmonic function $\log|f(s)| - k(\sigma)t - \epsilon t$ instead of $\log|f(s)| - \epsilon t$.

Lindelöf denotes† by $\mu(\sigma)$ the least upper bound of the numbers A such that $|\zeta(\sigma + it)|t^{-A}$ is bounded as $t \to \infty$. Otherwise stated, $\mu(\sigma)$ is characterized by the condition that $|\zeta(\sigma + it)|$ divided by $t^{\mu(\sigma)+\epsilon}$ is bounded as $t \to \infty$ if $\epsilon > 0$ but unbounded if $\epsilon < 0$. The above estimates show that $\mu(\sigma) = 0$ for $\sigma \geq 1$ and that $\mu(\sigma) = \frac{1}{2} - \sigma$ for $\sigma \leq 0$. Formula (1) shows that $\mu(\sigma)$ satisfies the functional equation $\mu(\sigma) = \mu(1 - \sigma) + \frac{1}{2} - \sigma$. The modified Lindelöf's theorem shows that $\mu(\sigma) \leq \frac{1}{2} - \frac{1}{2}\sigma$ for $0 \leq \sigma \leq 1$ and, more generally, it shows that $\mu(\sigma)$ is convex downward in the sense that any segment of the graph of μ lies below the line joining its endpoints. This implies that $\mu(\sigma) \geq 0$ for $\sigma < 1$ and that if $\mu(\sigma_0) = 0$ for some $\sigma_0 < 1$, then necessarily $\mu(\sigma) = 0$ for $\sigma_0 < \sigma < 1$.

The so-called‡ *Lindelöf hypothesis* is that $\mu(\sigma)$ is the simplest§ function which has all the above properties, namely, the function which is zero for $\sigma \geq \frac{1}{2}$ and $\frac{1}{2} - \sigma$ for $\sigma \leq \frac{1}{2}$. By the convexity of μ the Lindelöf hypothesis is equivalent to the hypothesis that $\mu(\frac{1}{2}) = 0$. It is shown in Section 9.4 that the Lindelöf hypothesis implies and is implied by a condition on the location of the roots ρ which is weaker than the Riemann hypothesis, so the Lindelöf hypothesis is, in Titchmarsh's phrase, less drastic than the Riemann hypothesis. Nonetheless, its proof appears to be no easier and it has never been proved or disproved.

†This μ is not, of course, related in any way to the Möbius function $\mu(n)$ defined in Section 5.6.

‡Actually Lindelöf conjectured that $|\zeta(s)|$ is *bounded* on $\text{Re}\, s = \sigma$ for $\sigma > \frac{1}{2}$, a conjecture which was shown above to be false.

§Another possibility is $\mu(\sigma) = \frac{1}{2} - \frac{1}{2}\sigma$ on $0 \leq \sigma \leq 1$. However, it is known (see Section 9.8) that $\mu(\frac{1}{2}) < \frac{1}{4}$, so this possibility is excluded.

9.3 THE THREE CIRCLES THEOREM

In 1912 Littlewood introduced a new technique into the study of the growth of ζ when he published in a brief note [L12] some new estimates obtained using the theorem now known as the "three circles" theorem. Littlewood attributes the theorem to no one, saying it was discovered independently by several authors, but Bohr and Landau [B7] say the theorem was first published by Hadamard in 1896 (although Hadamard published no proof) and it is now commonly known as "Hadamard's three circles theorem." In any case, the theorem consists of the following simple observations.

Let $f(s)$ be defined and analytic on an annulus $D = \{r_1 \le |s - s_0| \le r_3\}$. Given an upper bound M_3 for the modulus of f on the outer circle $|s - s_0| = r_3$ and an upper bound M_1 for the modulus of f on the inner circle $|s - s_0| = r_1$ the problem is to find an upper bound M_2 for the modulus of f on a concentric circle $|s - s_0| = r_2$ inside the annulus. The method is to consider the harmonic function $\log |f(s)|$ on the annulus D and to compare it to a harmonic function which is identically $\log M_1$ on the inner circle and identically $\log M_3$ on the outer circle. Such a function is easily found by applying linear interpolation and the fact that $a \log |s - s_0| + b$ is a two-parameter family of harmonic functions constant on circles $|s - s_0| = \text{const}$ (note the analogy with the modified Lindelöf theorem in the preceding section). This leads to consideration of the harmonic function

$$H(s) = \frac{(\log M_3 - \log M_1) \log |s - s_0| + \log M_1 \log r_3 - \log M_3 \log r_1}{\log r_3 - \log r_1}.$$

Since $H(s) \ge \log |f(s)|$ on ∂D and since both are harmonic [except that $\log |f(s)|$ may have singularities at zeros of f where it is $-\infty$], the same inequality holds throughout D, which for $|s - s_0| = r_2$ gives

$$\frac{(\log M_3 - \log M_1) \log r_2 + \log M_1 \log r_3 - \log M_3 \log r_1}{\log r_3 - \log r_1} \ge \log M_2$$

which simplifies to

$$\log M_1 \log \frac{r_3}{r_2} + \log M_3 \log \frac{r_2}{r_1} \ge \log M_2 \log \frac{r_3}{r_1}$$

or

$$M_1^{\log (r_3/r_2)} M_3^{\log (r_2/r_1)} \ge M_2^{\log (r_3/r_1)}$$

which is the desired estimate. Otherwise stated, $M_2 \le M_1{}^{\alpha} M_3{}^{\beta}$, where $\alpha + \beta = 1$ and $\alpha : \beta = \log (r_3/r_2) : \log (r_2/r_1)$, in which form M_2 appears as a sort of mean value between M_1 and M_3.

As a simple application of this theorem, Littlewood proved that *if the Riemann hypothesis is true, then for every* $\epsilon > 0$ *and* $\delta > 0$ *the function* $|\log \zeta(\sigma + it)|$ *is less than a constant times* $(\log t)^{2-2\sigma+\epsilon}$ *on the half-strip* $\{\frac{1}{2} + \delta \leq \sigma \leq 1, t \geq 2\}$, where $\log \zeta(s)$ is defined for $\operatorname{Re} s > \frac{1}{2}$ by virtue of the Riemann hypothesis. Since $\sigma > \frac{1}{2}$ implies $2 - 2\sigma + \epsilon < 1$ for ϵ sufficiently small, this shows that the Riemann hypothesis implies $\log |\zeta(\sigma + it)| = \operatorname{Re} \log \zeta(\sigma + it) \leq |\log \zeta(\sigma + it)| \leq K \log t (\log t)^{-\theta}$, where $\theta > 0$; hence for any $\epsilon' > 0$ it follows that $\log |\zeta(\sigma + it)| \leq \epsilon' \log t$ for all sufficiently large t; hence $|\zeta(s)| < t^{\epsilon'}$ on $\operatorname{Re} s = \sigma > \frac{1}{2}$ as $t \to \infty$—in short, *the Riemann hypothesis implies the Lindelöf hypothesis.* Since this is the consequence of Littlewood's theorem which is of principal† interest in this chapter and since it is subsumed in Backlund's proof (which took its inspiration from Littlewood) in the next section, the details of Littlewood's application of the three circles theorem will not be given here.

9.4 BACKLUND'S REFORMULATION OF THE LINDELÖF HYPOTHESIS

Backlund [B4] proved in 1918 that the *Lindelöf hypothesis is equivalent to the statement that for every* $\sigma > \frac{1}{2}$ *the number of roots in the rectangle* $\{T \leq \operatorname{Im} s \leq T + 1, \sigma \leq \operatorname{Re} s \leq 1\}$ *grows less rapidly than* $\log T$ *as* $T \to \infty$—more precisely, it is equivalent to the statement that for every $\epsilon > 0$ there is a T_0 such that the number of such zeros is less than $\epsilon \log T$ whenever $T \geq T_0$. It follows from this that the Riemann hypothesis implies the Lindelöf hypothesis because if the Riemann hypothesis is true, then there are never *any* zeros in the rectangle in question.

The implication in one direction is quite an easy consequence of Jensen's theorem. Consider a circle which passes through the points $\sigma + iT$ and $\sigma + i(T + 1)$ and which lies in $\operatorname{Re} s > \frac{1}{2}$. Let $\sigma_0 + i(T + \frac{1}{2}) = s_0$ be the center of this circle and let ρ be its radius. Finally, let r be the radius of the slightly larger circle concentric with this one and tangent to the line $\{\operatorname{Re} s = \frac{1}{2}\}$. Then by Jensen's theorem $\log |\zeta(s_0)| + \sum \log (r/|s_j - s_0|) \leq M$ where M is the maximum of $\log |\zeta(s)|$ on the larger circle and the sum on the left is over the zeros s_j of $\zeta(s)$ in the larger circle. If this sum is restricted to the zeros which lie in the rectangle $\{T \leq \operatorname{Im} s \leq T + 1, \sigma \leq \operatorname{Re} s \leq 1\}$, then it contains n terms, where n is the integer to be estimated, and each of them is at least $\log (r/\rho)$; hence

$$n \cdot \log (r/\rho) \leq M - \log |\zeta(s_0)|.$$

†The consequence which Littlewood was principally interested in, however, was that the Riemann hypothesis implies convergence throughout the halfplane $\operatorname{Re} s > \frac{1}{2}$ of the Dirichlet series $[\zeta(s)]^{-1} = \sum \mu(n) n^{-s}$ of Section 5.6. For a proof of this see Section 12.1.

Now $\log(r/\rho)$ is a positive number independent of T and†

$$-\log|\zeta(s_0)| = \log\frac{1}{|\zeta(s_0)|} = \log\left|\sum_n \mu(n)n^{-s_0}\right| \\ \leq \log\sum_n n^{-\sigma_0} = \log\zeta(\sigma_0)$$

for all t. If the Lindelöf hypothesis is true, then for every $\epsilon > 0$ there is a K such that $|\zeta(\sigma + it)| \leq Kt^\epsilon$ for $\frac{1}{2} \leq \sigma \leq \sigma_0 + r$; hence $M \leq \log[K(T + \frac{1}{2} + r)^\epsilon] = \epsilon\log(T + \frac{1}{2} + r) + \log K \leq 2\epsilon\log T$ for T sufficiently large; hence $n\log(r/\rho) \leq 3\epsilon\log T$ for T sufficiently large. Since $\epsilon > 0$ is arbitrary, this shows that the Lindelöf hypothesis implies n grows more slowly than $\log T$.

Backlund's proof of the converse uses the actual function which is used in the proof of Jensen's theorem in Section 2.2. Let $\sigma_0 > 1$ be fixed, let $s_0 = \sigma_0 + iT$, where, as before, T will go to infinity, let σ be fixed in the range $\frac{1}{2} < \sigma < 1$, let C be the circle with center s_0 tangent to the line $\operatorname{Re} s = \sigma$ so that the radius R of C is $\sigma_0 - \sigma$, let $s_1, s_2, \ldots, s_n$ be the zeros of ζ (counted with multiplicities) inside C, and let

(1) $$F(s) = \zeta(s)\prod_{\nu=1}^{n}\frac{R^2 - (\bar{s}_\nu - \bar{s}_0)(s - s_0)}{R(s - s_\nu)}.$$

(It will be assumed that ζ has no zeros on C, a condition which excludes at most a discrete set of T's.) The s_ν are contained in a finite number of rectangles of the form $\{T' \leq \operatorname{Im} s \leq T' + 1, \sigma \leq \operatorname{Re} s \leq 1\}$, and the number of these rectangles is independent of T; hence what is to be shown is that if for each σ the number n of these zeros grows more slowly than $\log T$, then the Lindelöf hypothesis must be true.

The first step is to consider $|\log F(s)|$ on a circle C_1 concentric with C lying entirely in the halfplane $\operatorname{Re} s > 1$, say the circle with center s_0 tangent to the line $\operatorname{Re} s = 1 + \Delta$, where $0 < \Delta < \sigma_0 - 1$. Let $\alpha_\nu(s)$ denote the νth factor in the product in (1) and consider $\log\alpha_\nu(s)$. Since α_ν is a fractional linear transformation which carries s_ν to ∞ and C to the unit circle, it carries C_1 to a circle which lies outside the unit circle and does not encompass it. Since this circle contains in its interior the point $\alpha_\nu(s_0) = R/(s_0 - s_\nu)$ which lies on the same ray from the origin as $\bar{s}_0 - \bar{s}_\nu$ and which therefore lies in the halfplane $\operatorname{Re}\alpha_\nu(s_0) > 0$, $\log\alpha_\nu(s)$ can be defined inside and on C_1 by the condition $|\operatorname{Im}\log\alpha_\nu(s_0)| < \frac{1}{2}\pi$. This gives a meaning to $\log F(s)$, namely, $\log F(s) = \log\zeta(s) + \sum_{\nu=1}^n \log\alpha_\nu(s)$, throughout the interior of C_1 and hence, by analytic continuation, throughout the interior of C where F is analytic and nonzero. Now $|\operatorname{Im}\log\alpha_\nu(s)| < 3\pi/2$ on C_1 because a circle which does not contain the origin cannot intersect both halves of the imaginary axis. On the other hand, $\operatorname{Re}\log\alpha_\nu(s) = \log|\alpha_\nu(s)|$ is positive on C_1 but less than $\log[(R^2 + RR_1)/R\Delta]$, where R_1 is the radius of C_1. Thus there is a bound $b \geq |\log\alpha_\nu(s)|$ for s on

†Note that σ_0 must be greater than one.

C_1 which is valid for all ν and all T. Since $|\log F(s)| \le |\log \zeta(s)| + nb$ and since $|\log \zeta(s)|$ is bounded on C_1 independently of T [namely, by $\log \zeta(1 + \Delta)$], this shows that *the given assumption on n implies that for every $\epsilon > 0$ there is a T_0 such that $|\log F(s)| < \epsilon \log T$ on C_1 whenever $T > T_0$*.

Consider next $\log |F(s)| = \operatorname{Re} \log F(s) = \log |\zeta(s)| + \sum_{\nu=1}^{n} \log |\alpha_\nu(s)|$ on C. Since α_ν is chosen in such a way that $|\alpha_\nu(s)| \equiv 1$ on C (see Section 2.2), this is $\log |\zeta(s)|$ which, since $|\zeta(s)| \le \text{const } t^2$, is less than a constant times $\log T$ as $T \to \infty$. On the other hand, $\log |F(s_0)|$ is bounded below because $\log |\zeta(s_0)| \ge -\log \zeta(\sigma_0)$ and $\log |\alpha_\nu(s_0)| = \log (R/|s_\nu - s_0|) \ge 0$; hence $\operatorname{Re} \log [F(s)/F(s_0)]$ is zero at the center of C and less than a constant times $\log T$ on C. Now by the lemma of Section 2.7 this implies that the *modulus* of $\log [F(s)/F(s_0)]$ is less than a constant times $\log T$ on a smaller circle. Specifically this lemma shows that on the circle C_3 concentric with C but slightly smaller, say with radius $R_3 = R - \eta$, where $\eta > 0$ is small, the modulus of $\log [F(s)/F(s_0)]$ is at most $2R_3(R - R_3)^{-1}$ times the maximum of $\operatorname{Re} \log [F(s)/F(s_0)]$ on C. Thus there is a constant K such that $|\log F(s)| < K \log T$ on C_3 for all sufficiently large T.

Finally let C_2 be a circle concentric with C_3 but slightly smaller still, say with radius $R_2 = R_3 - \eta = R - 2\eta$, and consider the modulus of $\log F(s)$ on C_2. By the three circles theorem this modulus is at most $(\epsilon \log T)^\alpha (K \log T)^\beta = \epsilon^\alpha K^\beta \log T$, where α and β are positive numbers independent of T which satisfy $\alpha + \beta = 1$. Since ϵ is arbitrarily small, so is $\epsilon^\alpha K^\beta$, and it follows that for any given $\delta > 0$ there is a T_0 such that $|\log F(s)| < \delta \log T$ inside and on C_2 whenever $T \ge T_0$. Since $\log |F(s)| \le |\log F(s)|$, this gives $|F(s)| < T^\delta$ and consequently, since $|\alpha_\nu(s)| > 1$ on C_2, $|\zeta(s)| < T^\delta$ inside and on C_2; hence $|\zeta(s)| < T^\delta$ throughout the strip $\sigma_0 - R_2 \le \operatorname{Re} s \le \sigma_0 + R_2$ once T is sufficiently large. Since $\sigma_0 - R_2 = \sigma + 2\eta$ is arbitrarily near $\frac{1}{2}$ and since δ is arbitrarily small, the Lindelöf hypothesis follows.

9.5 THE AVERAGE VALUE OF $S(t)$ IS ZERO

This section is devoted to the proof of Littlewood's theorem, mentioned in Section 8.2, that $\int_{t_1}^{t_2} S(t)\, dt$ grows no more rapidly than $\log t_2$ as $t_2 \to \infty$. The essential step is to show that $\int_{t_1}^{t_2} S(t)\, dt$ can be rewritten in the form

(1)
$$\int_{t_1}^{t_2} S(t)\, dt = \int_{1/2}^{\infty} \pi^{-1} \log |\zeta(\sigma + it_2)|\, d\sigma - \int_{1/2}^{\infty} \pi^{-1} \log |\zeta(\sigma + it_1)|\, d\sigma.$$

To prove this consider the rectangle $D = \{s: \frac{1}{2} \le \operatorname{Re} s \le K, t_1 \le \operatorname{Im} s \le t_2\}$, where K is a large constant. The function $\log \zeta(s)$ is well defined on the portion of D which lies to the right of $\operatorname{Re} s = 1$ and the function $S(t)$ is $\pi^{-1} \log \zeta(s)$ on the line $\operatorname{Re} s = \frac{1}{2}$ when $\log \zeta(s)$ is analytically continued along lines $\operatorname{Im} s = \text{const}$ which contain no zeros of $\zeta(s)$. Let D_ϵ denote the domain obtained by deleting from D all points whose imaginary parts lie within ϵ of the imaginary part of a zero of $\zeta(s)$. Then D_ϵ is a union of a finite number of rectangles and it contains no zeros of $\zeta(s)$, so $\pi^{-1} \log \zeta(s)$ is well defined throughout D_ϵ by analytic continuation and its imaginary part at points of ∂D_ϵ of the form $\frac{1}{2} + it$ is $S(t)$. Now by Cauchy's theorem

$$\int_{\partial D_\epsilon} \pi^{-1} \log \zeta(s)\, ds = 0.$$

Taking the real part of this equation gives

$$\int_{\partial D_\epsilon} \pi^{-1} \operatorname{Re} \log \zeta(s)\, d\sigma = \int_{\partial D_\epsilon} \pi^{-1} \operatorname{Im} \log \zeta(s)\, dt.$$

The integral on the left involves only the horizontal boundaries of D_ϵ. These consist of the top boundary $\operatorname{Im} s = t_2$, the bottom boundary $\operatorname{Im} s = t_1$, and the interior boundaries $\operatorname{Im} s = t \pm \epsilon$, where t is the imaginary part of a zero of $\zeta(s)$. Now the integral over a pair of interior boundaries can be written

$$\begin{aligned} &\int_{1/2}^{K} \pi^{-1} \operatorname{Re} \log \zeta(\sigma + it + i\epsilon)\, d\sigma \\ &\qquad - \int_{1/2}^{K} \pi^{-1} \operatorname{Re} \log \zeta(\sigma + it - i\epsilon)\, d\sigma \\ &= \int_{1/2}^{K} \pi^{-1} \log \left| \frac{\zeta(\sigma + it + i\epsilon)}{\zeta(\sigma + it - i\epsilon)} \right| d\sigma. \end{aligned}$$

Since $\log |\zeta(\sigma + it + i\epsilon)/\zeta(\sigma + it - i\epsilon)|$ approaches $\log 1 = 0$ uniformly as $\epsilon \downarrow 0$, the integrals over the interior boundaries cancel as $\epsilon \downarrow 0$. The integral on the right involves only the vertical boundaries of D_ϵ. These consist of the two line segments $\{\frac{1}{2} + it: t_1 \leq t \leq t_2\}$ and $\{K + it: t_1 \leq t \leq t_2\}$ with intervals of length 2ϵ deleted. Since deleting intervals of length 2ϵ from the domain of a convergent† integral and letting $\epsilon \downarrow 0$ does not change the value of the integral, and since $S(t)$ is Riemann integrable on $\{t_1 \leq t \leq t_2\}$ (it is continuous except for a finite number of jump discontinuities), the integral on the right approaches $-\int_{t_1}^{t_2} S(t)\, dt + \int_{t_1}^{t_2} \pi^{-1} \operatorname{Im} \log [\zeta(K + it)]\, dt$ as $\epsilon \downarrow 0$ and the equation

$$\begin{aligned} &\int_{1/2}^{K} \pi^{-1} \log |\zeta(\sigma + it_1)|\, d\sigma - \int_{1/2}^{K} \pi^{-1} \log |\zeta(\sigma + it_2)|\, d\sigma \\ &\qquad = -\int_{t_1}^{t_2} S(t)\, dt + \int_{t_1}^{t_2} \pi^{-1} \operatorname{Im} \log \zeta(K + it)\, dt \end{aligned}$$

†Im $\log \zeta$ is continuous on ∂D except for a finite number of jump discontinuities; hence it is Riemann integrable.

results. Thus to prove (1) it remains only to show that $\int_{t_1}^{t_2} \operatorname{Im} \log \zeta(K + it)\, dt$ approaches zero and $\int_{1/2}^{K} \operatorname{Re} \log \zeta(\sigma + it)\, d\sigma$ converges as $K \to \infty$. Since both the real and the imaginary part of a function are less in absolute value than its modulus, both of these statements follow from estimates of $|\log \zeta(K + it)| \leq \log \zeta(K)$ for large K. If, for given K, u is defined to be $u = 2^{-K} + 3^{-K} + 4^{-K} + \cdots$, then $0 < u \leq 2^{-K} + \int_2^\infty t^{-K}\, dt \leq 3 \cdot 2^{-K}$ for $K \geq 2$, and $\log \zeta(K) = \log (1 + u) < u \leq 3 \cdot 2^{-K}$, from which the desired conclusions follow immediately.

Thus Littlewood's estimate of $\int_{t_1}^{t_2} S(t)\, dt$ is equivalent to the statement that $|\int_{1/2}^{\infty} \log |\zeta(\sigma + it)|\, d\sigma|$ grows no faster than $\log t$ as $t \to \infty$. Since the estimate above gives $|\int_2^\infty \log |\zeta(\sigma + it)|\, d\sigma| \leq \int_2^\infty 3 \cdot 2^{-\sigma}\, d\sigma$, which is independent of t, it will suffice to show that $|\int_{1/2}^{2} \log |\zeta(\sigma + it)|\, d\sigma|$ grows no faster than $\log t$. The main step in the proof of this fact is an estimate of $|\log \zeta(s)|$ similar to Backlund's estimate in the preceding section.

Let C be the circle of radius $R = 2 + \delta$ ($\delta > 0$) with center at $s_0 = 2 + it$ (t large and variable), and let

$$F(s) = \zeta(s) \prod_{\nu=1}^{n} \frac{R^2 - (\bar{s}_\nu - \bar{s}_0)(s - s_0)}{R(s - s_\nu)} \tag{2}$$

where $s_1, s_2, \ldots, s_n$ are the zeros of ζ inside C. As before it will be assumed that there are no zeros on C, a condition which excludes only a discrete set of values of t and does not affect the validity of the conclusion of the argument. Then, as in the preceding section, $\log F(s)$ can be defined throughout the disk bounded by C, and at points s of C it satisfies $\operatorname{Re} \log [F(s)/F(s_0)] = \log |\zeta(s)| - \log |\zeta(s_0)| - \sum_{\nu=1}^{n} \log |R/(s_0 - s_\nu)| \leq \log |\zeta(s)| + \log \zeta(2)$ which is less than a constant times $\log t$ (see Section 6.7). Therefore by the lemma of Section 2.7 on the slightly smaller circle C_3 of radius $R_3 = 2$ about s_0, the modulus of $\log [F(s)/F(s_0)]$, and hence of $\log F(s)$, is less than a constant times $\log t$. Now

$$\begin{aligned} \left| \int_{1/2}^{2} \log |\zeta(\sigma + it)|\, d\sigma \right| &= \left| \operatorname{Re} \int_{1/2}^{2} \log \zeta(\sigma + it)\, d\sigma \right| \\ &\leq \int_{1/2}^{2} |\operatorname{Re} \log F(\sigma + it)|\, d\sigma \\ &\quad + \sum_{\nu=1}^{n} \left| \int_{1/2}^{2} \operatorname{Re} \log \alpha_\nu(\sigma + it)\, d\sigma \right|, \end{aligned}$$

where, as before, α_ν denotes the νth factor in definition (2) of $F(s)$. Now $|\operatorname{Re} \log F(\sigma + it)| \leq |\log F(\sigma + it)|$ has been shown to be less than a constant times $\log t$; hence the first term on the right is less than a constant times $\log t$. The second term on the right is a sum of n terms where, by von Mangoldt's theorem on the density of the roots (Section 3.4), n is less than a constant times $\log t$; so to prove the theorem, it will suffice to show that the terms of this sum $|\int_{1/2}^{2} \operatorname{Re} \log \alpha_\nu(\sigma + it)\, d\sigma|$ have an upper bound independent of

ν and t. Since the numerator of α_ν is bounded away from 0 and ∞ on the domain of integration, the log of its modulus is also; so this amounts essentially to finding a bound for $\int_{1/2}^{2} \log|\sigma + it - s_\nu|\, d\sigma$. Now $4 \geq |\sigma + it - s_\nu| \geq |\sigma - \sigma_\nu|$, where $\sigma_\nu = \text{Re}\, s_\nu$; so the integral in question is at most $\frac{3}{2}\log 4$ and at least $\int_{1/2}^{2} \log|\sigma - \sigma_\nu|\, d\sigma$. Although the integrand in this last integral is unbounded at $\sigma = \sigma_\nu$, the integral is convergent as an improper integral (or as a Lebesgue integral) and is easily shown to be bounded. This completes the proof that for every $t_1 > 0$ there is a $K > 0$ and a $T_0 > t_1$ such that $|\int_{t_1}^{t_2} S(t)\, dt| < K \log t_2$ whenever $t_2 \geq T_0$.

9.6 THE BOHR–LANDAU THEOREM

In 1914 Bohr and Landau [B8] proved a different sort of relationship between the growth of ζ and the location of its zeros. Roughly what they proved was that the fact that the *average* value of $|\zeta(s)|^2$ on lines $\text{Re}\, s = \sigma$ is bounded for $\sigma > \frac{1}{2}$, uniformly for $\sigma \geq \frac{1}{2} + \delta$, implies that *most* of the roots ρ lie in the range $\text{Re}\, s < \frac{1}{2} + \delta$ for any $\delta > 0$. [Actually Bohr and Landau deduced their conclusions about the roots ρ not from properties of ζ but from properties of the related function $1 - 2^{-s} + 3^{-s} - 4^{-s} + \cdots = (1 + 2^{-s} + 3^{-s} + \cdots) - 2(2^{-s} + 4^{-s} + 6^{-s} + \cdots) = (1 - 2^{1-s})\zeta(s)$, for which the needed facts about averages on lines $\text{Re}\, s = \sigma$ are easier to prove.] Specifically, it will be shown in the next section that *for every* $\sigma_0 > \frac{1}{2}$ *there exist* K, T_0 *such that* $(T - 1)^{-1} \int_1^T |\zeta(\sigma + it)|^2\, dt < K$ *whenever* $\sigma \geq \sigma_0$ and $T \geq T_0$. This section is devoted to Bohr and Landau's method of concluding from this that *for every* $\delta > 0$ *there is a* K' *such that the number of roots* ρ *in the range* $\{\text{Re}\, s \geq \frac{1}{2} + \delta, 0 \leq \text{Im}\, s \leq T\}$ *is less than* $K'T$ *for all* T. Since the total number $N(T)$ of roots in the range $\{0 \leq \text{Im}\, s \leq T\}$ is about $(T/2\pi)\log(T/2\pi)$ (see Section 6.7), this proves that the number of these roots to the right of the line $\text{Re}\, s = \frac{1}{2} + \delta$ divided by their total number approaches zero as $T \to \infty$. In short, *for any* $\delta > 0$ *all but an infinitesimal proportion of the roots* ρ *lie within* δ *of the line* $\text{Re}\, s = \frac{1}{2}$. In the sense that this is a statement about all but an infinitesimal proportion of the roots (one is tempted to say "almost all," but to avoid the misinterpretation that this might mean "all but a finite number," Littlewood's phrase "infinitesimal proportion" is better) it is to this day the strongest theorem on the location of the roots which substantiates the Riemann hypothesis.

Bohr and Landau's method of estimating the number of roots ρ to the right of $\text{Re}\, s = \frac{1}{2} + \delta$ given the above fact about the average of $|\zeta(s)|^2$ on $\text{Re}\, s = \sigma > \frac{1}{2}$ is as follows. Let $\delta > 0$ be given, let C be a circle through $\frac{1}{2} + \delta + it$ and $\frac{1}{2} + \delta + i(t + 1)$ which lies in the halfplane $\text{Re}\, s > \frac{1}{2}$, and let C^+ be a circle concentric with C, slightly larger than C but still contained in the halfplane $\operatorname{Re} s > \frac{1}{2}$. As in the proofs of this type given above, the choice of C, C^+, which is different for different values of t, is to differ merely by a vertical translation. Let r denote the radius of C, R the radius of C^+, $s_0 = \sigma_0 + i(t + \frac{1}{2})$ their common center, and n the number of zeros of ζ in the rectangle $\{\frac{1}{2} + \delta \le \operatorname{Re} s \le 1, t \le \operatorname{Im} s \le t + 1\}$. Since this rectangle is contained in C, Jensen's theorem gives

$$\log|\zeta(s_0)| + n \log \frac{R}{r} \le \frac{1}{2\pi}\int_0^{2\pi} \log|\zeta(s_0 + Re^{i\theta})|\,d\theta,$$

$$2\log|\zeta(s_0)| + 2n \log \frac{R}{r} \le \frac{1}{2\pi}\int_0^{2\pi} \log|\zeta(s_0 + Re^{i\theta})|^2\,d\theta.$$

Without loss of generality it can be assumed that $1 - 2^{-\sigma_0} - 3^{-\sigma_0} - \cdots > 0$ (increase σ_0 if necessary) so that there is an A with $|\zeta(s_0)| \ge A > 0$ for all t. The fact that the geometric mean of a function is less than or equal to its arithmetic mean gives

$$\frac{1}{2\pi}\int_0^{2\pi} \log|\zeta(s_0 + Re^{i\theta})|^2\,d\theta \le \log\left[\frac{1}{2\pi}\int_0^{2\pi} |\zeta(s_0 + Re^{i\theta})|^2\,d\theta\right].$$

Hence

$$A^2\left(\frac{R}{r}\right)^{2n} \le \frac{1}{2\pi}\int_0^{2\pi} |\zeta(s_0 + Re^{i\theta})|^2\,d\theta.$$

Moreover, the analogous inequality holds when R is replaced by any radius ρ between $(r + R)/2$ and R. Multiplying this inequality by $\rho\,d\rho$ and integrating then gives

$$A^2 r^{-2n}\int_{(r+R)/2}^{R} \rho^{2n+1}\,d\rho \le \frac{1}{2\pi}\iint |\zeta(s_0 + \rho e^{i\theta})|^2\,\rho\,d\rho\,d\theta.$$

The right side is a constant times the integral of $|\zeta(s)|^2$ over the annulus $\{(r + R)/2 \le |s - s_0| \le R\}$ which is less than the integral of $|\zeta(s)|^2$ over the entire disk bounded by C^+. The left side is greater than

$$A^2 r^{-2n}\left(\frac{r+R}{2}\right)^{2n+1}\left(\frac{R-r}{2}\right) = A^2\left(\frac{R^2 - r^2}{4}\right)\left(1 + \frac{R-r}{2r}\right)^{2n}$$
$$> A^2\left(\frac{R^2 - r^2}{4}\right)\cdot 2n\left(\frac{R-r}{2r}\right),$$

that is, greater than a constant times n. Hence n is less than a constant times the integral of $|\zeta(s)|^2$ over the disk bounded by C^+, the constant depending on A, r, R and hence on δ but not on t. Ignoring the range of t for which C^+ includes the singularity $s = 1$ of ζ and adding the inequality just obtained over all integer values of t above this range and below $T + 1$ shows (since the overlapping C^+'s include any given point at most $2R$ times) that the total number of zeros with imaginary parts less than a given integer is less than a constant plus a constant times an integral of $|\zeta(s)|^2$ over a strip of the form

$\{\sigma_0 - R \le \operatorname{Re} s \le \sigma_0 + R, 1 \le \operatorname{Im} s \le T + R\}$ which, by the property of the averages of $|\zeta(s)|^2$ to be proved in the next section, is less than a constant times T, as was to be shown.

Of course the actual constant obtained by this method of estimation is absurdly large, particularly in view of the fact that according to the Riemann hypothesis it should be zero. Another technique for proving the same estimate with a much smaller constant was given by Littlewood in 1924 [L14].

9.7 THE AVERAGE OF $|\zeta(s)|^2$

For $\sigma > 1$ it is easy to prove that the average $(T-1)^{-1}\int_1^T |\zeta(\sigma+it)|^2\,dt$ of $|\zeta(s)|^2$ on $\operatorname{Re} s = \sigma$ approaches $\zeta(2\sigma)$ as $T \to \infty$ and that the approach to the limit is uniform for $\sigma \ge \sigma_0 \ge 1$. First of all, since $\sigma \ne 1$ and since $|\zeta(\sigma+it)| = |\zeta(\sigma-it)|$, the average can be written in the more natural form $\lim_{T\to\infty}(1/2T)\int_{-T}^T |\zeta(\sigma+it)|^2\,dt$. Then using

$$|\zeta(\sigma+it)|^2 = \zeta(\sigma+it)\zeta(\sigma-it) = \sum_m \sum_n \frac{1}{n^{\sigma+it}} \cdot \frac{1}{m^{\sigma+it}}$$

and noting that this double series is uniformly convergent (it is dominated by $\sum\sum n^{-\sigma}m^{-\sigma}$), so it can be integrated termwise, gives

$$\lim_{T\to\infty} \frac{1}{2T}\int_{-T}^{T} |\zeta(\sigma+it)|^2\,dt = \lim_{T\to\infty} \sum_m \sum_n \frac{1}{n^\sigma}\cdot\frac{1}{m^\sigma}\,\frac{1}{2T}\int_{-T}^{T}\left(\frac{m}{n}\right)^{it} dt.$$

If $m = n$, the coefficient of $n^{-\sigma}m^{-\sigma} = n^{-2\sigma}$ is identically 1, whereas if $n \ne m$, it is $2\sin[T\log(n/m)]/2T\log(n/m)$. Since $(\sin h)/h$ is bounded, the limit of the sums is the sum of the limits which is just $\sum n^{-2\sigma} = \zeta(2\sigma)$ and this is true uniformly in σ, provided σ is bounded away from 1.

Since $\zeta(2\sigma)$ makes sense all the way to $\sigma = \frac{1}{2}$, it is natural to ask whether it is not still true that the average of $|\zeta(s)|^2$ on $\operatorname{Re} s = \sigma$ is $\zeta(2\sigma)$ for $\sigma > \frac{1}{2}$. The method of proof will of course have to be drastically modified because $\sum n^{-\sigma}$ no longer converges when $\sigma \le 1$, but the theorem is still true. This theorem appears in Landau's 1908 *Handbuch* [L3], but the central idea of the proof which follows is from Hardy and Littlewood [H6].

The proof for $\sigma > 1$ and the form of the theorem strongly suggest that the divergent series $\sum n^{-\sigma-it}$ will play a role in the proof. If, as in the estimation of $\zeta(1+it)$ in Section 9.2, one uses just the terms of this series in which $n < t$, then the remainder $R(\sigma, t)$ in $\zeta(\sigma+it) = \sum_{n<t} n^{-\sigma-it} + R(\sigma, t)$ must be estimated. Now the Euler–Maclaurin formula for $\zeta(\sigma+it)$ gives

$$\begin{aligned} R(\sigma, t) &= \zeta(\sigma+it) - \sum_{n<t} n^{-\sigma-it} \\ &= \sum_{t\le n<N} n^{-s} + \frac{N^{1-s}}{s-1} + \frac{1}{2}N^{-s} - s\int_N^\infty \bar{B}_1(x)x^{-s-1}\,dx, \end{aligned}$$

where $s = \sigma + it$ and where N is any integer larger than t. If N is the first integer larger than t, then the first three terms on the right have modulus less than a constant times $t^{-\sigma}$ as $t \to \infty$. Backlund's method of estimating the remaining term shows easily that it is less than a constant times $t^{1-\sigma}$, but this method ignores the cancellations in the integral due to the rapid oscillation of x^{it} for large t and the following more refined estimate shows that the fourth term, like the first three, is less than a constant times $t^{-\sigma}$ so that the same is true of $R(\sigma, t)$ itself.

Hardy and Littlewood estimate the integral $\int_N^\infty \bar{B}_1(x)x^{-s-1}\,dx$ by observing that the Fourier series of $\bar{B}_1(x)$, namely, $\bar{B}_1(x) = -\sum_{n=1}^\infty (\sin 2\pi nx)/n\pi$ is boundedly convergent† so that the termwise integration

$$\int_N^\infty \bar{B}_1(x)x^{-s-1}\,dx = -\sum_{n=1}^\infty \frac{1}{n\pi}\int_N^\infty (\sin 2\pi nx)x^{-s-1}\,dx$$

is justified by the Lebesgue dominated convergence theorem.‡ Now

$$\begin{aligned}\frac{1}{n\pi}&\int_N^\infty (\sin 2\pi nx)x^{-s-1}\,dx\\ &= \frac{1}{2in\pi}\int_N^\infty (e^{2\pi inx} - e^{-2\pi inx})x^{-\sigma-1-it}\,dx\\ &= \frac{1}{2\pi in}\int_N^\infty e^{iw(x)}x^{-\sigma-1}\,dx - \frac{1}{2\pi in}\int_N^\infty e^{-iv(x)}x^{-\sigma-1}\,dx,\end{aligned}$$

†Let $S_N(x)$ denote the sum of the first N terms. Then for $0 < x < \frac{1}{2}$ the difference $\bar{B}_1(x) - S_N(x)$ can be written in the form

$$\begin{aligned}x - \tfrac{1}{2} &+ \sum_{n=1}^N (\pi n)^{-1}\sin 2\pi nx\\ &= \int_0^x \Big(1 + \sum_{n=1}^N 2\cos 2\pi nt\Big)\,dt - \frac{1}{2}\\ &= \int_0^x \Big(\sum_{n=-N}^N e^{2\pi int}\Big)\,dt - \frac{1}{2}\\ &= \int_0^x \frac{e^{2\pi iNt+\pi it} - e^{-2\pi iNt-\pi it}}{e^{\pi it} - e^{-\pi it}}\,dt - \frac{1}{2}\\ &= \int_0^x \frac{\sin(2N+1)\pi t}{\sin \pi t}\,dt - \frac{1}{2}\\ &= \int_0^x \frac{\sin(2N+1)\pi t}{\pi t}\,dt + \int_0^x \left(\frac{1}{\sin \pi t} - \frac{1}{\pi t}\right)\sin(2N+1)\pi t\,dt - \frac{1}{2}\\ &= \frac{1}{\pi}\int_0^{(2N+1)\pi x} \frac{\sin u}{u}\,du + \int_0^x f(t)\sin(2N+1)\pi t\,dt - \frac{1}{2}\end{aligned}$$

where $f(t) = (\sin \pi t)^{-1} - (\pi t)^{-1} = (\pi t - \sin \pi t)/\pi t \sin \pi t$ is analytic for $|t| < 1$ and therefore bounded for $|t| \le \frac{1}{2}$, say by K. Then since $\int_0^h \sin u/u\,du$ is bounded (its maximum occurs at $h = \pi$), it follows that $\bar{B}_1(x) - S_N(x)$ has modulus less than $\pi^{-1}\int_0^\pi u^{-1}\sin u\,du + \frac{1}{2}K + \frac{1}{2}$ for $0 < x \le \frac{1}{2}$ and for *all* N. Since $S_N(0) \equiv 0$, $S_N(-x) \equiv -S_N(x)$, the same bound holds for $-\frac{1}{2} \le x \le \frac{1}{2}$ and, since $\bar{B}_1(s)$ is bounded, this implies the desired result.

‡See, for example, Edwards [E1, p. 437].

where $w(x) = 2\pi nx - t \log x$ and $v(x) = 2\pi nx + t \log x$. Since

$$\frac{dw}{dx} = 2\pi n - \frac{t}{x} > 2\pi n - \frac{N}{N}$$

is positive on $\{N < x < \infty\}$, x can be considered as a function of w and the first integral above can be expressed in terms of w

$$\lim_{K\to\infty} \frac{1}{2\pi i n} \int_N^K e^{iw(x)} x^{-\sigma-1}\, dx$$

$$= \lim_{K\to\infty} \frac{1}{2\pi n} \int_{w(N)}^{w(K)} (\sin w - i \cos w) x^{-\sigma-1} \frac{dw}{2\pi n - (t/x)}.$$

The real and imaginary parts of this integral are of the form $\int_A^B F(w) \sin w\, dw$ and $\int_A^B F(w) \cos w\, dw$, respectively, where F is a monotone decreasing positive function. Therefore they can be estimated using the following elementary lemma.

Lemma If $F(x)$ is positive and monotone nonincreasing on $\{A \le x \le B\}$, then $\int_A^B F(x) \sin x\, dx$ is at most $2F(A)$ in absolute value. The same estimate applies to $\int_A^B F(x) \cos x\, dx$.

Proof Let A' be an even multiple of π less than A and let B' be an odd multiple of π greater than B, say $2\mu\pi = A' \le A < B \le B' < 2\nu\pi + \pi$. Extend F to $\{A' \le x \le B'\}$ by defining it to be constantly $F(A)$ between A' and A and constantly $F(B)$ between B and B'. Then

$$\begin{aligned}
\int_{A'}^{B'} F(x) \sin x\, dx &= F(A) \int_{A'}^{A} \sin x\, dx + \int_A^B F(x) \sin x\, dx \\
&\quad + F(B) \int_B^{B'} \sin x\, dx \\
&= F(A)(-\cos A + 1) + \int_A^B F(x) \sin x\, dx \\
&\quad + F(B)(1 + \cos B) \\
&\ge \int_A^B F(x) \sin x\, dx.
\end{aligned}$$

On the other hand,

$$\begin{aligned}
\int_{A'}^{B'} F(x) \sin x\, dx &= \int_{2\mu\pi}^{2\mu\pi+\pi} F(x) \sin x\, dx + \int_{2\mu\pi+\pi}^{2\mu\pi+2\pi} F(x) \sin x\, dx + \cdots \\
&\quad + \int_{2\nu\pi}^{2\nu\pi+\pi} F(x) \sin x\, dx \\
&\le F(2\mu\pi) \int_{2\mu\pi}^{2\mu\pi+\pi} \sin x\, dx + F(2\mu\pi + 2\pi) \int_{2\mu\pi+\pi}^{2\mu\pi+2\pi} \sin x\, dx \\
&\quad + F(2\mu\pi + 2\pi) \int_{2\mu\pi+2\pi}^{2\mu\pi+3\pi} \sin x\, dx + \cdots \\
&\quad + F(2\nu\pi) \int_{2\nu\pi}^{2\nu\pi+\pi} \sin x\, dx \\
&= F(2\mu\pi) \int_{2\mu\pi}^{2\mu\pi+\pi} \sin x\, dx = 2F(A)
\end{aligned}$$

which proves $\int_A^B F(x)\sin x\,dx \le 2F(A)$. The proof of the lemma is now easily completed by using analogous arguments to find a lower estimate of $\int_A^B F(x)\sin x\,dx$ and upper and lower estimates of $\int_A^B F(x)\cos x\,dx$.

Applying the lemma to the integrals above (the real and imaginary parts of the integral involving v as well as the integral involving w) shows easily that $|(n\pi)^{-1}\int_N^\infty (\sin 2\pi nx)x^{-s-1}\,dx|$ is less than a constant times $n^{-2}N^{-\sigma-1}$. Summing these estimates over n shows that $|\int_N^\infty \bar{B}_1(x)x^{-s-1}\,dx|$ is less than a constant times $N^{-\sigma-1} \sim t^{-\sigma-1}$. Since $|s| \sim t$, this completes the proof that $R(\sigma, t)$ is less than a constant times $t^{-\sigma}$ as $t \to \infty$, and shows, moreover, that the estimate is uniform in σ.

With this estimate of $R(\sigma, t)$ the evaluation of the average of $|\zeta(\sigma+it)|^2$ is not too difficult. In the first place the average value of $|\sum_{n<t} n^{-\sigma-it}|^2$ is

$$\begin{aligned}
&\frac{1}{T-1}\int_1^T\Big(\sum_{m<t}\sum_{n<t} m^{-\sigma-it}n^{-\sigma+it}\Big)\,dt \\
&= \frac{1}{T-1}\int_1^T \sum_{n<t} n^{-2\sigma}\,dt + \frac{1}{T-1}\int_1^T \sum_{n<t}\sum_{m<n} m^{-\sigma}n^{-\sigma}\Big(\frac{n}{m}\Big)^{it}dt \\
&\quad + \frac{1}{T-1}\int_1^T \sum_{m<t}\sum_{n<m} m^{-\sigma}n^{-\sigma}\Big(\frac{n}{m}\Big)^{it}dt \\
&= \sum_{n<T}\frac{1}{T-1}\int_n^T n^{-2\sigma}\,dt + \sum_{n<T}\sum_{m<n}\frac{1}{T-1}\int_n^T m^{-\sigma}n^{-\sigma}\Big(\frac{n}{m}\Big)^{it}dt \\
&\quad + \sum_{m<T}\sum_{n<m}\frac{1}{T-1}\int_m^T m^{-\sigma}n^{-\sigma}\Big(\frac{n}{m}\Big)^{it}dt.
\end{aligned}$$

This first term is $\sum_{n<T}(T-n)/(T-1)n^{-2\sigma}$ which is essentially the Cesaro sum of the convergent series $\sum n^{-2\sigma} = \zeta(2\sigma)$ and which therefore approaches $\zeta(2\sigma)$ as $T \to \infty$. Since $\sum n^{-2\sigma}$ converges uniformly for $\sigma \ge \sigma_0$, it is elementary to show that this limit is approached uniformly in σ. Thus it is to be shown that the other terms in the average of $|\zeta(\sigma+it)|^2$ approach zero uniformly in σ. Each of the two remaining terms above has modulus at most

$$\sum_{n<T}\sum_{m<n}\frac{m^{-\sigma}n^{-\sigma}}{T-1}\left|\int_n^T e^{it\log(n/m)}\,dt\right| \le \sum_{n<T}\sum_{m<n}\frac{2m^{-\sigma}n^{-\sigma}}{(T-1)\log(n/m)}.$$

This sum is monotone decreasing as σ increases, so in order to show that it is uniformly small for $\sigma \ge \sigma_0 > \frac{1}{2}$ when T is large, it suffices to show that it is small for any fixed value of $\sigma > \frac{1}{2}$ which, for convenience, can be assumed to be less than 1. Now the above sum can be split into two parts:

$$\sum_{n<T}\sum_{m<n/2}\frac{2m^{-\sigma}n^{-\sigma}}{(T-1)\log(n/m)} + \sum_{n/2\le m<n<T}\sum\frac{2m^{-\sigma}n^{-\sigma}}{(T-1)\log(n/m)}$$

which with $r = n - m$, $\log(n/m) = -\log[1-(r/n)] > r/n$ in the second part shows it is less than

$$\frac{2}{(T-1)\log 2}\sum_{n<T}\sum_{m<T} m^{-\sigma}n^{-\sigma} + \sum_{n<T}\sum_{r\le n/2}\frac{2(n-r)^{-\sigma}n^{-\sigma}}{(T-1)(r/n)}$$
$$\le \frac{2}{(T-1)\log 2}\Big(\sum_{n<T} n^{-\sigma}\Big)^2 + \frac{2}{T-1}\sum_{n<T} n^{1-2\sigma}\sum_{r\le n/2}\frac{[1-(r/n)]^{-\sigma}}{r}$$
$$\le \frac{2}{(T-1)\log 2}\Big(\sum_{n<T} n^{-\sigma}\Big)^2 + \frac{2\cdot 2^{\sigma}}{T-1}\sum_{n<T} n^{1-2\sigma}\sum_{r\le n/2} r^{-1}$$
$$\sim \frac{2}{(T-1)\log 2}\Big(\int_1^T t^{-\sigma}\,dt\Big)^2 + \frac{2\cdot 2^{\sigma}}{T-1}\int_1^T t^{1-2\sigma}\log\frac{t}{2}\,dt$$
$$< \text{const}\,\frac{T^{2-2\sigma}}{T-1} + \text{const}\,\frac{T^{2-2\sigma+\epsilon}}{T-1}$$

which, because $\sigma > \frac{1}{2}$, approaches zero as $T \to \infty$. Finally $(T-1)^{-1} \times \int_1^T |\zeta(\sigma+it)|^2\,dt$ differs from $(T-1)^{-1}\int_1^T|\sum_{n<t} n^{-\sigma-it}|^2\,dt$ by an amount whose modulus is at most

$$\frac{1}{T-1}\left|2\,\mathrm{Re}\int_1^T\Big(\sum_{n<t} n^{-\sigma-it}\Big)\overline{R(\sigma,t)}\,dt\right| + \int_1^T |R(\sigma,t)|^2\,dt$$
$$\le \frac{2}{T-1}\Big(\int_1^T\Big|\sum_{n<t} n^{-\sigma-it}\Big|^2 dt\Big)^{1/2}\Big(\int_1^T |R(\sigma,t)|^2\,dt\Big)^{1/2}$$
$$+\frac{1}{T-1}\int_1^T |R(\sigma,t)|^2\,dt$$

by the Schwarz inequality. Since the average value of $|R(\sigma,t)|^2$ approaches zero as $T \to \infty$, this approaches $2\cdot[\zeta(2\sigma)]^{1/2}\cdot\sqrt{0} + 0 = 0$ as $T \to \infty$ and the proof is complete. Specifically what has been shown is that *given* $\sigma_0 > \frac{1}{2}$ *and given* $\epsilon > 0$, *there is a* T_0 *such that*

$$\left|\zeta(2\sigma) - \frac{1}{T-1}\int_1^T |\zeta(\sigma+it)|^2\,dt\right| < \epsilon$$

whenever $\sigma \ge \sigma_0$, $T \ge T_0$. This of course implies the statement needed for the Bohr–Landau theorem, namely, that $(T-1)^{-1}\int_1^T |\zeta(\sigma+it)|^2\,dt$ is uniformly bounded for $\sigma \ge \sigma_0 > \frac{1}{2}$, $T \ge T_0$.

9.8 FURTHER RESULTS. LANDAU'S NOTATION *o*, *O*

This section is devoted to the statements, without proof, of various refinements and extensions of the theorems proved above. For fuller accounts, with proofs and references to the primary sources, see either of the books [T3] or [T8] of Titchmarsh, which were the source of much of the material of this chapter.

In describing these theorems it will be useful to introduce the following notation of Landau [L3, p.61]: The notation "$f(x) = O(g(x))$ as $x \to \infty$" means that there is a constant K and a value x_0 of x such that $|f(x)| < Kg(x)$ whenever $x \geq x_0$. In words, "the modulus of f grows no faster than a constant times g as $x \to \infty$." Here f may be complex valued, but g is real and positive. The notation "$f(x) = o(g(x))$ as $x \to \infty$" means that for every $\epsilon > 0$ there is a value x_0 of x such that $|f(x)| < \epsilon g(x)$ whenever $x \geq x_0$. In words, "the modulus of f grows more slowly than g as $x \to \infty$." There are various obvious extensions of this notation such as "$f(x) = O(g(x))$ as $x \to 0$," which means there exist K, x_0 such that $|f(x)| < Kg(x)$ whenever $0 < |x| < x_0$ or "$f(x) = F(x) + O(g(x))$ as $x \to \infty$," which means "$f(x) - F(x) = O(g(x))$ as $x \to \infty$," etc., which will also be used. In this notation the prime number theorem can be stated $\pi(x) = \mathrm{Li}(x) + o(\mathrm{Li}(x)) = \mathrm{Li}(x) + o(x/\log x)$, de la Vallée Poussin's estimate of the error can be written "$\pi(x) = \mathrm{Li}(x) + o(x \exp[-(c \log x)^{1/2}])$ for some $c > 0$," and the Lindelöf hypothesis can be written "$\zeta(\frac{1}{2} + it) = o(t^{\epsilon})$ for every $\epsilon > 0$," etc.

It was proved in Section 9.2 that $\zeta(1 + it) = O(\log t)$. Weyl [W4] in 1921 proved that, in fact, $\zeta(1 + it) = O(\log t/\log\log t)$ and thus, in particular, $\zeta(1 + it) = o(\log t)$. Weyl's proof was based on a new method of evaluating the "exponential sums" $\sum_{a \leq n \leq b} n^{-1-it} = \sum_{a \leq n \leq b} \exp[-(1 + it)\log n]$ which occur in the Euler–Maclaurin formula for $\zeta(1 + it)$. Weyl's method was improved upon by Vinogradoff who proved $\zeta(1 + it) = O([\log t \log\log t]^{3/4})$. These investigations of $\zeta(1 + it)$ led to improvements in de la Vallée Poussin's estimate (Section 5.2) of the amount by which the real part β of a root $\rho = \beta + i\gamma$ must be less than one. Littlewood in 1922 proved that $\beta < 1 - (K \log\log\gamma/\log\gamma)$ and Vinogradov and Korolov in 1958 proved that $\beta < 1 - K(\log t)^{-a}$ for any $a > \frac{2}{3}$. This was improved slightly by Richert (see notes to Walfisz' book [W1]) who proved $\beta < 1 - K(\log t)^{-2/3} (\log\log t)^{-1/3}$. Each of these improvements gives a corresponding improvement of de la Vallée Poussin's estimate of the error in the prime number theorem; for example, Richert's estimate above gives $\pi(x) = \mathrm{Li}(x) + O(x \exp[-c(\log x)^{3/5}(\log\log x)^{-1/5}])$ for some $c > 0$ (see Walfisz [W1]). As is probably clear from the mere statement of the results, the methods used in proving these facts are very demanding. Nonetheless, the improvement over de la Vallée Poussin's result is very slight in comparison with the estimate $\pi(x) = \mathrm{Li}(x) + O(x^{1/2} \log x)$ which would follow from the Riemann hypothesis (see Section 5.5).

It was also proved in Section 9.2 that $\zeta(\frac{1}{2} + it) = O(t^{1/4} \log t)$. Actually the Riemann–Siegel formula (with error estimate) shows that

$$|\zeta(\tfrac{1}{2} + it)| = |Z(t)| = \Big|2 \sum_{2\pi n^2 < t} n^{-1/2} \cos[\vartheta(t) - t \log n]\Big| + O(t^{-1/4})$$

$$= O\left(\int_1^{(t/2\pi)^{1/2}} t^{-1/2}\, dt\right) + O(t^{-1/4}) = O(t^{1/4}).$$

Hardy and Littlewood applied Weyl's method to the estimation of the exponential sum which is the main term of the Riemann–Siegel formula† to prove that this estimate can be improved to $\zeta(\frac{1}{2} + it) = O(t^{1/6}(\log t)^{3/2})$. This has been improved to $\zeta(\frac{1}{2} + it) = o(t^{1/6})$ and slightly beyond, for example, to $\zeta(\frac{1}{2} + it) = O(t^{19/116})$. Note that this implies $\mu(\frac{1}{2}) < \frac{1}{6}$ which, by the convexity of Lindelöf's function μ, implies improvements on the bound $\mu(\sigma) \leq \frac{1}{2} - \frac{1}{2}\sigma$ throughout $\{0 < \sigma < 1\}$.

Since the average value of $|\zeta(s)|^2$ on $\operatorname{Re} s = \sigma$ is $\zeta(2\sigma)$ for $\sigma > \frac{1}{2}$ and since $\zeta(2\sigma) \to \infty$ as $\sigma \downarrow \frac{1}{2}$, it is to be expected that the average value of $|\zeta(s)|^2$ on $\operatorname{Re} s = \frac{1}{2}$ is infinite. Hardy and Littlewood proved not only that this is true but that

$$\frac{1}{2T}\int_{-T}^{T}\left|\zeta\left(\frac{1}{2} + it\right)\right|^2 dt \sim \log T$$

in the sense that the relative error approaches zero as $T \to \infty$. Similar results hold for the average values of $|\zeta(s)|^4$ on $\operatorname{Re} s = \sigma$. For $\sigma > \frac{1}{2}$ this average is $[\zeta(2\sigma)]^4/\zeta(4\sigma)$ and for $\sigma = \frac{1}{2}$

$$\frac{1}{2T}\int_{-T}^{T}\left|\zeta\left(\frac{1}{2} + it\right)\right|^4 dt \sim \frac{1}{2\pi^2}(\log T)^4.$$

The averages of $|\zeta(s)|^{2k}$ for integers $k > 2$ are much more difficult and very little is known about them.

If the Lindelöf hypothesis is true, then of course many of the estimates can be improved. For example, Cramér showed that the Lindelöf hypothesis implies $S(t) = o(\log t)$. This in turn implies that the number of roots ρ with imaginary parts between T and $T + 1$ is approximately $(1/2\pi) \log T$ and that the relative error in this approximation approaches zero as $T \to \infty$. Littlewood showed that the Lindelöf hypothesis implies that $\int_0^T S(t)\, dt = o(\log T)$. As for averages of $|\zeta(s)|^{2k}$, Hardy and Littlewood showed that the Lindelöf hypothesis is *equivalent* to the statement that for all positive integers k

$$\frac{1}{2T}\int_{-T}^{T}\left|\zeta\left(\frac{1}{2} + it\right)\right|^{2k} dt = o(T^{\epsilon}) \qquad \text{for all} \quad \epsilon > 0.$$

If the Riemann hypothesis is true, then $\zeta(1 + it) = O(\log \log t)$ and $\zeta(\frac{1}{2} + it) = O(\exp(c \log t/\log \log t))$ for some c. (Note that the last estimate implies the Lindelöf hypothesis.) If the Riemann hypothesis is true, then the estimates of $S(t)$ can be strengthened to $S(t) = O(\log t/\log \log t)$ and $\int_0^T S(t)\, dt = O(\log T/\log \log T)$. On the other hand, the Riemann hypothesis also implies a *lower* bound on the rate of growth of $S(t)$. Namely, it was shown by Bohr and Landau that if the Riemann hypothesis is true, then for every $\epsilon > 0$ the inequalities $S(t) > (\log t)^{(1/2)-\epsilon}$ and $S(t) < -(\log t)^{(1/2)-\epsilon}$ have arbitrarily

†This occurred before the publication of the Riemann–Siegel formula. Hardy and Littlewood derived the main term of this formula independently as a special case of their "approximate functional equation."

large solutions t so that, in particular, $S(t)$ is not bounded either above or below. It is not known whether $S(t)$ is in fact unbounded.* This theorem of Bohr–Landau does not assume the full Riemann hypothesis but only that the number of roots off the line is *finite*, in which form it is the basic tool in Titchmarsh's proof that there must be infinitely many exceptions to Gram's law and in the proof of Section 8.4 that there must be infinitely many exceptions to Rosser's rule.

Since the proof of this theorem is not readily accessible in the literature and since it plays such an important part in Section 8.4, it seems worthwhile to include a proof here. The following proof of the weaker theorem that *if there are only a finite number of exceptions to the Riemann hypothesis, then $S(t)$ cannot be bounded below* is taken from a 1911 paper of Landau [L3a]. It uses the 1910 theorem of Bohr [B6a] which states that $\zeta(s)$ is unbounded on $\{\text{Re } s > 1, \text{Im } s \geq 1\}$.

Suppose there is a q such that there are no exceptions to the Riemann hypothesis in the halfplane $\{\text{Im } s \geq q\}$ and suppose that $S(t)$ is bounded below. Then a contradiction to the known fact that $\zeta(s)$ is unbounded on $\{\text{Re } s > 1, \text{Im } s \geq 1\}$ can be derived as follows. Consider $\text{Im} \log \zeta(s)$ on the half-strip $D = \{\frac{1}{2} \leq \text{Re } s \leq 1 + \delta, \text{Im } s \geq q\}$. Since $S(t) = \pi^{-1} \text{Im} \log \zeta(\frac{1}{2} + it)$ at all points where $\zeta(\frac{1}{2} + it) \neq 0$, the assumption on $S(t)$ implies that $\text{Im} \log \zeta(\frac{1}{2} + it)$ is bounded below on the left side of D except at the points where it is not defined. Let D_0 be the domain obtained from D by deleting very small semicircular neighborhoods of the roots ρ on its boundary. Since $\zeta(s) = (s - \rho)^n g(s)$, $\text{Im} \log \zeta(s) = n \text{Im} \log (s - \rho) + \text{Im} \log g(s)$ where $\text{Im} \log g(s)$ is continuous and bounded in a neighborhood of ρ, and since $\text{Im} \log (s - \rho)$ along any semicircle in $\{\text{Re } s \geq \frac{1}{2}\}$ assumes its smallest value on $\text{Re } s = \frac{1}{2}$ where $S(t)$ is bounded below, D_0 can be determined in such a way that $\text{Im} \log \zeta(s)$ is bounded below on the left side of D_0, the bound being independent of the size of the deleted semicircles. On the bottom side $\text{Im } s = q$ of D_0, $\text{Im} \log \zeta(s)$ is continuous and therefore bounded. On the right side $\text{Re } s = 1 + \delta$ of D_0, the inequality $\text{Im} \log \zeta(s) \leq |\log \zeta(s)| \leq \log \zeta(1 + \delta)$ shows that $\text{Im} \log \zeta(s)$ is bounded below. Finally, the estimates of Section 6.7 show that $\text{Im} \log \zeta(\sigma + it) \leq \text{const} \log M \leq \text{const} \log t$ for large t when $\frac{1}{2} \leq \sigma \leq 1 + \delta$ ($\frac{1}{2} + it$ not a root ρ). Thus there is a constant K such that $\text{Im} \log \zeta(s) \geq -K$ for all s on the boundary of D_0, and for any $\epsilon > 0$ the inequality $\text{Im} \log \zeta(s) + \epsilon t \geq -K$ holds for all $s = \sigma + it$ on the boundary of the domain $\{s \in D_0 : \text{Im } s \leq Q\}$ for all sufficiently large Q. Since $\text{Im} \log \zeta(s) + \epsilon t$ is harmonic on this domain, it follows that $\text{Im} \log \zeta(s) + \epsilon t \geq -K$ throughout D_0. Since ϵ was arbitrary, $\text{Im} \log \zeta(s) \geq -K$ throughout D_0. Since the deleted circles were arbitrary, $\text{Im} \log \zeta(s) \geq -K$ in the interior of D. Consider now the smaller strip $D_1 = \{s: (1/2) + (\delta/2) \leq \text{Re } s \leq 1 + (\delta/2), \text{Im } s \geq q + 1\}$. The lemma of Section 2.7 for deducing a bound on $|f|$ from a one-sided bound on its real or imaginary part then implies that $\log \zeta(s)$ is bounded on D_1. But then $|\zeta(s)| = e^{\text{Re} \log \zeta(s)}$ would be bounded on D_1, which it is not. This contradiction proves the theorem.

* *Note added in second printing:* This statement is an error. Selberg proved that $S(t)$ is unbounded (Contributions to the theory of the Riemann zeta-function, *Arch. for Math. og Naturv.*, B, **48** (1946), no. 5).

Chapter 10

Fourier Analysis

10.1 INVARIANT OPERATORS ON R^+ AND THEIR TRANSFORMS

One of the basic ideas in Riemann's original paper is, as Chapter 1 shows, the idea of Fourier analysis. This chapter is devoted to the formulation of a more modern approach to Fourier analysis which I believe sheds some light on the meaning of the zeta function and on its relation to the distribution of primes.

The approach to Fourier analysis I have in mind is that in which the fundamental object of study is the algebra of invariant operators on the functions on a group. The group in this instance is the multiplicative group of positive real numbers, which will be denoted R^+. Consider the vector space V of all complex-valued functions on R^+ and consider linear operators $V \longrightarrow V$. The simplest such operators are the translation operators, which are defined as follows. For each positive real number u the translation operator T_u is the operator which carries a function f in V to the function whose value at x is the value of f at ux, in symbols

$$T_u \colon f(x) \mapsto f(ux),$$

where $f(x)$ is a generic element of V and where the "barred arrow" $\mapsto$ is used to denote the effect of the operator in question on a generic element of its domain. Translation operators $V \longrightarrow V$ are defined for all of V. However, many of the most interesting operators are defined only on subspaces of V—for example, only for smooth functions or functions which vanish at ∞—and the term "operator" should not be taken to imply that the domain is necessarily all of V.

An operator $V \longrightarrow V$ is said to be *invariant* if it commutes with all translation operators. More precisely, a linear operator $L\colon V \longrightarrow V$ is said to be invariant if its domain is invariant under the action of the translation operators T_u and if $T_uLf = LT_uf$ for all $u \in R^+$ and all f in the domain of L. The simplest examples of invariant operators are the translation operators themselves, which are invariant by virtue of the fact that the group R^+ is commutative. Then superpositions of translation operators are also invariant, for example, the summation operator $f(x) \mapsto \sum_{n=1}^{\infty} f(nx)$ or integral operators $f(x) \mapsto \int_0^\infty f(ux)F(u)\,du$. (The exact domain of definition of these operators will not be important in what follows since functions will merely be assumed to satisfy whatever conditions are needed. For the summation operator the domain might be taken to be the set of functions $f \in V$ which satisfy $\lim_{x\to\infty} x^{1+\epsilon}|f(x)| = 0$ for some $\epsilon > 0$. For an integral operator in which $F(u)$ is continuous and $|F(u)|$ is bounded, the domain might be taken to be those continuous functions $f \in V$ which satisfy $\lim_{x\to\infty} x^{1+\epsilon}|f(x)| = 0$ and $\lim_{x\to 0} x^{1-\epsilon}|f(x)| = 0$ for some $\epsilon > 0$, etc.) A different sort of invariant operator is the differential operator $\lim_{h\to 1}(T_h - T_1)/(h-1)$ which carries $f(x)$ to $xf'(x)$ provided f is differentiable.

For any complex number s the one-dimensional subspace of V generated by the function† $f(x) = x^{-s}$ has the property of being invariant under all invariant operators; otherwise stated, an invariant operator carries $f(x) = x^{-s}$ to a multiple of itself. This can be proved as follows. For fixed u, T_u multiplies f by u^{-s}. Hence, since L is linear, LT_uf is u^{-s} times Lf. On the other hand, $LT_uf = T_uLf$, so T_u multiplies Lf by u^{-s} for every u. But this means $Lf(u) = (T_uLf)(1) = u^{-s}(Lf)(1)$, so Lf is $(Lf)(1)$ times f as was to be shown. For example, the summation operator $f(x) \mapsto \sum_1^\infty f(nx)$ carries $f(x) = x^{-s}$ to $\zeta(s)f$ (provided $\operatorname{Re} s > 1$), the integral operator $f(x) \mapsto \int_0^\infty f(ux)F(u)\,du$ carries it to $\int_0^\infty u^{-s}F(u)\,du$ times f (provided F and s are such that this integral converges—that is, f is in the domain of the operator), and the differential operator $f(x) \mapsto xf'(x)$ carries it to $(-s)$ times f. The fundamental idea of Fourier analysis is to *analyze* invariant operators—literally to take them apart—and to study them in terms of their actions on these one-dimensional invariant subspaces.

The *transform* of an invariant operator is the function whose domain is the set of complex numbers s such that the function $f(x) = x^{-s}$ lies in the domain of the operator and whose value for such an s is the factor by which the operator multiplies $f(x) = x^{-s}$. Thus the entire function $s \mapsto u^{-s}$ is the transform of the translation operator T_u, the zeta function $s \mapsto \zeta(s)$ for Re

†The reason for taking x^{-s} as the basic function rather than x^s or x^{is} is to put the equations in the most familiar form and, in particular, to make $\operatorname{Re} s = \frac{1}{2}$ the line of symmetry in the functional equation.

$s > 1$ is the transform of the summation operator $f(x) \mapsto \sum_{n=1}^{\infty} f(nx)$, and the function $s \mapsto -s$ is the transform of the differential operator $f(x) \mapsto xf'(x)$. The basic formula $\log \zeta(s) = \int_0^\infty x^{-s}\, dJ(x)$ for $\operatorname{Re} s > 1$ of Riemann's paper [(3) of Section 1.11] is the statement that $\log \zeta(s)$ for $\operatorname{Re} s > 1$ is the transform of the invariant operator $f(x) \mapsto \int_0^\infty f(ux)\, dJ(u)$. Similarly, the formula $-\zeta'(s)/\zeta(s) = \int_0^\infty x^{-s}\, d\psi(x)$ which was the starting point of the derivation of von Mangoldt's formula for $\psi(x)$ (see Section 3.1) is the statement that $-\zeta'(s)/\zeta(s)$ is the transform of $f(x) \mapsto \int_0^\infty f(ux)\, d\psi(u)$. The technique by which Riemann derived his formula for $J(x)$ and von Mangoldt his formula for $\psi(x)$ is a technique of *inversion*, of going from the transform to the operator, which might well be called Fourier *synthesis*—putting the operator back together again when its effect on the invariant subspaces is known.

10.2 ADJOINTS AND THEIR TRANSFORMS

In order to define the adjoint of an invariant operator (on the vector space V of complex-valued functions on the multiplicative group R^+ of positive real numbers), it is necessary to define an inner product on the vector space V. The natural definition would be $\langle f, g\rangle = \int_0^\infty f(x)\overline{g(x)}\, d\mu(x)$, where $d\mu$ is the invariant measure on the group R^+, namely, $d\mu(x) = d \log x = x^{-1}\, dx$. However, the functional equation $\xi(s) = \xi(1-s)$ involves an inner product on V which is natural with respect to the additive structure of R^+ rather than the multiplicative structure, namely, the inner product

$$\langle f, g\rangle = \int_0^\infty f(x)\overline{g(x)}\, dx.$$

This inner product is defined on a rather small subset of V—for example, the inner product of $f_1(x) = x^{-s_1}$ and $f_2(x) = x^{-s_2}$ is undefined for any pair of complex numbers s_1, s_2—but it suffices for the definition of a *formal adjoint* L^* of an operator L as an operator such that $\langle Lf, g\rangle = \langle f, L^*g\rangle$ whenever both sides are defined. Again, touchy points regarding the domains of definition of L and L^* can be avoided by restricting consideration to particular cases.

For example, the adjoint of the translation operator T_u is easily found by a change of variable in the integral

$$\langle T_u f, g\rangle = \int_0^\infty f(ux)\overline{g(x)}\, dx = \int_0^\infty f(y)\overline{g(y/u)}u^{-1}\, dy$$

to be the operator $g(x) \mapsto u^{-1}g(u^{-1}x)$, that is, the operator $u^{-1}T_{u^{-1}}$. Since the adjoint of a sum is the sum of the adjoints, this implies the adjoint of the summation operator $f(x) \mapsto \sum_{n=1}^{\infty} f(nx)$ is the operator $f(x) \mapsto \sum_{n=1}^{\infty} n^{-1}f(n^{-1}x)$

and the adjoint of the integral operator $f(x) \mapsto \int_0^\infty f(ux)F(u)\,du$ is

$$f(x) \mapsto \int_0^\infty u^{-1}f(u^{-1}x)\overline{F(u)}\,du = \int_0^\infty f(vx)v^{-1}\overline{F(v^{-1})}\,dv.$$

The adjoint of the differential operator $f(x) \mapsto xf'(x)$ is found by integration by parts,

$$\int_0^\infty xf'(x)\overline{g(x)}\,dx = xf(x)\overline{g(x)}\Big|_0^\infty - \int_0^\infty f(x)\frac{d}{dx}[x\overline{g(x)}]\,dx,$$

to be the operator $g(x) \mapsto -d[xg(x)]/dx$.

Now in terms of transforms this operation is, from the above examples, clearly related to the substitution $s \mapsto 1 - s$ since they show that an operator whose transform is u^{-s} has an adjoint whose transform is $u^{-(1-s)}$, that an operator whose transform is $\zeta(s)$ (Re $s > 1$) has an adjoint whose transform is $\zeta(1 - s)$ (Re $s < 0$), that an operator whose transform is $\int_0^\infty u^{-s}F(u)\,du$ has adjoint whose transform is $\int_0^\infty u^{-(1-s)}\overline{F(u)}\,du$, and that an operator whose transform is $-s$ has an adjoint whose transform is $-(1 - s)$. The general rule is that an operator whose transform is $\phi(s)$ has an adjoint whose transform is $\overline{\phi(1 - \bar{s})}$. This can be thought of as "conjugate transpose," where the "transpose" operation is $\phi(s) \mapsto \phi(1 - \bar{s})$. If ϕ is analytic and real on the real axis, then by the reflection principle $\phi(\bar{s}) = \overline{\phi(s)}$, so in this case $\overline{\phi(1 - \bar{s})}$ can be written simply $\phi(1 - s)$.

In this way the functional equation $\xi(s) = \xi(1 - s)$ seems to be saying that some operator is self-adjoint. A specific sense in which this is true is described in the following section.

10.3 A SELF-ADJOINT OPERATOR WITH TRANSFORM $\xi(s)$

Riemann's second proof of the functional equation (see Section 1.7) depends on the functional equation $1 + 2\psi(x) = x^{-1/2}[1 + 2\psi(x^{-1})]$ from the theory of theta functions. With $x = u^2$ and $G(u) = 1 + 2\psi(u^2) = \sum_{-\infty}^{\infty} \exp(-\pi n^2 u^2)$, this equation takes the simple form $G(u) = u^{-1}G(u^{-1})$, which implies, by Section 10.2, that the invariant operator

$$f(x) \mapsto \int_0^\infty f(ux)G(u)\,du \tag{1}$$

is formally self-adjoint. This operator has no transform at all; that is, the integral

$$\int_0^\infty u^{-s}G(u)\,du \tag{2}$$

does not converge for any s. However, as will be shown in this section, it can be modified in such a way as to be made convergent, and when this is done the formal self-adjointness of this operator is in essence equivalent to the functional equation of the zeta function.

The integral (2) would converge at ∞ if the constant term of $G(u) = 1 + 2\sum_{n=1}^{\infty} \exp(-\pi n^2 x^2)$ were absent. The constant term can be eliminated by differentiating G or, better, by applying the invariant operator $G(u) \mapsto uG'(u)$. By the definition of "adjoint" this is formally the same as applying the adjoint $f(u) \mapsto -d[uf(u)]/du$ to the first factor in the integrand of (2), which multiplies the first factor by $s - 1$. To preserve self-adjointness, it is natural then to apply $G(u) \mapsto -d[uG(u)]/du$ to the second factor or $f(u) \mapsto uf'(u)$ to the first, which multiplies the first factor by $-s$. Thus, formally,

$$\text{(3)} \qquad \int_0^\infty u^{-s}\left[\left(-\frac{d}{du}u^2\frac{d}{du}\right)G(u)\right]du = \int_0^\infty \left[\left(-u\frac{d^2}{du^2}u\right)u^{-s}\right]G(u)\,du$$
$$= s(1-s)\int_0^\infty u^{-s}G(u)\,du.$$

The right side is $s(1-s)$ times the (formal) transform of a real self-adjoint operator and therefore is in some sense invariant under the substitution $s \longrightarrow 1 - s$. However, the left side is in fact a well-defined analytic function of s for all s. To see this, let

$$H(u) = \frac{d}{du}\left[u^2\frac{d}{du}G(u)\right]$$

so that the integral in question is $-\int_0^\infty u^{-s}H(u)\,du$. Then

$$H(u) = \frac{d}{du}\left[u^2\frac{d}{du}\{G(u) - 1\}\right]$$

goes to zero faster than any power of u as $u \longrightarrow \infty$, and the integral $-\int_0^\infty u^{-s}H(u)\,du$ therefore converges at ∞. On the other hand, applying $(d/du)u^2(d/du)$ to both sides of the functional equation $G(u) = (1/u)G(1/u)$ gives two expressions,

$$\text{(4)} \qquad 2uG'(u) + u^2G''(u) = \frac{2}{u^2}G'\left(\frac{1}{u}\right) + \frac{1}{u^3}G''\left(\frac{1}{u}\right),$$

for $H(u)$ and hence gives $H(u) = (1/u)H(1/u)$; that is, H satisfies the same functional equation as G. This implies that $H(u)$ goes to zero faster than any power of u as $u \downarrow 0$ and hence that the integral $-\int_0^\infty u^{-s}H(u)\,du$ converges at 0. Therefore the left side of (3) is an entire function of s which, because of (3), would be expected to be invariant under $s \longrightarrow 1 - s$. That this is the case follows immediately from $H(x) = x^{-1}H(x^{-1})$ which gives

$$-\int_0^\infty u^{-s}H(u)\,du = \int_\infty^0 u^{1-s}H(u)\,d\log u$$
$$= -\int_0^\infty v^{s-1}H\left(\frac{1}{v}\right)d\log v = -\int_0^\infty u^{s-1}H(u)\,du.$$

Since, as will now be shown, $\int_0^\infty u^{-s}H(u)\,du = 2\xi(1-s)$, this proves the functional equation $\xi(s) = \xi(1-s)$.

Let s be a negative real number. Then

$$\begin{aligned}\int_0^\infty u^{-s}H(u)\,du &= \int_0^\infty u^{-s}\left(\frac{d}{du}u^2\frac{d}{du}G(u)\right)du \\ &= \int_0^\infty u^{-s}\left(\frac{d}{du}u^2\frac{d}{du}[G(u)-1]\right)du \\ &= -\int_0^\infty \left(\frac{d}{du}u^{-s}\right)\left(u^2\frac{d}{du}[G(u)-1]\right)du\end{aligned}$$

(because $G(u) - 1$ and all its derivatives vanish more rapidly than any power of u as $u \to \infty$ and because, therefore, $G(u) - u^{-1} = u^{-1}[G(u^{-1}) - 1]$ and all its derivatives vanish more rapidly than any power of u as $u \downarrow 0$, which implies that $u^2\,d[G(u)-1]/du = u^2\,d[G(u)-u^{-1}]/du + u^2\,d[u^{-1}-1]/du$ is bounded as $u \downarrow 0$)

$$\begin{aligned}&= s\int_0^\infty u^{1-s}\frac{d}{du}[G(u)-1]\,du \\ &= -s\int_0^\infty \left(\frac{d}{du}u^{1-s}\right)[G(u)-1]\,du\end{aligned}$$

[because $G(u) - 1$ vanishes more rapidly than any power of u as $u \to \infty$, while $G(u) - 1 = G(u) - u^{-1} + u^{-1} - 1$ grows no more rapidly than u^{-1} as $u \downarrow 0$]

$$\begin{aligned}&= s(s-1)\int_0^\infty u^{-s}[G(u)-1]\,du \\ &= s(s-1)\int_0^\infty u^{-s}2\sum_{n=1}^\infty e^{-\pi n^2u^2}\,du \\ &= s(s-1)\sum_{n=1}^\infty\int_0^\infty u^{1-s}e^{-\pi n^2u^2}2\,d\log u\end{aligned}$$

(the interchange being valid by absolute convergence)

$$\begin{aligned}&= s(s-1)\sum_{n=1}^\infty\int_0^\infty\left(\frac{v}{\pi n^2}\right)^{(1-s)/2}e^{-v}\,d\log v \\ &= s(s-1)\pi^{(s-1)/2}\sum_{n=1}^\infty\frac{1}{n^{1-s}}\int_0^\infty e^{-v}v^{(1-s)/2}v^{-1}\,dv \\ &= s(s-1)\pi^{(s-1)/2}\zeta(1-s)\,\Pi\left(\frac{1-s}{2}-1\right) \\ &= (-s)2\Pi\left(\frac{1-s}{2}\right)\pi^{-(1-s)/2}\zeta(1-s) = 2\xi(1-s).\end{aligned}$$

Therefore, by analytic continuation, the same equation holds for all s. Note that (3) then states that formally

$$\int_0^\infty u^{-s}G(u)\,du = \frac{2\xi(s)}{s(s-1)};$$

that is, the function $2\xi(s)/s(s-1)$ is formally the transform of the operator (1), but, since this operator has no transform, some "convergence factor" such as the replacement of G by H is necessary. For a different method of giving meaning to the idea that $2\xi(s)/s(s-1)$ is the transform of the self-adjoint operator $f(x) \mapsto \int_0^\infty f(ux)G(u)\,du$, see Section 10.5.

In summary, the functional equation $\xi(s) = \xi(1-s)$ can be deduced from the functional equation $G(u) = u^{-1}G(u^{-1})$ as follows. First show that the function $H(u) = (d/du)u^2(d/du)G(u)$ satisfies the same functional equation as G does, that is, $H(u) = u^{-1}H(u^{-1})$. This is immediate from (4). Then show that $\int_0^\infty u^{-s}H(u)\,du$ converges for all s, and that for negative real s it is $2\xi(1-s)$. This shows that $\xi(1-s)$ is an entire function and, because $H(u) = u^{-1}H(u^{-1})$, that it is invariant under $s \to 1-s$.

10.4 THE FUNCTIONAL EQUATION

It was shown in the preceding section that the functional equation $\xi(s) = \xi(1-s)$ can be deduced simply and naturally from the fact that the function $G(u) = \sum_{-\infty}^{\infty} \exp(-\pi n^2 u^2)$ satisfies the functional equation $G(u) = u^{-1}G(u^{-1})$ or, what is the same, from the fact that the operator $f(x) \mapsto \int_0^\infty f(ux)G(u)\,du$ is formally self-adjoint in the sense of Section 10.2. Thus in order to understand the functional equation of ξ, it is natural to study the functional equation of G.

The functional equation of G results immediately from two formulas—the Poisson summation formula and the formula

(1) $$e^{-\pi u^2} = \int_{-\infty}^{\infty} e^{-\pi x^2} e^{2\pi i x u}\,dx.$$

Consider Poisson summation first.

Let the *Fourier transform* of a complex-valued function $f(x)$ on the real line be defined by

$$\hat{f}(u) = \int_{-\infty}^{\infty} f(x) e^{2\pi i x u}\,dx,$$

a definition which differs from the usual definition by a factor of 2π in the exponential. Then the *Poisson summation formula* states that under suitable conditions on f (involving its smoothness and its vanishing at ∞), the sum of the Fourier transform is the sum of the function; that is,

(2) $$\sum_{n=-\infty}^{\infty} \hat{f}(n) = \sum_{n=-\infty}^{\infty} f(n).$$

This fact follows very easily from the theory of Fourier series when one considers the "periodified" function $F(x) = \sum_{n=-\infty}^{\infty} f(x+n)$ (assuming f vanishes rapidly enough at infinity for this sum to converge for all x between 0 and 1). This function F is periodic with period 1 and therefore (again under suitable assumptions about f) by the theory of Fourier series can be expanded as a series $\sum_{-\infty}^{\infty} a_n e^{2\pi i x n}$ in which the coefficients are given by

$$a_n = \int_0^1 F(x) e^{-2\pi i n x}\,dx = \int_0^1 \sum_{m=-\infty}^{\infty} f(x+m) e^{-2\pi i n x}\,dx$$

$$= \sum_{m=-\infty}^{\infty} \int_m^{m+1} f(y) e^{-2\pi i n (y-m)}\,dy = \int_{-\infty}^{\infty} f(y) e^{-2\pi i n y}\,dy = \hat{f}(-n).$$

Therefore, setting $x = 0$ in $F(x) = \sum a_n e^{2\pi i n x}$ gives the Poisson summation formula (2). Since the above operations are all valid in the case of the function $f(x) = \exp(-\pi x^2 u^{-2})$ for any positive u, this together with (1) gives

$$G(u) = \sum_{n=-\infty}^{\infty} e^{-\pi n^2 u^2} = \sum_{n=-\infty}^{\infty} \int_{-\infty}^{\infty} e^{-\pi x^2} e^{2\pi i x n u}\,dx$$

$$= \sum_{n=-\infty}^{\infty} \int_{-\infty}^{\infty} e^{-\pi (y/u)^2} e^{2\pi i n y} u^{-1}\,dy$$

$$= u^{-1} \sum_{n=-\infty}^{\infty} \hat{f}(n) = u^{-1} \sum_{n=-\infty}^{\infty} e^{-\pi n^2 u^{-2}} = u^{-1} G(u^{-1})$$

as was to be shown.

Finally, consider the proof of (1). In the special case $u = 0$ this is the formula

$$1 = \int_{-\infty}^{\infty} e^{-\pi x^2}\,dx \tag{3}$$

which is one of the basic formulas of calculus, being essentially equivalent to the formula $\Pi(-\frac{1}{2}) = \pi^{1/2}$, to the fact that the constant in Stirling's formula is $\frac{1}{2}\log 2\pi$ (see Section 6.3), or to Wallis' product for π. It can be proved simply by

$$\left(\int_{-\infty}^{\infty} e^{-\pi x^2}\,dx\right)^2 = \int_{-\infty}^{\infty}\int_{-\infty}^{\infty} e^{-\pi x^2} e^{-\pi y^2}\,dx\,dy$$

$$= \int_0^{2\pi}\int_0^{\infty} e^{-\pi r^2} r\,dr\,d\theta = \int_0^{2\pi} \frac{1}{2\pi}\,d\theta = 1$$

which, since the integral must be positive, proves (3). But then the change of variable $x = y - iu$ and Cauchy's theorem gives

$$1 = \int_{-\infty+iu}^{\infty+iu} e^{-\pi(y-iu)^2}\,dy = \int_{-\infty}^{\infty} e^{-\pi(y-iu)^2}\,dy$$

which is the desired formula (1).

This completes the proof that $G(u) = u^{-1}G(u^{-1})$. The structure of this proof can be interpreted in the following way. Let W be a complex vector space with inner product and let M be a linear transformation of W. A linear transformation $A\colon W \longrightarrow W$ is said to be *self-reciprocal relative to* M if A^*M

$= A$, where A^* denotes the adjoint of A. *If A and B are both self-reciprocal relative to M and if A and B commute, then A^*B is self-adjoint* because $A^*B = (A^*M)^*B = M^*AB = M^*BA = (B^*M)^*A = B^*A = (A^*B)^*$. This is roughly the situation in the above proof, with W equal to the vector space of complex-valued functions on the entire real line R, with the inner product equal to $\langle f, g\rangle = \int_{-\infty}^{\infty} f(x)\overline{g(x)}\,dx$, with M equal to the operator $f(x) \mapsto \int_{-\infty}^{\infty} f(ux)e^{2\pi iu}\,du$ resembling Fourier transform, with A equal to the summation operator $f(x) \mapsto \sum_{n=-\infty}^{\infty} f(nx)$, and with B equal to the operator $f(x) \mapsto \int_{-\infty}^{\infty} f(ux) \exp(-\pi u^2)\,du$. Then A and B are formally self-reciprocal relative to M and A^*B is the operator $f(x) \mapsto \int_{-\infty}^{\infty} f(ux)G(u)\,du$; so the above theorem states that $\int_{-\infty}^{\infty} f(ux)G(u)\,du$ is self-adjoint. If V is identified with the subspace of W consisting of even functions (doubling the inner product on V), this implies $G(u) = u^{-1}G(u^{-1})$ as desired. Of course this is only roughly true and certain problems about domains of definition have to be ignored; however, since a rigorous proof of $G(u) = u^{-1}G(u^{-1})$ has already been given and since the only issue here is to understand the structure of the theorem, these problems will be passed over.

Consider first the statement that A is self-reciprocal relative to M, that is, $A^*M = A$. Now A is invariant relative to the multiplicative structure of R [it is a superposition of multiplicative "translation "operators $f(x) \mapsto f(nx)$]; hence, formally, so is A^* invariant [this is the part of the proof which is only formally correct because the adjoint of the "translation" $f(x) \mapsto f(0x)$ is not defined because it involves division by 0]. Now an important characteristic of invariant operators is that if they are applied to $f(ux)$ considered as a function of x for fixed u, the result is the same as if they are applied to $f(ux)$ considered as a function of u for fixed x because both of these are equal to the operator applied to $f(y)$ as a function of y evaluated at $y = ux$; this follows from $L[f(ux)]$ (as a function of x) $= (L \circ T_u)f = T_u Lf$ (by invariance) $=(Lf)(ux)$. Therefore, to apply A^* to

$$(Mf)(x) = \int_{-\infty}^{\infty} f(ux)e^{2\pi iu}\,du$$

is the same as to apply A^* to $f(ux)$ in this integral considered as a function of u for fixed x. But by the definition of A^* (because A is real), this is the same as to apply A to the second factor $e^{2\pi iu}$ of the integrand. Thus

$$(A^*Mf)(x) = \sum_{n=-\infty}^{\infty} \int_{-\infty}^{\infty} f(ux)e^{2\pi inu}\,du.$$

But with $F(u) = f(ux)$ the Poisson summation formula shows that this is (formally) equal to

$$\sum_{n=-\infty}^{\infty} \int_{-\infty}^{\infty} F(u)e^{2\pi inu}\,du = \sum_{-\infty}^{\infty} \hat{F}(n) = \sum_{-\infty}^{\infty} F(n) = \sum_{-\infty}^{\infty} f(nx)$$
$$= (Af)(x).$$

Hence $A^*M = A$ as was to be shown. Similarly B^*M is formally equal to the result of applying B to the second factor of Mf; that is,

$$(B^*Mf)(x) = \int_{-\infty}^{\infty} f(ux)\left[\int_{-\infty}^{\infty} e^{2\pi ivu}e^{-\pi v^2}\,dv\right] du$$
$$= \int_{-\infty}^{\infty} f(ux)e^{-\pi u^2}\,du = (Bf)(x)$$

by (1); hence $B^*M = B$. Finally, by the same argument, A^*B is formally equal to the result of applying A to the second factor in the integral Bf; that is,

$$(A^*Bf)(x) = \sum_{n=-\infty}^{\infty} \int_{-\infty}^{\infty} f(ux)e^{-\pi n^2u^2}\,du = \int_{-\infty}^{\infty} f(ux)G(u)\,du$$

as was to be shown.

10.5 $2\xi(s)/s(s-1)$ AS A TRANSFORM

The statement that $2\xi(s)/s(s-1)$ is formally the transform of $f(x) \mapsto \int_0^\infty f(ux)G(u)\,du$ can be given substance as follows. A continuous analog of Euler's ludicrous formula

$$\sum_{-\infty}^{\infty} x^n = (1 + x + x^2 + \cdots) + (x^{-1} + x^{-2} + \cdots)$$
$$= \frac{1}{1-x} + \frac{1}{x-1} = 0$$

is

(1) $$\int_0^\infty x^{-s}\,dx = \int_0^1 x^{-s}\,dx + \int_1^\infty x^{-s}\,dx$$
$$= \frac{1}{1-s} - \frac{1}{1-s} = 0.$$

This is of course nonsense because the values of s for which the above integrals converge are mutually exclusive—the first integral being convergent for $\operatorname{Re} s < 1$ and the second being convergent for $\operatorname{Re} s > 1$—but it does suggest that the formal transform of $f(x) \mapsto \int_0^\infty f(ux)\,dx$ is zero and hence that the formal transform of $f(x) \mapsto \int_0^\infty f(ux)[G(u) - 1]\,du$ ought to be $2\xi(s)/s(s-1)$. However, since this operator actually has a transform for $\operatorname{Re} s < 0$, this suggests the correct formula

$$\int_0^\infty u^{-s}[G(u) - 1]\,du = \frac{2\xi(s)}{s(s-1)} \qquad (\operatorname{Re} s < 0)$$

which was proved in Section 10.3. Setting $u = v^{-1}$ in this formula gives

$$\int_0^\infty v^{s-1}\left[G(v) - \frac{1}{v}\right] dv = \frac{2\xi(s)}{s(s-1)} \qquad (\operatorname{Re} s < 0)$$

and, hence,

$$\int_0^\infty u^{-s}\left[G(u) - \frac{1}{u}\right] du = \frac{2\xi(s)}{s(s-1)} \qquad (\operatorname{Re} s > 1)$$

[using $\xi(s) = \xi(1-s)$]. This too can be interpreted as saying that $2\xi(s)/s(s-1)$ is formally the transform of $f(x) \mapsto \int_0^\infty f(ux)G(u)\,du$ because the formal transform of $f(x) \mapsto \int_0^\infty f(ux)u^{-1}\,du$ is zero by (1). Now since

$$\int_0^\infty u^{-s}\left[G(u) - 1 - \frac{1}{u}\right] du$$

is convergent for s in the critical strip $\{0 < \operatorname{Re} s < 1\}$, the same considerations lead one to expect that the value of this integral in the strip where it converges will also be $2\xi(s)/(s-1)$. That this is actually the case can be proved by considering the function

$$\int_1^\infty u^{-s}[G(u) - 1]\,du + \int_0^1 u^{-s}\left[G(u) - 1 - \frac{1}{u}\right] du - \frac{1}{s}$$

which is defined and analytic throughout the halfplane $\{\operatorname{Re} s < 1\}$ except for the pole at $s = 0$. Since this function agrees with

$$\int_0^\infty u^{-s}[G(u) - 1]\,du \qquad \text{on} \quad \{\operatorname{Re} s < 0\}$$

and with

$$\int_0^\infty u^{-s}\left[G(u) - 1 - \frac{1}{u}\right] du \qquad \text{on} \quad \{0 < \operatorname{Re} s < 1\}$$

it follows that these two functions are analytic continuations of one another. Thus

$$\frac{2\xi(s)}{s(s-1)} = \begin{cases} \displaystyle\int_0^\infty u^{-s}[G(u) - 1]\,du & (\operatorname{Re} s < 0), \\[2ex] \displaystyle\int_0^\infty u^{-s}\left[G(u) - 1 - \frac{1}{u}\right] du & (0 < \operatorname{Re} s < 1), \\[2ex] \displaystyle\int_0^\infty u^{-s}\left[G(u) - \frac{1}{u}\right] du & (1 < \operatorname{Re} s), \end{cases}$$

are all three literally true in the stipulated ranges of s and all three say that formally $2\xi(s)/(s-1)$ is the transform of $f(x) \mapsto \int_0^\infty f(ux)G(u)\,du$.

10.6 FOURIER INVERSION

The problem of Fourier inversion or Fourier synthesis is the problem of finding an invariant operator when its transform—that is, its effect on the one-dimensional invariant subspaces—is known. Riemann's technique of accomplishing this by changes of variable in Fourier's theorem (see Section

1.12) gives more generally

$$\Phi(s) = \int_0^\infty u^{-s} \frac{\phi(u)}{u}\, du \tag{1}$$

$$\Longleftrightarrow \phi(u) = \frac{1}{2\pi i} \int_{a-i\infty}^{a+i\infty} \Phi(s) u^s\, ds$$

provided ϕ and/or Φ satisfy suitable conditions. That is, under suitable conditions, an operator with the given transform $\Phi(s)$ can be found by defining

$$\phi(u) = \frac{1}{2\pi i} \int_{a-i\infty}^{a+i\infty} \Phi(s) u^s\, du$$

and taking the operator to be $f(x) \mapsto \int_0^\infty f(ux)\phi(u)\, d \log u$. In actual practice this formula is perhaps best understood as a heuristic one because the problems of proving convergence and of proving the applicability of Fourier's theorem are substantial.

In this context the proof of von Mangoldt's formula for ψ in Chapter 3 can be described as follows. If $\phi(u)$ is defined to be 0 for $u < 1$ and 1 for $u > 1$, then $\Phi(s)$ is $1/s$ for Re $s > 0$ and the basic integral formula of Section 3.3 is the statement that the inversion formula (1) is valid in this case provided $a > 0$. This result can be generalized by changing ϕ to

$$\phi(u) = \begin{cases} 0 & \text{for} \quad u < y, \\ 1 & \text{for} \quad u > y, \end{cases}$$

which changes $\Phi(s)$ to y^{-s}/s. Then the inversion formula (1) is still valid. On the other hand $\Phi(s)$ can be changed from $1/s$ to $1/(s - \alpha)$ which changes ϕ to

$$\frac{1}{2\pi i} \int_{a-i\infty}^{a+i\infty} \frac{u^s\, ds}{s - \alpha} = \frac{1}{2\pi i} \int_{a-i\infty-\alpha}^{a+i\infty+\alpha} \frac{u^{z+\alpha}\, dz}{z}$$

which, when $a > \alpha$, gives $\phi(u) = u^\alpha$ for $u > 1$ and zero for $u < 1$, and the inversion (1) is valid provided $a >$ Re α. (If α is real, this is immediate. If α is not real, then the final estimate of Section 3.3 must be applied to the evaluation of the conditionally convergent integral from $a - i\infty - \alpha$ to $a + i\infty - \alpha$ in order to show it is the same as the integral from $a -$ Re $\alpha - i\infty$ to $a -$ Re $\alpha + i\infty$.) Now

$$-\frac{\zeta'(s)}{\zeta(s)} = \int_0^\infty u^{-s}\, d\psi(u) = -\int_0^\infty \psi(u)\, d(u^{-s}) = s \int_0^\infty u^{-s} \frac{\psi(u)}{u}\, du,$$

so when $\phi(u) = \psi(u)$, the transform $\Phi(s)$ is $-\zeta'(s)/s\zeta(s)$. But this function can be written in two different ways as superpositions of functions with known inverse transforms, namely, as

$$-\frac{\zeta'(s)}{s\zeta(s)} = \frac{1}{s} \int_0^\infty u^{-s}\, d\psi(u) = \sum_{n=1}^\infty \Lambda(n) \frac{n^{-s}}{s}$$

and as

$$-\frac{\zeta'(s)}{s\zeta(s)} = \frac{1}{s-1} - \sum_\rho \frac{1}{\rho(s-\rho)} + \sum_n \frac{1}{2n(s+2n)} - \frac{c}{s},$$

where $c = \zeta'(0)/\zeta(0)$ [see (7) of Section 3.2]. Hence termwise application of the inversion (1) gives a $\phi(u)$ which, on the one hand, is equal to

$$\sum_{n=1}^{\infty} \Lambda(n)\phi_n(u) = \psi(u),$$

where $\phi_n(u)$ is the function which is 1 for $u < n$ and 0 for $u > n$, and, on the other hand, is equal to

$$u - \sum \frac{u^\rho}{\rho} + \sum \frac{u^{-2n}}{2n} - c$$

for $u > 1$ and 0 for $u < 1$. The proof of von Mangoldt's formula is simply the proof that these termwise inversions are valid.

10.7 PARSEVAL'S EQUATION

One of the basic theorems of Fourier analysis is Parseval's theorem, which states that under suitable conditions Fourier transform is a unitary transformation; that is, if $\hat{f}$ is the Fourier transform of f

$$\hat{f}(u) = \int_{-\infty}^{\infty} f(x)e^{2\pi ixu}\, dx,$$

then

(1) $$\int_{-\infty}^{\infty} |f(x)|^2\, dx = \int_{-\infty}^{\infty} |\hat{f}(x)|^2\, dx.$$

This statement of the theorem deals of course with Fourier transforms relative to the additive structure of the real numbers, but it can easily be translated into a theorem concerning Fourier transforms on the multiplicative group R^+ as follows.

If $\Phi(s)$ is the transform of the invariant operator $f(x) \mapsto \int_0^\infty f(ux)\phi(u)\, du$, that is, if

$$\Phi(s) = \int_0^\infty u^{-s}\phi(u)\, du,$$

then

$$\begin{aligned}\Phi(a+it) &= \int_0^\infty u^{-a-it}\phi(u)\, du \\ &= \int_0^\infty e^{-it\log u}u^{1-a}\phi(u)\, d\log u \\ &= \int_{-\infty}^{\infty} e^{-itv}e^{(1-a)v}\phi(e^v)\, dv,\end{aligned}$$

$$\Phi(a - 2\pi ix) = \int_{-\infty}^{\infty} e^{(1-a)v}\phi(e^v)e^{2\pi ixv}\, dv;$$

so (1) with $f(x) = e^{(1-a)x}\phi(e^x)$ and consequently with $\hat{f}(u) = \Phi(a - 2\pi iu)$ gives

$$\int_{-\infty}^{\infty} |\Phi(a - 2\pi ix)|^2\, dx = \int_{-\infty}^{\infty} e^{2(1-a)x}|\phi(e^x)|^2\, dx,$$

$$\frac{-1}{2\pi i}\int_{a+2\pi i\infty}^{a-2\pi i\infty} |\Phi(s)|^2\, ds = \int_{0}^{\infty} e^{2(1-a)\log u}|\phi(e^{\log u})|^2\, d\log u,$$

$$\frac{1}{2\pi i}\int_{a-i\infty}^{a+i\infty} |\Phi(s)|^2\, ds = \int_{0}^{\infty} u^{1-2a}|\phi(u)|^2\, du,$$

and in particular

$$\frac{1}{2\pi i}\int_{(1/2)-i\infty}^{(1/2)+i\infty} |\Phi(s)|^2\, dx = \int_{0}^{\infty} |\phi(u)|^2\, du.$$

This theorem is used in Chapter 11 in the study of zeros of $\xi(s)$ on the line $\operatorname{Re} s = \frac{1}{2}$. For a proof of the theorem in the needed cases see Bochner [B5] or Titchmarsh [T7].

10.8 THE VALUES OF $\zeta(-n)$

The zeta function can be evaluated at negative integers as follows. Consider the operator whose transform is $(1 - s)\zeta(s)$, namely, the composition of $f(x) \mapsto d[xf(x)]/dx$ with the operator $f(x) \mapsto \sum_{n=1}^{\infty} f(nx)$. Since $(1 - s)\zeta(s)$ is an entire function, this operator is defined, at least formally, for all functions of the form $f(x) = x^{-s}$. Consider the effect of this operator on $f(x) = e^{-x}$. This can be found in two different ways as follows.

On the one hand $f(x) = \sum_{n=0}^{\infty} (-1)^n x^n/n!$, so if the operator whose transform is $(1 - s)\zeta(s)$ is applied termwise to this series, one finds the function

$$\sum_{n=0}^{\infty} \frac{(-1)^n}{n!}(1 + n)\zeta(-n)x^n$$

as the resulting function because the operator multiplies $x^{-(-n)}$ by $[1 - (-n)]$ $\zeta(-n)$. On the other hand the summation operator carries $f(x) = e^{-x}$ to

$$e^{-x} + e^{-2x} + e^{-3x} + \cdots = \frac{1}{e^x - 1},$$

and its composition with $f(x) \mapsto d[xf(x)]/dx$ carries it to

$$\frac{d}{dx}\frac{x}{e^x - 1} = \frac{d}{dx}\sum_{n=0}^{\infty}\frac{B_n x^n}{n!} = \sum_{n=1}^{\infty}\frac{nB_n x^{n-1}}{n!} = \sum_{n=0}^{\infty}\frac{B_{n+1}x^n}{n!}$$

by the definition of the Bernoulli numbers [see (1) of Section 1.5]. Thus, equating the coefficients of x^n in the two expressions, one finds

$$(-1)^n(n + 1)\zeta(-n) = B_{n+1},$$

$$\zeta(-n) = (-1)^n[B_{n+1}/(n + 1)] \qquad (n = 0, 1, 2, \ldots)$$

which agrees with the value found in Section 1.5.

However, it must be admitted that this argument is very far from being a rigorous proof. For one thing, the series expansion of $x(e^x - 1)^{-1}$ in terms of Bernoulli numbers is valid only for $|x| < 2\pi$. Actually the evaluation of $\zeta(-n)$ using Riemann's integral as in Section 1.5 can be regarded as a method of making mathematical sense out of the above nonsense.

10.9 MÖBIUS INVERSION

The Möbius inversion formula is simply the inverse transform of the Euler product formula

$$\zeta(s)\prod_p\left(1-\frac{1}{p^s}\right)=1$$

in that it states that the summation operator $f(x)\mapsto\sum f(nx)$ with transform $\zeta(s)$ can be inverted by composing it with the operators $f(x)\mapsto f(x)-f(px)$ with transform $1-p^{-s}$, where p ranges over all primes. If $\sum f(nx)$ converges absolutely (and in particular if f is zero for all sufficiently large x), this follows easily from the fact that after a finite number of steps the above operations reduce $\sum f(nx)$ to $\sum f(kx)$, where k ranges over all integers not divisible by any of the primes that have been used; since this means that the first k past $k = 1$ is very large, it implies that $\sum f(kx)$ approaches $f(x)$ as more and more primes are used (see note, Section 1.17).

The inverse transform of the expanded product

$$\prod_p\left(1-\frac{1}{p^s}\right)=\sum_{n=1}^{\infty}\frac{\mu(n)}{n^s}$$

states that the composition of the operators $f(x)\mapsto f(x)-f(px)$ over all primes p can also be written in the form $f(x)\mapsto\sum_{n=1}^{\infty}\mu(n)f(nx)$. [Here $\mu(n)$ is zero unless n is a product of distinct prime factors, is 1 if n is a product of an even number of distinct prime factors, and is -1 if n is a product of an odd number of distinct prime factors.] This too is very easily proved in the case where $\sum f(nx)$ is absolutely convergent. Thus the Möbius inversion formula can be written in the form

$$g(x)=\sum_{n=1}^{\infty}f(nx)\Longleftrightarrow f(x)=\sum_{n=1}^{\infty}\mu(n)g(nx)$$

provided $\sum f(nx)$ and $\sum g(nx)$ both converge absolutely. Yet another statement of it is

$$f(x)=\sum_{n=1}^{\infty}\sum_{m=1}^{\infty}\mu(n)f(mnx)$$

under suitable conditions on f. For example, if $\sum\sum f(mnx)$ converges absolutely, then the double series can be rearranged,

$$f(x)=\sum_{N=1}^{\infty}\left[\sum_{m|N}\mu(m)\right]f(Nx), \tag{1}$$

and the inversion formula is equivalent to the identity

$$\sum_{m|N} \mu(m) = \begin{cases} 1 & \text{if} \quad N = 1, \\ 0 & \text{otherwise,} \end{cases}$$

which can be proved by applying Möbius inversion in the form (1) to the function $f(x)$ which is 1 for $x \leq 1$ and 0 for $f(x) > 1$, and then setting $x = 1$, $x = \frac{1}{2}$, $x = \frac{1}{3}$, etc. In Chapter 12 the slightly different statement

$$g(x) = \sum_{n=1}^{\infty} f\left(\frac{x}{n}\right) \Longleftrightarrow f(x) = \sum_{n=1}^{\infty} \mu(n) g\left(\frac{x}{n}\right)$$

of Möbius inversion will be needed.

10.10 RAMANUJAN'S FORMULA

Hardy, in his book [H4] on Ramanujan, states that Ramanujan "was especially fond and made continual use" of the formula

$$\int_0^\infty x^{s-1}[\phi(0) - x\phi(1) + x^2\phi(2) - \cdots]\, dx = \frac{\pi}{\sin \pi s}\phi(-s).$$

In the discussion that follows it will be convenient to recast this in the equivalent form

$$(1) \qquad \int_0^\infty x^{-s}\left[\phi(1) - x\phi(2) + \frac{x^2}{2!}\phi(3) - \cdots + (-1)^n \frac{x^n}{n!}\phi(n+1) + \cdots\right] dx = \Pi(-s)\phi(s)$$

in which s has been replaced by $1 - s$, $\phi(s)$ by $\phi(s+1)/\Pi(s)$, and $\pi(\sin \pi s)^{-1}$ by $\Pi(-s)\Pi(s-1)$ [see (6) of Section 1.3]. In this form Ramanujan's formula can be deduced from

$$(2) \qquad \int_0^\infty x^{-s} e^{-x}\, dx = \Pi(-s)$$

by observing that application of an operator with transform $\phi(s)$ to the first factor x^{-s} of the integrand on the one hand multiplies the integral by $\phi(s)$ but on the other hand is the same as application of an operator with transform $\phi(1 - \bar{s})$ (the conjugate of the adjoint) to the second factor; since $e^{-x} = \sum_{n=0}^{\infty} (-1)^n x^n/n!$ and since an operator with transform $\phi(1 - \bar{s})$ multiplies $x^n = x^{-(-n)}$ by $\phi(1 - (-n)) = \phi(n+1)$, this gives formula (1).

This heuristic argument is of course not a proof, but it does show how the formula can be proved for certain functions $\phi(s)$. For example, if $\phi(s) = a^{-s}$ for $a > 0$, then the integral on the left side of (1) is simply $\int_0^\infty x^{-s} e^{-x/a} a^{-1}\, dx = \int_0^\infty (ay)^{-s} e^{-y}\, dy = \phi(s)\Pi(-s)$ as was to be shown. Note that the integral is convergent only for Re $s < 1$ and that the right side of (1) gives an analytic continuation of the value of the integral past the pole at $s = 1$.

As a second example consider the binomial coefficient function $\phi(s) = \binom{-s}{n}$ for a fixed integer $n \geq 0$. This is the transform of the operator $f(x) \mapsto x^n/n!\, d^n f(x)/dx^n$ and the right side of (1) can be expressed as an integral $1/n! \int_0^\infty [x^n\, d^n x^{-s}/dx^n]e^{-x}\, dx$ for Re $s < 1$. Integration by parts is valid for Re $s < 1$ and puts this integral in the form $(-1)^n/n! \int_0^\infty x^{-s}\, (d^n[x^n e^{-x}]/dx^n)\, dx$. Since by simple termwise operations on power series

$$\begin{aligned}
&\frac{(-1)^n}{n!}\frac{d^n}{dx^n}[x^n e^{-x}] \\
&\quad= \frac{(-1)^n}{n!}\frac{d^n}{dx^n}\sum_{m=0}^{\infty}\frac{(-1)^m x^{m+n}}{m!} \\
&\quad= \sum_{m=0}^{\infty}\frac{(-1)^n(m+n)(m+n-1)\cdots(m+1)}{n!}\cdot\frac{(-1)^m x^m}{m!} \\
&\quad= \sum_{m=0}^{\infty}\binom{-m-1}{n}\frac{(-1)^m x^m}{m!},
\end{aligned}$$

this puts the integral in the desired form and shows that Ramanujan's formula (1) is valid for this ϕ. Note that the integral again converges only for Re $s < 1$ and that the right side of (1) gives the analytic continuation of the value of the integral past the pole at $s = 1$. Since it is linear in ϕ, Ramanujan's formula is true in the same sense for any linear combination of the polynomials $\phi(s) = \binom{-s}{n}$ and hence for any polynomial $\phi(s)$.

As a third example consider the case $\phi(s) = \Pi(s-1)$. In this case the series in the integral is $\sum_{n=0}^{\infty}\phi(n+1)(-x)^n/n! = \sum_{n=0}^{\infty}(-x)^n = (1+x)^{-1}$, so the integral is $\int_0^\infty x^{-s}(1+x)^{-1}\, dx$, which is convergent for $0 < \text{Re } s < 1$. Ramanujan's formula says that the value of this integral should be $\Pi(-s)\Pi(s-1)$, a fact which is easily proved by applying the operator $f(x) \mapsto \int_0^\infty f(ux)e^{-u}\, du$ with transform $\Pi(-s)$ to the second factor of the integrand in $\Pi(-s) = \int_0^\infty x^{-s}e^{-x}\, dx$ and evaluating the result in two ways to find

$$\begin{aligned}
\int_0^\infty x^{-s}\left(\int_0^\infty e^{-ux}e^{-u}\, du\right) dx &= \int_0^\infty x^{-s}\left.\frac{e^{-u(x+1)}}{-(x+1)}\right|_{u=0}^{u=\infty} dx \\
&= \int_0^\infty x^{-s}(x+1)^{-1}\, dx
\end{aligned}$$

and

$$\begin{aligned}
&\int_0^\infty x^{-s}\left(\int_0^\infty e^{-ux}e^{-u}\, du\right) dx \\
&\quad= \int_0^\infty e^{-u}\left(\int_0^\infty x^{1-s}e^{-ux}\, d\log x\right) du \\
&\quad= \int_0^\infty e^{-u}\left(u^{s-1}\int_0^\infty y^{1-s}e^{-y}\, d\log y\right) du \\
&\quad= \Pi(s-1)\Pi(-s)
\end{aligned}$$

which proves the desired equation.

Ramanujan's formula fails in the case of the function $\phi(s) = \sin \pi s$ because in this case the integral is identically zero, but $\Pi(-s)\phi(s)$ is not. However, if the formula is regarded as a method of extending a given function $\phi(1), \phi(2), \phi(3), \ldots$ defined at positive integers to an analytic function $\phi(s)$ defined for all s (or as many s as possible), then the formula works very well, extending the function $0 = \phi(1) = \phi(2) = \phi(3) = \cdots$ to the function $\phi(s) \equiv 0$ rather than to the more complicated function $\phi(s) = \sin \pi s$.

Now if $\phi(s)$ is an analytic function, then $\Pi(-s)\phi(s)$ has poles at positive integers $s = n$, and the residues of these poles are

$$\begin{aligned}\lim_{s\to n}(s-n)\Pi(-s)\phi(s) &= \lim_{s\to n}(s-n)\frac{\Pi(n-s)\phi(s)}{(n-s)(n-1-s)\cdots(1-s)} \\ &= -1\frac{\Pi(0)\phi(n)}{(-1)(-2)\cdots(1-n)} \\ &= (-1)^n\frac{\phi(n)}{(n-1)!}\end{aligned}$$

and, conversely, if $F(s)$ is a function with simple poles of residue $(-1)^n[\phi(n)/(n-1)!]$ at positive integers n, then $F(s)/\Pi(-s) = \phi(s)$ defines a function with values $\phi(n)$ at positive integers. Thus Ramanujan's formula can be regarded as the statement that the analytic function defined by

$$\int_0^\infty x^{-s}\left[\phi(1) - \phi(2)x + \frac{\phi(3)x^2}{2!} - \cdots\right] dx$$

has an analytic continuation [if $\phi(1) \neq 0$, this integral does not converge for $\operatorname{Re} s > 1$] with poles at $s = 1, 2, 3, \ldots$ with residues $-\phi(1), \phi(2), -\phi(3)/2!, \ldots$. In this way Ramanujan's formula becomes—provided the series $\sum_{n=0}^\infty \phi(n+1)(-x)^n/n!$ has a sum which is $O(x^{-\alpha})$ for some positive α as $x \to \infty$—a special case of the following theorem.

Theorem Let $\Phi(x)$ be a continuous function on the positive real axis which satisfies $\Phi(x) = O(x^{-\alpha})$ for some $\alpha > 0$ as $x \to \infty$ and which has an asymptotic expansion $\Phi(x) \sim \sum_{n=0}^\infty a_n x^n$ as $x \downarrow 0$. Then the analytic function $F(s)$ defined in the strip $\{1 - \alpha < \operatorname{Re} s < 1\}$ by the integral $F(s) = \int_0^\infty x^{-s}\Phi(x)\,dx$ has an analytic continuation to the entire halfplane $\{1 - \alpha < \operatorname{Re} s\}$ with no singularities other than simple poles at positive integers n, and the residue of the pole at n is $-a_{n-1}$.

Proof The integral

$$F_n(s) = \int_0^\infty x^{-s}[\Phi(x) - a_0 - a_1 x - \cdots - a_{n-1}x^{n-1}]\,dx$$

is convergent at ∞ if $\operatorname{Re} s > n$ and convergent at 0 if $\operatorname{Re} s < n + 1$ (because the integrand is like $x^{-s}a_n x^n$ as $x \longrightarrow 0$), so it defines an analytic function $F_n(s)$ in the strip $\{n < \operatorname{Re} s < n + 1\}$. In the strip $\{n - 1 < \operatorname{Re} s < n + 1\}$ the function

$$\int_0^1 x^{-s}[\Phi(x) - a_0 - \cdots - a_{n-1}x^{n-1}]\,dx + \frac{a_{n-1}}{n-s}$$
$$+ \int_1^{\infty} x^{-s}[\Phi(x) - a_0 - \cdots - a_{n-2}x^{n-2}]\,dx$$

is defined and analytic, except for a simple pole with residue $-a_{n-1}$ at $s = n$, and agrees with $F_{n-1}(s)$ and $F_n(s)$ in their respective strips of definition. Thus the functions $F_n(s)$ are all analytic continuations of each other and in the same way $F_1(s)$ is an analytic continuation of $F(s)$. Since the analytic function they all define has the stated properties, this proves the theorem. [If Φ is analytic on a neighborhood of the positive real axis and analytic at 0, then the theorem can also be proved quite easily using Riemann's method of Section 1.4 of considering the integral $\int_{+\infty}^{+\infty} (-x)^{-s}\Phi(x)\,dx$.]

Applying a trivial modification of this theorem to Abel's formula

$$\Pi(-s)\zeta(1-s) = \int_0^{\infty} x^{-s}\left(\frac{1}{e^x - 1}\right) dx \qquad (\operatorname{Re} s < 0) \tag{3}$$

[(1) of Section 1.4] shows that since

$$(e^x - 1)^{-1} = \frac{1}{x} - \frac{1}{2} + \frac{B_2 x}{2!} + \frac{B_4 x^3}{4!} - \cdots,$$

the function $\Pi(-s)\zeta(1 - s)$ has an analytic continuation to the entire complex plane with simple poles at $0, 1, 2, \ldots$ having residues $-1, \frac{1}{2}, -B_2/2, 0, -B_4/4!, \ldots, -B_n/n!, \ldots$. Thus $\zeta(1 - s)$ has an analytic continuation which has a simple pole with residue -1 at $s = 0$ but which is analytic at all positive integers n and has the value

$$-\frac{B_n}{n!} \cdot \frac{(n-1)!}{(-1)^n} = (-1)^{n-1}\frac{B_n}{n}$$

at n. Thus the theorem very easily gives the analytic continuation of ζ and its values at $0, -1, -2, \ldots$.

Consider now the application of Ramanujan's formula to guess the values of ζ on the basis of the values

$$\zeta(2n) = \frac{(2\pi)^{2n}(-1)^{n+1}B_{2n}}{2(2n)!} \qquad (n = 1, 2, \ldots)$$

found by Euler [(2) of Section 1.5]. Setting $\phi(s) = \zeta(2s)$ in Ramanujan's formula leads to a series which cannot be summed in any obvious way, but setting $\phi(s) = \Pi(s-1)\zeta(2s)$ leads to the series

$$\sum_{n=1}^{\infty} \phi(n)\frac{(-1)^{n-1}x^{n-1}}{(n-1)!} = \frac{1}{x}\sum_{n=1}^{\infty} \frac{(2\pi)^{2n}(-1)^{n+1}B_{2n}}{2(2n)!}(-1)^{n-1}x^n$$
$$= \frac{1}{2x}\sum_{m=2}^{\infty} \frac{(2\pi x^{1/2})^m B_m}{m!}$$

(because $B_3 = B_5 = \cdots = 0$) which has the sum

$$\frac{1}{2x}\left(\frac{2\pi x^{1/2}}{\exp(2\pi x^{1/2}) - 1} - 1 + \frac{2\pi x^{1/2}}{2}\right).$$

Thus Ramanujan's formula would give

(4) $$\int_0^{\infty} x^{-s}\frac{1}{2x}\left(\frac{2\pi x^{1/2}}{\exp(2\pi x^{1/2}) - 1} - 1 + \frac{2\pi x^{1/2}}{2}\right) dx = \Pi(-s)\Pi(s-1)\zeta(2s)$$

if it were true in this case. The theorem proved above shows that this formula gives the correct values of $\zeta(2n)$ $(n = 1, 2, 3, \ldots)$ but does not show that it gives the correct value of $\zeta(s)$ for other values of s. However, the integral on the left can be rewritten

$$\int_0^{\infty} \left(\frac{y}{2\pi}\right)^{-2s}\left(\frac{y}{e^y - 1} - 1 + \frac{y}{2}\right) d\log y$$
$$= (2\pi)^{2s}\int_0^{\infty} y^{-2s}\left(\frac{1}{e^y - 1} - \frac{1}{y} + \frac{1}{2}\right) dy$$

which shows—by the principle of Section 10.5—that it has the analytic continuation

(5) $$(2\pi)^{2s}\int_0^{\infty} y^{-2s}\left(\frac{1}{e^y - 1}\right) dy$$

and therefore, by Abel's formula (3), that it is $(2\pi)^{2s}\Pi(-2s)\zeta(1-2s)$. The method of Section 10.5 and of the proof of the theorem above can be used to prove very easily that (5) is indeed an analytic continuation of the integral in (4) and therefore to prove that *Ramanujan's formula in the case* (4) *is equivalent to*

$$(2\pi)^{2s}\Pi(-2s)\zeta(1-2s) = \Pi(-s)\Pi(s-1)\zeta(2s)$$

which is the functional equation of the zeta function [see (4) of Section 1.6]. Thus Ramanujan's formula does hold in this case even though the theorem above does not suffice to prove it.

As a final example, consider the case $\phi(s) = \Pi(s)\zeta(1-s)$. In this case the series is $\sum_{n=0}^{\infty} (n+1)!\zeta(-n)(-1)^n x^n/n! = \sum_{n=0}^{\infty} B_{n+1}x^n$ and Ramanujan's formula takes the form

(6) $$\int_0^{\infty} x^{-s}(-\tfrac{1}{2} + B_2 x + B_4 x^3 + \cdots)\, dx = \Pi(-s)\Pi(s)\zeta(1-s).$$

This formula is meaningless as it stands because the power series is divergent for all $x \neq 0$. However, it is an asymptotic expansion for small x of a function which can be identified and the integral can be made meaningful as follows. By Stirling's formula

$$\log \Pi(x) \sim \left(x + \frac{1}{2}\right) \log x - x + \frac{1}{2} \log 2\pi + \frac{B_2}{2x} + \frac{B_4}{4 \cdot 3x^3} + \cdots,$$

$$\frac{\Pi'(x)}{\Pi(x)} \sim \log x + \frac{1}{2x} - \frac{B_2}{2x^2} - \frac{B_4}{4x^4} - \frac{B_6}{6x^6} - \cdots,$$

$$\frac{d}{dx}\left[\frac{\Pi'(x)}{\Pi(x)} - \log x\right] \sim -\frac{1}{2x^2} + \frac{B_2}{x^3} + \frac{B_4}{x^5} + \frac{B_6}{x^7} + \cdots.$$

On the other hand the change of variable $y = x^{-1}$ in the integral of (6) puts it in the form

$$\int_0^\infty y^{s-1}\left(\frac{1}{2} + B_2 y^{-1} + B_4 y^{-3} + \cdots\right) d \log y$$
$$= \int_0^\infty y^s \left(-\frac{1}{2y^2} + \frac{B_2}{y^3} + \frac{B_4}{y^5} + \cdots\right) dy$$

so (6) suggests the equation

$$\int_0^\infty y^s \frac{d}{dy}\left(\frac{\Pi'(y)}{\Pi(y)} - \log y\right) dy = \Pi(-s)\Pi(s)\zeta(1-s)$$

which by integration by parts [the function $\Pi'(y)/\Pi(y) - \log y$ is asymptotic to $(2y)^{-1}$ as $y \to \infty$ and asymptotic to $-\log y$ as $y \to 0$, so the integral is convergent for $0 < \operatorname{Re} s < 1$] is equivalent to

$$-s \int_0^\infty y^{s-1} \left(\frac{\Pi'(y)}{\Pi(y)} - \log y\right) dy = \Pi(-s)\Pi(s)\zeta(1-s)$$

or, with $s \to 1 - s$, equivalent to

$$\int_0^\infty y^{-s}\left(\log y - \frac{\Pi'(y)}{\Pi(y)}\right) dy = \frac{\pi}{\sin \pi s}\zeta(s), \tag{7}$$

a formula which is in fact true for all s in the strip $\{0 < \operatorname{Re} s < 1\}$ where the integral converges; see Titchmarsh [T8, formula (2.9.2)].

A stronger version of Ramanujan's formula (1) than the one embodied in the theorem above can be proved using Fourier inversion. Since this involves the behavior of $\phi(s)$ on lines $\operatorname{Re} s = \text{const}$ in the complex plane, it necessarily involves considering ϕ as a function of a complex variable and therefore, as Hardy observes, it lay outside the range of Ramanujan's ideas and techniques in a very essential way. For example, Fourier inversion can be used to prove the following theorem.

Theorem Let $F(s)$ be analytic in a halfplane $\{\operatorname{Re} s > 1 - \alpha\}$ (for $\alpha > 0$) except for simple poles at $s = 1, 2, 3, \ldots$ with residues $-a_0, -a_1, -a_2, \ldots$, respectively. Suppose, moreover, that the growth of $F(s)$ in the complex plane satisfies suitable conditions and in particular that $F(s) \longrightarrow 0$ very rapidly as $\operatorname{Im} s \longrightarrow \pm\infty$ along lines $\operatorname{Re} s = \text{const}$. Then $F(s)$ in the strip $\{1 - \alpha < \operatorname{Re} s < 1\}$ can be represented in the form $F(s) = \int_0^\infty x^{-s}\Phi(x)\,dx$, where $\Phi(x)$ is analytic for $x > 0$, where $\Phi(x) = O(x^{-\alpha+\epsilon})$ as $x \longrightarrow \infty$ (for every $\epsilon > 0$), and where $\Phi(x) \sim \sum_{n=0}^\infty a_n x^n$ is an asymptotic expansion of $\Phi(x)$ as $x \downarrow 0$. If, moreover, $F(s)$ does not grow too rapidly as $\operatorname{Re} s \longrightarrow \infty$, then $\Phi(x)$ is analytic at 0 and, consequently, has $\sum a_n x^n$ as its power series expansion near 0.

Proof The idea of the proof is simply the Fourier inversion formula

$$F(s) = \int_0^\infty x^{-s}\Phi(x)\,dx \Longleftrightarrow \Phi(x) = \frac{1}{2\pi i}\int_{c-i\infty}^{c+i\infty} F(s)x^{s-1}\,ds$$

[in (1) of Section 10.6 set $\phi(u) = u\Phi(u)$ and set $\Phi(s) = F(s)$]. Let c lie between $1 - \alpha$ and 1, and let the formula on the right define Φ. The assumption that $F(s) \longrightarrow 0$ rapidly as $s \longrightarrow c \pm i\infty$ guarantees that $\Phi(x)$ is then defined and analytic for all x in the slit plane and that $\Phi(x)$ is independent of the choice of c. Fourier inversion shows that $F(s)$ has the desired representation, and it remains only to show that $\Phi(x)$ has the stated properties as $x \longrightarrow \infty$ and as $x \downarrow 0$. Since $\Phi(x)$ is a superposition of functions x^{s-1} ($\operatorname{Re} s = c$) all of which are $O(x^{c-1+\epsilon})$ as $x \longrightarrow \infty$, the same is true—by passage to the limit under the integral sign—of $\Phi(x)$. Since c is any number greater than $1 - \alpha$, this gives $\Phi(x) = O(x^{-\alpha+\epsilon})$ as $x \longrightarrow \infty$. The integral of $(1/2\pi i)F(s)x^{s-1}\,ds$ over the boundary of the strip $\{c \leq \operatorname{Re} s \leq n + \frac{1}{2}\}$ is on the one hand the sum of the residues in the strip—which is $-a_0 - a_1 x - \cdots - a_{n-1}x^{n-1}$—and is on the other hand $1/2\pi i \int_{n+1/2-i\infty}^{n+1/2+i\infty} F(s)x^{s-1}\,ds - \Phi(x)$. This shows that $\Phi(x) - a_0 - a_1 x - \cdots - a_{n-1}x^{n-1}$ is a superposition of functions which are $O(x^{n-(1/2)})$ as $x \longrightarrow 0$, hence by passage to the limit under the integral sign that $\Phi(x) \sim a_0 + a_1 x + a_2 x^2 + \cdots$ is an asymptotic expansion. To prove that it actually converges to $\Phi(x)$ for small x, it suffices to prove that $\lim_{n\to\infty} \int_{n+1/2-i\infty}^{n+1/2+i\infty} F(s)\, e^{s\log x}\,ds$ is zero; since $e^{s\log x} \longrightarrow 0$ very rapidly for small x, this will be true if the decrease of $F(s)$ for $\operatorname{Im} s \longrightarrow \pm\infty$ is uniform as $\operatorname{Re} s \longrightarrow \infty$ and if $|F(s)|$ does not grow rapidly as $\operatorname{Re} s \longrightarrow \infty$.

In the case $F(s) = \pi/\sin \pi s$ the elementary estimate $F(c + it) \leq \text{const}\; e^{-\pi|t|}$ as $t \longrightarrow \infty$ (uniform in c) shows that the theorem applies and gives $\pi/\sin \pi s = \int_0^\infty x^{-s}[1 - x + x^2 - x^3 + \cdots]\,dx = \int_0^\infty x^{-s}(1 + x)^{-1}\,dx = \Pi(-s)\Pi(s - 1)$ (see above), that is, the theorem gives the product formula for the sine (6) of Section 1.3. In the case $F(s) = \Pi(-s)\Pi(s - 1)\zeta(2s) = (\pi/\sin \pi s)\zeta(2s)$, the fact that $\zeta(2s)$ is bounded for $\operatorname{Re} s > \frac{3}{4}$ shows that the theorem applies to give formula (4) and hence the functional equation of the zeta function [given the formula for the values of $\zeta(2), \zeta(4), \zeta(6), \ldots$]. Formula (7) does not quite come under the theorem as stated because $(\pi/\sin \pi s)\,\zeta(s)$ has a double pole at $s = 1$. However, the same methods can be applied to prove that

$$\frac{\pi}{\sin \pi s}\zeta(s) = \int_0^\infty x^{-s}\left(\log x + \gamma + \sum_{n=1}^\infty \zeta(n+1)(-x)^n\right) dx,$$

so (7) is equivalent to the elementary formula $-\Pi'(x)/\Pi(x) = \gamma + \sum_{n=1}^\infty \zeta(n+1)(-x)^n$; this formula can be proved by taking the logarithmic derivative of Euler's formula (4) of Section 1.3 and using $(x+n)^{-1} = n^{-1}\sum_{m=0}^\infty (-x/n)^m$.

Chapter 11

Zeros on the Line

11.1 HARDY'S THEOREM

In 1914 Hardy [H3] proved that *there are infinitely many roots* ρ *of* $\xi(s) = 0$ *on the line* $\operatorname{Re} s = \frac{1}{2}$. Except for the numerical work of Gram and Backlund, this was the first concrete result concerning zeros on the line. As was stated in Section 1.9, Hardy and Littlewood [H6] later proved—in 1921—that *the number of roots on the line segment from* $\frac{1}{2}$ *to* $\frac{1}{2} + iT$ *is at least* KT for some positive constant K and all sufficiently large T, and still later—in 1942—Selberg [S1] proved that the *number of such roots is at least* $KT \log T$ for some positive constant K and all sufficiently large T. This chapter is devoted to the proofs of these three theorems. Note that each of them supersedes the preceding one so that logically it would suffice to prove just Selberg's estimate. However, each of the three proofs is essentially an elaboration of the preceding one, so it is natural—both logically and historically—to prove all three. The proofs given here follow those of Titchmarsh's book [T8]. Although the basic ideas of these proofs are essentially the same as in the originals, Titchmarsh has simplified and clarified them considerably.

The idea of the proof of Hardy's theorem is to apply Fourier inversion (see Section 10.6) to one of the expressions of ξ as a transform (see Section 10.5), say

$$\frac{2\xi(s)}{s(s-1)} = \int_0^\infty u^{-s}\left[G(u) - 1 - \frac{1}{u}\right] du \qquad (0 < \operatorname{Re} s < 1),$$

to find a formula such as

$$(1) \qquad G(x) - 1 - \frac{1}{x} = \frac{1}{2\pi i}\int_{a-i\infty}^{a+i\infty} \frac{2\xi(s)}{s(s-1)} x^{s-1}\, ds \qquad (0 < a < 1).$$

The applicability of Fourier inversion in this case follows from the most basic theorems of the theory of Fourier integrals—see, for example, Taylor [T2]. With $a = \frac{1}{2}$ the right side of this equation is an integral involving the function $\xi(\frac{1}{2} + it)$ to be studied and the left side is a function about which a great deal was known in the nineteenth century. (It is essentially the function ψ which occurs in Riemann's second proof of the functional equation—see Section 1.7.) The idea of Hardy's proof is to use information about the function on the left to draw conclusions about the integrand on the right.

The function $G(x) = \sum_{-\infty}^{\infty} \exp(-\pi n^2 x^2)$ is defined whenever $\operatorname{Re} x^2 > 1$, which means that it is defined not only for positive real x as in (1) but also for complex values of x in the wedge $\{-\pi/4 < \operatorname{Im} \log x < \pi/4\}$. In this wedge it is of course an analytic function of the complex variable x. However, it very definitely has singularities on the boundary of the wedge, and in fact these singularities of (1) for complex x are what Hardy's proof uses. Specifically, *$G(x)$ and all its derivatives approach zero as x approaches $i^{1/2}$* $(= e^{i\pi/4})$. This fact about G, which originally was discovered in connection with the theory of θ-functions, can easily be proved as follows:

$$\begin{aligned} G(x) &= \sum_{-\infty}^{\infty} e^{-\pi n^2 x^2} = \sum_{-\infty}^{\infty} e^{-\pi n^2 i} e^{-\pi n^2 (x^2 - i)} = \sum_{-\infty}^{\infty} (-1)^n e^{-\pi n^2 (x^2 - i)} \\ &= -G((x^2 - i)^{1/2}) + 2G(2(x^2 - i)^{1/2}). \end{aligned}$$

The functional equation $G(x) = x^{-1}G(x^{-1})$ then gives

$$\begin{aligned} G(x) &= -\frac{1}{(x^2 - i)^{1/2}} G\left(\frac{1}{(x^2 - i)^{1/2}}\right) + \frac{2}{2(x^2 - i)^{1/2}} G\left(\frac{1}{2(x^2 - i)^{1/2}}\right) \\ &= \frac{1}{(x^2 - i)^{1/2}} \left[\sum_{-\infty}^{\infty} e^{-\pi n^2 2^{-2} (x^2 - i)^{-1}} - \sum_{-\infty}^{\infty} e^{-\pi n^2 (x^2 - i)^{-1}}\right] \\ &= \frac{1}{(x^2 - i)^{1/2}} \sum_{n = \text{odd}} e^{-\pi n^2 2^{-2} (x^2 - i)^{-1}}. \end{aligned}$$

Since $e^{-1/u}$ approaches zero as $u \downarrow 0$ more rapidly than any power of u does, this shows that $G(x)$ and all its derivatives approach zero as x approaches $i^{1/2}$ from within the wedge $\{-\pi/4 < \operatorname{Im} \log x < \pi/4\}$, say along the circle $|x| = 1$.

Consider now the integral on the right side of (1) for complex values of x. It will converge provided $\xi(a + it)$ goes to zero rapidly enough as $t \to \pm\infty$. Now $|\xi(s)| = |\Pi(s/2)\pi^{-s/2}(s - 1)\zeta(s)|$ for $s = a + it$ is easily estimated; $|\zeta(s)|$ grows less rapidly than a constant times t^2 as $t \to \pm\infty$ (see Section 6.7), $|s - 1|$ grows like $|t|$, $|\pi^{-s/2}|$ is constant, and $|\Pi(s/2)|$ grows like e^B where

$$\begin{aligned} B &= \operatorname{Re} \log \Pi\left(\frac{a + it}{2}\right) \\ &= \operatorname{Re}\left\{\left(\frac{a + it + 1}{2}\right) \log\left(\frac{a + it}{2}\right) - \frac{a + it}{2} + \cdots\right\} \\ &= \operatorname{Re}\left\{\left(\frac{a + it + 1}{2}\right) \log\left(\frac{it}{2}\right) + \cdots\right\} \end{aligned}$$

$$= \frac{a+1}{2}\log\frac{|t|}{2} - \frac{t}{2}\operatorname{Im}\log\left(\frac{it}{2}\right) + \cdots$$
$$= \frac{a+1}{2}\log|t| - \frac{|t|\pi}{4} + \cdots,$$

where the omitted terms remain bounded as $t \to \pm\infty$. Thus $|\xi(a+it)|$ is less than a power of $|t|$ times $e^{-|t|\pi/4}$ as $t \to \pm\infty$. Since the factor $x^{s-1} = x^{a-1}x^{it} = \text{const } e^{it\log x}$ grows like $\exp(\pm|t||\operatorname{Im}\log x|)$ as $t \to \pm\infty$, this shows that the decrease of $\xi(s)$ overwhelms the increase of x^{s-1} in the integral (1) provided $|\operatorname{Im}\log x| < \pi/4$ and hence that *the integral* (1) *converges throughout the wedge* $\{-\pi/4 < \operatorname{Im}\log x < \pi/4\}$. By analytic continuation, then, formula (1) remains valid throughout this wedge.

Formula (1) takes a simpler form if the operator $x(d^2/dx^2)x$ is applied to both sides to give

$$H(x) = \frac{1}{2\pi i}\int_{a-i\infty}^{a+i\infty} 2\xi(s)x^{s-1}\,ds \qquad (0 < a < 1).$$

Clearly $H(x) = x(d^2/dx^2)x\,G(x)$ has, like G, the property that it and all its derivatives approach zero as $x \to i^{1/2}$. Moreover, the above estimates of the integrand justify termwise integration to give, when $a = \frac{1}{2}$,

$$H(x) = \frac{1}{\pi}\int_{-\infty}^{\infty}\xi\left(\frac{1}{2}+it\right)x^{-1/2}x^{it}\,dt,$$
$$x^{1/2}H(x) = \frac{1}{\pi}\int_{-\infty}^{\infty}\xi\left(\frac{1}{2}+it\right)\sum_{0}^{\infty}\frac{(it\log x)^n}{n!}\,dt$$
$$= \sum_{0}^{\infty} c_n(i\log x)^n,$$

where

$$c_n = \frac{1}{\pi n!}\int_{-\infty}^{\infty}\xi\left(\frac{1}{2}+it\right)t^n\,dt.$$

The integrals c_n are zero for odd n by the symmetry of ξ. If Hardy's theorem were *false*, that is, if there were only a finite number of zeros of $\xi(\frac{1}{2}+it)$, then $\xi(\frac{1}{2}+it)$ would have the same sign for all large t and one would expect—because of the high weight it places on large values of t—that c_{2n} would have this same sign for all sufficiently large n. This can be proved simply by observing that if $\xi(\frac{1}{2}+it)$ is positive for $t \geq T$, then

$$\begin{aligned}\Pi(2n)\pi c_{2n} &= 2\int_0^{\infty}\xi(\tfrac{1}{2}+it)t^{2n}\,dt\\ &\geq 2\int_0^{T+2}\xi(\tfrac{1}{2}+it)t^{2n}\,dt\\ &\geq 2\left\{-\int_0^{T}|\xi(\tfrac{1}{2}+it)|\,T^{2n}\,dt\right.\\ &\qquad\left.+\int_{T+1}^{T+2}\xi(\tfrac{1}{2}+it)(T+1)^{2n}\,dt\right\}\\ &\geq \text{const }(T+1)^{2n} - \text{const }(T)^{2n}\end{aligned}$$

is positive for all sufficiently large n and similarly that if $\xi(\frac{1}{2}+it)$ is negative for $t \geq T$, then c_{2n} is negative for all sufficiently large n. Thus if the above formula for $x^{1/2}H(x)$ is differentiated sufficiently many times with respect to $i \log x$, then the right side becomes an even power series in which all terms have the same sign. Thus if $x \longrightarrow i^{1/2}$ upward along the circle $|x| = 1$, it follows that $i \log x \downarrow -\pi/4$ through real values and the value of this even power series cannot approach zero. On the other hand it must approach zero because to differentiate with respect to $i \log x$ is the same as to apply $ix(d/dx)$, and doing this any number of times carries $x^{1/2}H(x)$ to a function which approaches zero as $x \longrightarrow i^{1/2}$. This contradiction proves Hardy's theorem.

Another proof of Hardy's theorem which is worthy of mention is that of Titchmarsh [T4]. Titchmarsh showed that in using the Riemann–Siegel formula† at Gram points g_n *on average* only the first term

$$Z(g_n) = 2\cos\vartheta(g_n) + \cdots = (-1)^n \cdot 2 + \cdots$$

counts. More specifically, he proved that

$$\lim_{N\to\infty} \frac{1}{N}\sum_{n=1}^{N} Z(g_{2n}) = 2, \qquad \lim_{N\to\infty} \frac{1}{N}\sum_{n=1}^{N} Z(g_{2n+1}) = -2.$$

This of course proves that Z must change sign infinitely often and hence proves Hardy's theorem. It also proves that on the average Gram's law is true in the strong sense that

$$\lim_{N\to\infty} \frac{1}{N}\sum_{n=1}^{N} \operatorname{Re}\zeta\left(\frac{1}{2} + ig_n\right) = 2$$

since $\zeta(\frac{1}{2}+ig_n) = \operatorname{Re}\zeta(\frac{1}{2}+ig_n) = (-1)^n Z(g_n)$.

11.2 THERE ARE AT LEAST *KT* ZEROS ON THE LINE

The proof of the fact that there are positive constants K, T_0 such that the number of roots ρ on the line segment from $\frac{1}{2}$ to $\frac{1}{2}+iT$ is at least KT whenever $T \geq T_0$ begins, as did the proof of the preceding section, with the formula

$$\frac{2\xi(s)}{s(s-1)} = \int_0^\infty u^{-s}\left[G(u) - 1 - \frac{1}{u}\right] du$$

(valid for s in the so-called critical strip $0 < \operatorname{Re} s < 1$). The proof of the preceding section depended on the fact that the integral

$$\frac{1}{2\pi i}\int_{1/2-i\infty}^{1/2+i\infty} \frac{2\xi(s)}{s(s-1)} x^{s-1}\, ds$$

†However, as with the Hardy–Littlewood estimates of $\zeta(\frac{1}{2}+it)$ described in Section 9.8, this work preceded publication of the Riemann–Siegel formula and was based instead on the so-called approximate functional equation.

approaches zero very rapidly as x approaches $i^{1/2}$ from below along the circle $|x| = 1$. The present proof depends on a *local* study of this integral, that is, on a study of the integral over finite intervals of the line $\operatorname{Re} s = \frac{1}{2}$. Let s denote the midpoint of the interval under consideration, let $2k$ denote its length, and let

$$I_{x,k}(s) = \frac{1}{2\pi i}\int_{s-ik}^{s+ik} \frac{2\xi(v)}{v(v-1)} x^{v-1}\,dv. \tag{1}$$

Then $I_{x,k}(s)$ can be rewritten in the form

$$\begin{aligned}
&\frac{1}{2\pi i}\int_{s-ik}^{s+ik}\int_0^\infty \left(\frac{u}{x}\right)^{-v}\left[G(u) - 1 - \frac{1}{u}\right]x^{-1}\,du\,dv \\
&= \frac{1}{2\pi i}\int_{s-ik}^{s+ik}\int_0^{\infty/x} w^{-v}\left[G(xw) - 1 - \frac{1}{xw}\right]dw\,dv \\
&= \frac{1}{2\pi i}\int_{s-ik}^{s+ik}\int_0^{\infty} w^{-v}\left[G(xw) - 1 - \frac{1}{xw}\right]dw\,dv \\
&= \int_0^\infty \left(\frac{1}{2\pi i}\int_{s-ik}^{s+ik} w^{-v}\,dv\right)\left[G(xw) - 1 - \frac{1}{xw}\right]dw \\
&= \frac{1}{\pi}\int_0^\infty \frac{w^{-s}\sin(k\log w)}{\log w}\left[G(xw) - 1 - \frac{1}{xw}\right]dw
\end{aligned}$$

[where use is made of the fact that $G(u) - 1$ approaches zero very rapidly as u goes to infinity along any ray $u = xw$ in the wedge $-(\pi/4) \le \operatorname{Im}\log x \le (\pi/4)$, $w =$ real]. This expresses $I_{x,k}(s)$ as the transform of an operator and shows, by virtue of Parseval's theorem (Section 10.7), that

$$\begin{aligned}
&\frac{1}{2\pi i}\int_{(1/2)-i\infty}^{(1/2)+i\infty} |I_{x,k}(s)|^2\,ds \\
&= \frac{1}{\pi^2}\int_0^\infty \left|\frac{\sin(k\log w)}{\log w}\right|^2 \left|G(xw) - 1 - \frac{1}{xw}\right|^2 dw.
\end{aligned} \tag{2}$$

The idea of the proof is to use this formula with explicit estimates of the function G to find an upper bound for $\int |I|^2\,ds$ and to show that for suitable choices of x, k it is much smaller than it could be if $\xi(\frac{1}{2} + it)$ did not change sign frequently.

The first step, therefore, is to derive an upper estimate of $\int |I|^2\,ds$. Note first that the symmetries of G and $x^{-1} = \bar{x}$ imply that the integral on the right side of (2) is equal to twice the integral from 1 to ∞. [The factor $\sin(k\log w)/\log w$ is unchanged under $w \to w^{-1}$. The factor $G(xw) - 1 - (xw)^{-1}$ becomes $G(1/\bar{x}w) - 1 - \bar{x}w = \bar{x}w[(\bar{x}w)^{-1}G(1/\bar{x}w) - (\bar{x}w)^{-1} - 1] = \bar{x}w[G(\bar{x}w) - 1 - (\bar{x}w)^{-1}]$ under $w \to w^{-1}$, so the square of its modulus is multiplied by w^2. Since dw becomes $-dw/w^2$, it follows that the integral from 0 to 1 becomes the integral from 1 to ∞.] Now

$$\left|\frac{\sin ky}{y}\right| \le \begin{cases} k & \text{for}\quad 0 \le y < \pi/k, \\ y^{-1} & \text{for}\quad \pi/k \le y < \infty; \end{cases}$$

so this implies that

$$(3)\qquad \frac{1}{2\pi i}\int_{(1/2)-i\infty}^{(1/2)+i\infty}|I_{x,k}(s)|^2\,ds \le \frac{2k^2}{\pi^2}\int_1^{e^{\pi/k}}\left|G(xw)-1-\frac{1}{xw}\right|^2 dw$$
$$+\frac{2}{\pi^2}\int_{e^{\pi/k}}^{\infty}(\log w)^{-2}\left|G(xw)-1-\frac{1}{xw}\right|^2 dw.$$

The two integrals on the right can be estimated using the explicit formula for $G(u)-1$. The first integral will be considered first, after which the second integral is easily estimated by the same techniques.

The parallelogram law $2|A|^2+2|B|^2=|A+B|^2+|A-B|^2\ge |A+B|^2$ shows that the first integral on the right side of (3) is at most

$$\frac{4k^2}{\pi^2}\int_1^{e^{\pi/k}}|G(xw)-1|^2\,dw+\frac{4k^2}{\pi^2}\int_1^{e^{\pi/k}}\frac{dw}{w^2}.$$

The second integral here is simply $4k^2\pi^{-2}[1-e^{-\pi/k}]$, so it is of the order of magnitude of k^2 and it will suffice to estimate the first integral. Since $G(u)-1 = 2\sum_{n=1}^{\infty}\exp(-\pi n^2u^2)$, this integral can be written in the form

$$\frac{16k^2}{\pi^2}\int_1^{e^{\pi/k}}\sum_{n=1}^{\infty}\sum_{m=1}^{\infty}e^{-\pi n^2x^2w^2}e^{-\pi m^2x^{-2}w^2}\,dw.$$

Let $x=e^{i\pi/4}e^{-i\delta}$ so that $x^2=\sin 2\delta+i\cos 2\delta$ and this integral becomes

$$(4)\qquad \frac{16k^2}{\pi^2}\int_1^{e^{\pi/k}}\sum_{n=1}^{\infty}\sum_{m=1}^{\infty}e^{-\pi(n^2+m^2)w^2\sin 2\delta}e^{-i\pi(n^2-m^2)w^2\cos 2\delta}\,dw.$$

The double sum converges absolutely and can therefore be rearranged as three sums, one in which $m>n$, one in which $m=n$, and one in which $m<n$. The integral of the terms with $m=n$ is easily estimated by using

$$\sum_{n=1}^{\infty}e^{-\pi(2n^2)w^2\sin 2\delta}=\tfrac{1}{2}\{G[w(2\sin 2\delta)^{1/2}]-1\}.$$

Since $u[G(u)-1]$ is bounded both as $u\to\infty$ and as $u\to 0$, there is a constant K such that $G(u)-1<Ku^{-1}$ for all positive u and the terms of (4) with $m=n$ contribute at most

$$\frac{16k^2}{\pi^2}\int_1^{e^{\pi/k}}\frac{1}{2}\,\frac{K}{w(2\sin 2\delta)^{1/2}}\,dw=\frac{4\sqrt{2}\,Kk^2}{\pi^2(\sin 2\delta)^{1/2}}\log(e^{\pi/k})$$

which for small values of δ is less than a constant times $k\delta^{-1/2}$.

It will now be shown that the remaining terms $m\ne n$ of (4) are much smaller than $k\delta^{-1/2}$. The terms with $m>n$ are the complex conjugates of those with $m<n$, so it will suffice to estimate the latter. Termwise integration is easily justified so the quantity to be estimated is equal to the sum over all (m,n) with $m<n$ of

$$(5)\qquad \frac{16k^2}{\pi^2}\int_1^{e^{\pi/k}}e^{-\pi(n^2+m^2)w^2\sin 2\delta}e^{-i\pi(n^2-m^2)w^2\cos 2\delta}\,dw.$$

The real part of this integral is

$$\frac{16k^2}{\pi^2}\int_1^{e^{\pi/k}} f(w)\cos V(w)\,dw,$$

where $f(w) = \exp[-\pi(n^2 + m^2)w^2 \sin 2\delta]$ and $V(w) = \pi(n^2 - m^2)w^2 \cos 2\delta$. Now $\cos 2\delta$ is positive for small δ, so $V(w)$ is a monotone increasing function of w, and this integral can be written in terms of the variable V as

$$\frac{16k^2}{\pi^2}\int_{V(1)}^{V(e^{\pi/k})} \frac{f}{V'}\cos V\,dV,$$

where f and V' are functions of V by composition with the inverse function $V \longrightarrow w$. Since f is decreasing and V' is increasing, the lemma of Section 9.7 says that this integral is at most

$$\frac{16k^2}{\pi^2}\cdot 2\frac{f(1)}{V'(1)} = \frac{32k^2}{\pi^2}\,\frac{e^{-\pi(n^2+m^2)\sin 2\delta}}{2\pi(n^2 - m^2)\cos 2\delta}.$$

A similar estimate applies to the imaginary part and hence to the modulus of the integral (5). It follows that for small δ the total modulus of the terms to be estimated is at most a constant times

$$\sum_{n=1}^{\infty}\sum_{m<n}\frac{k^2e^{-\pi(n^2+m^2)\sin 2\delta}}{(n^2 - m^2)} \le \sum_{n=1}^{\infty}\sum_{m<n}\frac{k^2e^{-\pi n^2\sin 2\delta}}{(n+m)(n-m)}$$
$$\le \sum_{n=1}^{\infty}\frac{k^2e^{-\pi n^2\sin 2\delta}}{n}\sum_{m<n}\frac{1}{n-m}.$$

Now $\sum_{m<n} 1/(n - m) = \sum_{m<n} 1/m$ is less than a constant times $\log n$, so the quantity to be estimated is less than a constant times

$$k^2\sum_{n=1}^{\infty} e^{-\pi n^2 \sin 2\delta}\frac{\log n}{n}.$$

The function $\exp(-\pi u^2 \sin 2\delta)(\log u)/u$ is decreasing for $u > e$, so this sum is at most k^2 times

$$\frac{e^{-\pi 4\sin 2\delta}\log 2}{2} + \frac{e^{-\pi 9 \sin 2\delta}\log 3}{3} + \int_3^{\infty} e^{-\pi u^2 \sin 2\delta}\log u\, d\log u$$
$$\le \text{const} + \int_3^{(\sin 2\delta)^{-1/2}} e^{-\pi u^2 \sin 2\delta}\log u\, d\log u$$
$$+ \int_1^{\infty} e^{-\pi v^2}\log\left(\frac{v}{(\sin 2\delta)^{1/2}}\right) d\log v$$
$$\le \text{const} + \frac{1}{2}(\log u)^2\Big|_{u=3}^{(\sin 2\delta)^{-1/2}}$$
$$+ \frac{1}{2}\log\left(\frac{1}{\sin 2\delta}\right)\int_1^{\infty}\exp^{-\pi v^2} d\log v + \text{const}$$
$$\le \text{const} + \text{const}\left(\log\frac{1}{\sin 2\delta}\right)^2 + \text{const}\left(\log\frac{1}{\sin 2\delta}\right).$$

Given any $\epsilon > 0$ this is much less than $\epsilon\delta^{-1/2}$ for all sufficiently small δ. Putting all these estimates together then shows that *there is a constant K_1 such that for every $\epsilon > 0$ the first integral on the right side of* (3) *has modulus less than $\epsilon k^2\delta^{-1/2} + K_1 k\delta^{-1/2}$ for all sufficiently small positive δ.*

Analogous arguments prove that the same estimate applies to the second integral on the right side of (3). Briefly,

$$\frac{2}{\pi^2}\int_{e^{\pi/k}}^{\infty}(\log w)^{-2}\left|\frac{1}{xw}\right|^2 dw \le \frac{2}{\pi^2}(\log e^{\pi/k})^{-2}\int_{e^{\pi/k}}^{\infty} w^{-2}\,dw$$
$$\le \text{const}\, k^2 \le \epsilon k^2\delta^{-1/2}.$$

When $|G(xw) - 1|^2$ is written as a double sum over n and m, the total of the terms with $m = n$ is at most a constant times

$$\int_{e^{\pi/k}}^{\infty}(\log w)^{-2}\sum_{n=1}^{\infty} e^{-\pi n^2 2w^2 \sin 2\delta}\,dw$$
$$= \int_{e^{\pi/k}}^{\infty}(\log w)^{-2}[G(w(2\sin 2\delta)^{1/2}) - 1]\,dw$$
$$\le K\int_{e^{\pi/k}}^{\infty}(\log w)^{-2}\frac{1}{w(2\sin 2\delta)^{1/2}}\,dw$$
$$\le \frac{K}{(2\sin 2\delta)^{1/2}}\int_{e^{\pi/k}}^{\infty}(\log w)^{-2}\,d\log w$$
$$= \frac{K}{(2\sin 2\delta)^{1/2}}(\log e^{\pi/k})^{-1} \le \text{const}\, k\delta^{-1/2}$$

for all sufficiently small δ. Finally, the terms with $m \neq n$ are at most a constant times

$$\sum_{n=1}^{\infty}\sum_{m<n}\int_{e^{\pi/k}}^{\infty}(\log w)^{-2} e^{-\pi(n^2+m^2)w^2\sin 2\delta} e^{-i\pi(n^2-m^2)w^2\cos 2\delta}\,dw$$
$$\le \text{const}\sum_{n=1}^{\infty}\sum_{m<n}\frac{e^{-\pi(n^2+m^2)\sin 2\delta}}{(\log e^{\pi/k})^2(n^2 - m^2)}$$

which by the same sequence of estimates as before is less than $\epsilon k^2\delta^{-1/2}$ for all sufficiently small δ.

In what follows it will be convenient to consider $x = e^{-i\pi/4}e^{i\delta}$ rather than $x = e^{i\pi/4}e^{-i\delta}$ (because then the significant values of the integral occur for positive values of t—see below). Since this replaces $I_{x,k}(s)$ by its complex conjugate, the same estimates apply and what has been proved is that *if $I_{x,k}(s)$ is defined as in* (1) *with $x = e^{-i\pi/4}e^{i\delta}$, then there is a constant K' such that given $\epsilon > 0$ the inequality*

$$\frac{1}{2\pi i}\int_{(1/2)-i\infty}^{(1/2)+i\infty}|I_{x,k}(s)|^2\,ds < \frac{K'k + \epsilon k^2}{\delta^{1/2}}$$

holds for all sufficiently small positive values of δ ($k > 0$ being arbitrary).

Later in the proof this estimate will be used in combination with the Schwarz inequality to obtain an estimate of the integral of $|I|$. First, however, some estimates will be made of the value of $|I|$ implied by the assumption that $\xi(\frac{1}{2}+it)$ does not change sign in its domain of integration. Since

$$I_{x,k}\left(\frac{1}{2}+it\right) = \frac{1}{2\pi}\int_{t-k}^{t+k}\frac{2\xi(\frac{1}{2}+iu)}{-u^2-\frac{1}{4}}x^{-1/2}x^{iu}\,du$$
$$= -x^{-1/2}\frac{1}{\pi}\int_{t-k}^{t+k}\frac{\xi(\frac{1}{2}+iu)}{u^2+\frac{1}{4}}e^{\pi u/4}e^{-u\delta}\,du$$

(where $x = e^{-i\pi/4}e^{i\delta}$), the integral $J(t) = J_{x,k}(t)$ defined by

$$J_{x,k}(t) = \frac{1}{\pi}\int_{t-k}^{t+k}\frac{|\xi(\frac{1}{2}+iu)|}{u^2+\frac{1}{4}}e^{\pi u/4}e^{-u\delta}\,du$$

has the property that $|J(t)| \geq |I(\frac{1}{2}+it)|$ for all t and $|J(t)| = |I(\frac{1}{2}+it)|$ whenever the interval of integration of $I(\frac{1}{2}+it)$ contains no roots ρ. The basic idea of the proof is to show that in a suitable sense $J(t)$ is much larger than $I(\frac{1}{2}+it)$ on the average. Thus estimates of $J(t)$ from below are required.

It was shown in the preceding section that $|\Pi(s/2)|$ for $s = \frac{1}{2}+it$ is e^B where $B = \frac{3}{4}\log|t| - |t|\pi/4 + \cdots$, the omitted terms remaining bounded as $|t| \to \infty$. Combining this with the formula $\xi(s) = \Pi(s/2)\pi^{-s/2}(s-1)\zeta(s)$ and obvious estimates of $|\pi^{-s/2}|$ and $|s-1|$ gives

$$\frac{|\xi(\frac{1}{2}+iu)|}{u^2+\frac{1}{4}}e^{u\pi/4} \geq \text{const}\,\frac{u^{3/4}e^{-u\pi/4}\pi^{-1/4}u\,|\zeta(\frac{1}{2}+iu)|\,e^{u\pi/4}}{u^2+\frac{1}{4}}$$
$$\geq \text{const}\,u^{-1/4}|\zeta(\tfrac{1}{2}+iu)|$$

for $u \geq 1$. Therefore, for large t,

$$J(t) \geq \text{const}\,(t+k)^{-1/4}e^{-(t+k)\delta}\int_{t-k}^{t+k}|\zeta(\tfrac{1}{2}+iu)|\,du$$

and to estimate $J(t)$ from below it will suffice to estimate $\int|\zeta|$ from below. This can be done using a technique very similar to the technique of Section 9.7, which is also due to Hardy–Littlewood.

It was shown in Section 9.7 that

$$\zeta(\tfrac{1}{2}+iv) = \sum_{n<v} n^{-(1/2)-iv} + R(v),$$

where $R(v)$ is less than a constant times $v^{-1/2}$ as $v \to \infty$. Thus for $t-k \leq v \leq t+k$

$$\zeta(\tfrac{1}{2}+iv) = \sum_{n<t} n^{-(1/2)-iv} + E(v) + R(v),$$

where $E(v)$ is plus or minus the sum of $n^{-1/2-iv}$ over all integers n between v and t. Since $E(v)$ consists of at most $k+1$ terms each of modulus at most $n^{-1/2} \leq (t-k)^{-1/2}$, this shows that for $t-k \leq v \leq t+k$

$$\Big|-\zeta(\tfrac{1}{2}+iv) + \sum_{n<t} n^{-(1/2)-iv}\Big| \leq (k+1)(t-k)^{-1/2} + \text{const}\,v^{-1/2}$$
$$\leq (k+\text{const})(t-k)^{-1/2}.$$

Assume $k \geq 1$, so the right side can be written as a constant times $k(t-k)^{-1/2}$. Then

$$\operatorname{Re}\{-\zeta(\tfrac{1}{2}+iv)+\sum_{n<t} n^{-(1/2)-iv}\} \leq \text{const}\, k(t-k)^{-1/2},$$

$$\begin{aligned}\int_{t-k}^{t+k} |\zeta(\tfrac{1}{2}+iv)|\,dv &\geq \int_{t-k}^{t+k} \operatorname{Re}\zeta(\tfrac{1}{2}+iv)\,dv \\ &\geq \int_{t-k}^{t+k} [\operatorname{Re}\sum_{n<t} n^{-(1/2)-iv} - \text{const}\, k(t-k)^{-1/2}]\,dv \\ &= 2k + \sum_{2\leq n<t} \operatorname{Re}\int_{t-k}^{t+k} n^{-(1/2)-iv}\,dv \\ &\quad - 2k(\text{const}\, k)(t-k)^{-1/2} \\ &= 2k + \operatorname{Re}\sum_{2\leq n<t} \frac{2\sin(k\log n)}{n^{(1/2)+it}\log n} \\ &\quad - \text{const}\, k^2(t-k)^{-1/2} \\ &\geq 2k - 2\left|\sum_{2\leq n<t} \frac{1}{n^{(1/2)+it}\log n}\right| \\ &\quad - \text{const}\, k^2(t-k)^{-1/2}.\end{aligned}$$

If t is much larger than k, the last term is insignificant compared to the first. Now although the middle term is not necessarily small, it is *on the average* smaller than the first term, so that on the average the first term $2k$ is a lower bound. Specifically, over any interval $A \leq t \leq B$ with $B > A \geq 1$, the integral

$$\int_A^B \sum_{2\leq n<t} \left|\frac{1}{n^{(1/2)+it}\log n}\right|^2 dt = \int_A^B \sum_{2\leq m<t,\, 2\leq n<t}\sum \frac{1}{n^{1/2}m^{1/2}\log n\log m}\left(\frac{m}{n}\right)^{it} dt$$

can be estimated as follows. The terms with $m = n$ contribute just $(B-A)$ times a partial sum of the series $\sum n^{-1}(\log n)^{-2}$. Since this series is convergent, its partial sums are bounded and this is at most a constant times $(B-A)$. Each of the terms† with $m \neq n$ is of the form

$$\frac{1}{n^{1/2}m^{1/2}\log n\log m}\int_b^B \left(\frac{m}{n}\right)^{it} dt,$$

where $b = \max(A, m, n)$; so regardless of the value of b its modulus is at most

$$\frac{2}{n^{1/2}m^{1/2}\log n\log m\,|\log(m/n)|}$$

and the total of the remaining terms has modulus at most

$$4\sum_{2\leq m<n<B}\sum \frac{1}{n^{1/2}m^{1/2}\log n\log m\log(n/m)}.$$

As in Section 9.7 divide this sum into two parts according to whether $m <$

†The sum is finite, so termwise integration is valid.

$\frac{1}{2}n$ or $m \geq \frac{1}{2}n$. The total of the terms with $m < \frac{1}{2}n$ is at most

$$4\sum\sum \frac{1}{n^{1/2}m^{1/2}\log n \log m \log 2} = \frac{4}{\log 2}\Big(\sum_{2\leq n<B} \frac{1}{n^{1/2}\log n}\Big)^2$$
$$\leq \frac{4}{\log 2}\Big(\int_1^B u^{-1/2}\,du\Big)^2 \leq \text{const } B.$$

To estimate the total of the terms with $m \geq \frac{1}{2}n$, set $r = n - m$ so that $\log(n/m) = -\log[1 - (r/n)] > r/n$ and the total is less than

$$4\sum_{3\leq n<B}\sum_{n/2\leq m<n} \frac{1}{n^{1/2}(n-r)^{1/2}\log n \log(n/2)(r/n)}$$
$$= 4\sum_{3\leq n<B} \frac{1}{\log n \log(n/2)} \sum_{1\leq r\leq n/2} \frac{1}{r[1-(r/n)]^{1/2}}$$
$$\leq 4\sqrt{2}\sum_{3\leq n<B} \frac{1}{\log n \log(n/2)} \sum_{1\leq r\leq n/2} \frac{1}{r}$$
$$\leq 4\sqrt{2}\sum_{3\leq n\leq B} \frac{\log(n/2)+1}{\log n \log(n/2)},$$

which, since the terms of the sum are bounded, is less than a constant times B too. This proves that

$$\int_A^B \Big|\sum_{2\leq n<t} \frac{1}{n^{(1/2)+it}\log n}\Big|^2 dt < K_2 B$$

for some positive constant K_2. Thus by the Schwarz inequality

$$\int_A^B \Big|\sum \frac{1}{n^{(1/2)+it}\log n}\Big| dt \leq \Big(\int_A^B 1^2\,dt\Big)^{1/2}\Big(\int_A^B \Big|\sum \frac{1}{n^{(1/2)+it}\log n}\Big|^2 dt\Big)^{1/2}$$
$$\leq (B-A)^{1/2}(K_2 B)^{1/2} \leq K_2^{1/2}B,$$

so the average order of magnitude of the middle term is bounded and, therefore, when k is sufficiently large, the first term $2k$ is on the average dominant.

Now let ν be the number of zeros of $\xi(\frac{1}{2}+it)$ in the interval $\{0 \leq t \leq B+k\}$. Let the entire real axis be divided into intervals of length k and for each of the ν zeros strike out the interval which contains it and the two intervals which adjoin this one. Let S be the subset of $\{A \leq t \leq B\}$ consisting of points which do not lie in the stricken intervals. Then the total length of the intervals of S is at least $B - A - 3\nu k$ since a length of at most $3k$ was stricken for each zero. On the other hand $|I(\frac{1}{2}+it)| = J(t)$ for all t in S (there is no zero between $t-k$ and $t+k$) so

$$\int_S |I(\tfrac{1}{2}+it)|\,dt = \int_S J(t)\,dt \geq \int_S \text{const }(B+k)^{-1/4}$$
$$\times e^{-(B+k)\delta}\int_{t-k}^{t+k} |\zeta(\tfrac{1}{2}+iu)|\,du$$
$$\geq \text{const }(B+k)^{-1/4}e^{-(B+k)\delta}$$
$$\times \int_S \Big[2k - 2\Big|\sum \frac{1}{n^{(1/2)+it}\log n}\Big| - \text{const } k^2(t-k)^{-1/2}\Big]dt$$
$$\geq \text{const }(B+k)^{-1/4}e^{-(B+k)\delta}$$
$$\times [2k(B-A-3\nu k) - \text{const } B - \text{const } k^2 B^{1/2}].$$

To simplify this, let $(B + k)\delta = 1$, which can be regarded as a choice of B given δ, k and let $B - A = \frac{1}{2}\delta^{-1}$, which can be regarded as a choice of A. (Note that $B > A \geq 1$ for $k \geq 1$ and δ small.) Then the above estimate becomes

$$\begin{aligned} \int_S |I(\tfrac{1}{2} + it)|\, dt & \\ &\geq \text{const}\, \delta^{1/4}[2k(\tfrac{1}{2}\delta^{-1} - 3\nu k) - \text{const}\, \delta^{-1} - \text{const}\, k^2\delta^{-1/2}] \\ &= K_1 k\delta^{-3/4} - K_2 k^2 \nu \delta^{1/4} - K_3\delta^{-3/4} - K_4 k^2 \delta^{-1/4}, \end{aligned}$$

where K_1, K_2, K_3, K_4 are positive constants. Note that the third term will be insignificant compared to the first if k is large enough and the fourth term will be insignificant compared to the first if δ is small enough. On the other hand

$$\begin{aligned} \int_S |I(\tfrac{1}{2} + it)|\, dt &\leq \int_A^B |I(\tfrac{1}{2} + it)|\, dt \\ &\leq \left(\int_A^B 1^2\, dt\right)^{1/2} \left[\int_A^B |I(\tfrac{1}{2} + it)|^2\, dt\right]^{1/2} \\ &\leq (B - A)^{1/2} \left[\frac{1}{i}\int_{(1/2)-i\infty}^{(1/2)+i\infty} |I(s)|^2\, ds\right]^{1/2} \\ &\leq \text{const}\, \delta^{-1/2}\left(\frac{K'k + \epsilon k^2}{\delta^{1/2}}\right)^{1/2} \\ &= K_5 \delta^{-3/4}(K'k + \epsilon k^2)^{1/2}. \end{aligned}$$

Thus

$$K_5\delta^{-3/4}(K'k + \epsilon k^2)^{1/2} \geq K_1 k\delta^{-3/4} - K_2 k^2 \nu\delta^{1/4} - K_3 \delta^{-3/4} - K_4 k^2 \delta^{-1/4},$$

$$\nu \geq \frac{K_1}{K_2} k^{-1}\delta^{-1} - \frac{K_3}{K_2} k^{-2}\delta^{-1} - \frac{K_4}{K_2}\delta^{-1/2} - \frac{K_5}{K_2}\delta^{-1}k^{-1}\left(\frac{K'}{k} + \epsilon\right)^{1/2}.$$

The coefficient of $\delta^{-1}k^{-1}$ on the right can be made positive by choosing ϵ sufficiently small and k sufficiently large. Therefore with this fixed value of k it has been shown that for all sufficiently small δ the number ν of roots ρ on the line segment from $\frac{1}{2}$ to $\frac{1}{2} + i\delta^{-1}$ is at least $K_6\delta^{-1} - K_7\delta^{-1/2}$ with $K_6 > 0$. Since the $\delta^{-1/2}$ term is insignificant for small δ this proves the theorem.

11.3 THERE ARE AT LEAST *KT* log *T* ZEROS ON THE LINE

The basic structure of Selberg's proof is the same as that of the Hardy–Littlewood proof in the preceding section, but the proof begins not with the transform equation

$$(1) \qquad \frac{2\xi(s)}{s(s-1)} = \int_0^\infty u^{-s}\left[G(u) - 1 - \frac{1}{u}\right] du \qquad (0 < \operatorname{Re} s < 1)$$

but with a transform equation in which the left side is $2\xi(s)[s(s - 1)]^{-1}\phi(s) \cdot\phi^*(s)$ with $\phi^*(s) = \overline{\phi(1 - \bar{s})}$ the "adjoint" of $\phi(s)$ and with $\phi(s)$ specially chosen. In essence $\phi(s)$ is chosen to be an approximation to $\zeta(s)^{-1/2}$. Loosely speaking, this has the effect of approximately canceling the zeros of $\xi(s)$ and smoothing it out in such a way that the estimates of $|I|$ can be sharpened. (See Selberg's 1946 paper [S2] for a discussion of the motivation for the choice of ϕ.)

Specifically $\phi(s)$ is defined as follows. The function $\zeta(s)^{-1/2}$ can for $\operatorname{Re} s > 1$ be written as the transform of an operator of the form $f(x) \mapsto \sum_{n=1}^{\infty} \alpha_n f(nx)$. For this it suffices to write

$$\zeta(s)^{-1/2} = \prod_p \left(1 - \frac{1}{p^s}\right)^{1/2}$$
$$= \prod_p \left(1 - \frac{1}{2} p^{-s} + \frac{(\frac{1}{2})(-\frac{1}{2})}{2} p^{-2s} - \cdots\right).$$

If this product is expanded, there is exactly one term in ν^{-s} for every positive integer ν and its coefficient is given explicitly by

$$(2) \qquad (-1)^{n_1}\binom{\frac{1}{2}}{n_1}(-1)^{n_2}\binom{\frac{1}{2}}{n_2}\cdots(-1)^{n_k}\binom{\frac{1}{2}}{n_k},$$

where $\nu = p_1^{n_1} p_2^{n_2} \cdots p_k^{n_k}$ is the prime factorization of ν. Let α_ν denote the coefficient (2). Then

$$|\alpha_\nu| \le \left|\binom{-\frac{1}{2}}{n_1}\binom{-\frac{1}{2}}{n_2}\cdots\binom{-\frac{1}{2}}{n_k}\right| \le 1,$$

so the series $\sum \alpha_n n^{-s}$ converges for $\operatorname{Re} s > 1$. Moreover the absolute convergence of the product for $\zeta(s)^{-1/2}$ shows that $\zeta(s)^{-1/2} = \sum \alpha_n n^{-s}$ for $\operatorname{Re} s > 1$. However, because $\zeta(s)$ has a simple pole at $s = 1$, this function $\zeta(s)^{-1/2}$ has a singularity at $s = 1$ and cannot be continued in any simple way over to the critical line $\operatorname{Re} s = \frac{1}{2}$. Selberg deals with this by using a sort of *convergence factor*, by introducing a large parameter X and setting

$$\beta_n = \begin{cases} \left(1 - \dfrac{\log n}{\log X}\right) \alpha_n, & n \le X, \\ 0, & n \ge X, \end{cases}$$

$$\phi(s) = \phi_X(s) = \sum_{n=1}^{\infty} \beta_n n^{-s}.$$

Since $\beta_n \sim \alpha_n$ for small values of n, the function $\phi(s)$ is in some sense an approximation to $\zeta(s)^{-1/2}$, at least in the halfplane $\operatorname{Re} s > 1$. On the other hand, the series defining ϕ is finite so $\phi(s)$ is defined and analytic for all s.

With this definition of $\phi(s)$ set

$$I(s) = \frac{1}{2\pi i} \int_{s-ik}^{s+ik} \frac{2\xi(s)}{s(s-1)} \phi(s)\phi^*(s) x^{s-1}\, ds$$

where $\phi^*(s) = \overline{\phi(1 - \bar{s})} = \phi(1 - s)$. Then I depends on three parameters,

k (half the length of the interval of integration), X [the large parameter measuring, roughly, the degree of approximation of $\phi(s)$ to $\zeta(s)^{-1/2}$], and x (a complex number on the unit circle $|x| = 1$ near $i^{-1/2}$ but above it, say $x = i^{-1/2}e^{i\delta}$, where δ is small and positive). The idea of the proof is to show that, when these parameters are suitably chosen, the modulus of $I(\frac{1}{2} + it)$ is on the average much less than

$$J(t) = \frac{1}{2\pi}\int_{t-k}^{t+k} \frac{2|\xi(\frac{1}{2} + iv)|}{v^2 + \frac{1}{4}} \left|\phi\left(\frac{1}{2} + iv\right)\right|^2 e^{\pi v/4} e^{-v\delta}\, dv.$$

Since the modulus of $I(\frac{1}{2} + it)$ is equal to $J(t)$ unless $\xi(\frac{1}{2} + iv)$ changes sign in the interval $\{t - k \leq v \leq t + k\}$, this will show that on the average it is to be expected that $\xi(\frac{1}{2} + iv)$ does change sign and therefore that there is very often a root in the interval.

As before, the first step in the estimation of $|I|$ is to write $I(s)$ as the transform of an operator and to apply the Parseval formula. First write $2\xi(s) \cdot [s(s-1)]^{-1}\phi(s)\phi^*(s)$ as the transform of an operator by composing the operator with transform (1) with the operators

$$f(x) \mapsto \sum_{n=1}^{\infty} \beta_n f(nx), \qquad f(x) \mapsto \sum_{n=1}^{\infty} \frac{\beta_n}{n} f\left(\frac{x}{n}\right)$$

with transforms $\phi(s)$, $\phi^*(s)$, respectively, to find

$$\frac{2\xi(s)}{s(s-1)}\phi(s)\phi^*(s) = \int_0^\infty u^{-s} \sum_{\mu=1}^{\infty}\sum_{\nu=1}^{\infty} \frac{\beta_\mu\beta_\nu}{\nu}\left[G\left(\frac{\mu u}{\nu}\right) - 1 - \frac{\nu}{\mu u}\right] du.$$

(The sums are actually finite, so there is no problem with termwise integration.) Put xu in place of u. Then the path of integration becomes the ray through x^{-1}, but the rapid vanishing of $G(xu) - 1$ as $u \to \infty$ on the real axis and the consequent (by the functional equation of G) rapid vanishing of $G(xu) - (xu)^{-1}$ as $u \downarrow 0$ makes it valid to replace this path of integration by the real axis. Since $G(u) - 1 = 2\sum_{n=1}^{\infty} \exp(-\pi n^2u^2)$ and since

$$\sum_{\mu=1}^{\infty}\sum_{\nu=1}^{\infty} \frac{\beta_\mu\beta_\nu}{\mu} = \left(\sum_{\mu=1}^{\infty}\frac{\beta_\mu}{\mu}\right)\left(\sum_{\nu=1}^{\infty}\beta_\nu\right) = \phi(1)\phi(0),$$

this puts the formula in the form

$$\text{(3)} \qquad \frac{2\xi(s)}{s(s-1)}\phi(s)\phi^*(s)x^{s-1} = \int_0^\infty u^{-s}\left[\Psi(xu) - \frac{\phi(1)\phi(0)}{xu}\right] du,$$

where

$$\Psi(xu) = 2\sum_{n=1}^{\infty}\sum_{\mu=1}^{\infty}\sum_{\nu=1}^{\infty} e^{-\pi n^2\mu^2x^2u^2/\nu^2}\frac{\beta_\mu\beta_\nu}{\nu}.$$

Then integration ds from $s - ik$ to $s + ik$ on both sides gives

$$I(s) = \int_0^\infty u^{-s}\frac{\sin(k\log u)}{\pi\log u}\left[\Psi(xu) - \frac{\phi(1)\phi(0)}{xu}\right] du;$$

so by the Parseval formula

$$(4)\quad \frac{1}{2\pi i}\int_{(1/2)-i\infty}^{(1/2)+i\infty} |I(s)|^2\,ds = \int_0^\infty \left(\frac{\sin(k\log u)}{\pi\log u}\right)^2 \left|\Psi(xu) - \frac{\phi(1)\phi(0)}{xu}\right|^2 du.$$

The left side of this equation is to be estimated by estimating the right side using the explicit formula for Ψ. Rather than estimating the right side directly, however, Selberg bases his estimates of it on an estimate of the integral

$$W(z,\theta) = \int_z^\infty |\Psi(xu)|^2 u^{-\theta}\,du$$

instead. This estimate, which is proved in Section 11.4 below, is the following.

Lemma There exist constants K and δ_0 such that

$$W(z,\theta) \le \frac{K}{\delta^{1/2}\theta z^\theta \log X}$$

holds for all δ in the range $0 < \delta \le \delta_0$ (where δ enters the definition of W because $x = e^{-i\pi/4}e^{i\delta}$) provided the other parameters satisfy the restrictions $0 < \theta \le \frac{1}{2}$, $1 \le z \le \delta^{-1/15}$, and $1 \le X \le \delta^{-1/15}$ (where X enters the definition of W because the β's depend on X).

This lemma will be used not only in estimating $\int |I|^2$ but also in estimating $\int |J|^2$. Consider first the estimation of $\int |I|^2$. As in the previous case the symmetry of the integrand implies that the right side of (4) can be replaced by twice the integral from one to infinity. If this interval of integration is subdivided at the point h (in the previous argument h was $e^{\pi/k}$) and the usual inequality for $x^{-1}\sin x$ is used, then

$$\begin{aligned}
\frac{1}{2\pi i}\int_{(1/2)-i\infty}^{(1/2)+i\infty} |I(s)|^2\,ds &\le \frac{2k^2}{\pi^2}\int_1^h \left|\Psi(xu) - \frac{\phi(0)\phi(1)}{xu}\right|^2 du \\
&\quad + \frac{2}{\pi^2}\int_h^\infty (\log u)^{-2}\left|\Psi(xu) - \frac{\phi(0)\phi(1)}{xu}\right|^2 du \\
&\le \frac{4k^2}{\pi^2}\int_1^h |\Psi(xu)|^2\,du + \frac{4k^2}{\pi^2}|\phi(1)\phi(0)|^2\left(1 - \frac{1}{h}\right) \\
&\quad + \frac{4}{\pi^2}\int_h^\infty (\log u)^{-2}|\Psi(xu)|^2\,du \\
&\quad + \frac{4}{\pi^2}|\phi(1)\phi(0)|^2\int_h^\infty \frac{du}{u^2(\log u)^2}.
\end{aligned}$$

Assume $h \ge e$. (Later h will go to infinity.) The second and fourth terms combined are at most

$$\begin{aligned}
&|\phi(1)\phi(0)|^2\frac{4}{\pi^2}\left[k^2\left(1 - \frac{1}{h}\right) + \frac{1}{(\log h)^2}\int_h^\infty \frac{du}{u^2}\right] \\
&\le \left|\sum_\mu \frac{\beta_\mu}{\mu}\right|^2 |\textstyle\sum_\nu \beta_\nu|^2\frac{4}{\pi^2}\left(k^2 + \frac{1}{(\log h)^2}\right)
\end{aligned}$$

which, since $|\beta_j| \le 1$ and $\beta_j = 0$ for $j \ge X$, is at most a constant times $(\log X)^2 X^2[k^2 + (\log h)^{-2}]$. Therefore

$$\frac{1}{2\pi i}\int_{(1/2)+i\infty}^{(1/2)+i\infty} |I(s)|^2\, ds \le \frac{4k^2}{\pi^2}\int_1^h |\Psi(xu)|^2\, du + \frac{4}{\pi^2}\int_h^\infty (\log u)^{-2}|\Psi(xu)|^2\, du$$
$$+ K_1 k^2 X^2 (\log X)^2 + K_1 X^2 \frac{(\log X)^2}{(\log h)^2}.$$

The lemma can be used to estimate the two integrals on the right. Let K, δ_0 be as in the lemma, let $0 < \theta \le \frac{1}{2}$, $1 \le h \le \delta^{-1/15}$, and let $1 \le X \le \delta^{-1/15}$. Then the first integral above is at most

$$\frac{4k^2}{\pi^2}\int_1^h u^\theta u^{-\theta} |\Psi(xu)|^2\, du$$
$$\le \frac{4k^2}{\pi^2}\int_1^h z^\theta\left[-\frac{\partial W}{\partial z}\right] dz$$
$$= -\frac{4k^2}{\pi^2} z^\theta W(z,\theta)\Big|_1^h + \frac{4k^2\theta}{\pi^2}\int_1^h z^{\theta-1}W(z,\theta)\, dz$$
$$\le \frac{4k^2}{\pi^2}\frac{K}{\delta^{1/2}\theta\log X} + \frac{4k^2\theta}{\pi^2}\int_1^h \frac{Kz^{\theta-1}}{\delta^{1/2}\theta z^\theta \log X}\, dz$$
$$= \frac{4k^2 K}{\pi^2\delta^{1/2}\log X}\left(\frac{1}{\theta} + \log h\right).$$

If $\theta = \frac{1}{2}$ and h is very large, the second term dominates and the integral is at most a constant times $k^2\delta^{-1/2}(\log X)^{-1}(\log h)$. To estimate the second integral, use the identity

$$\int_0^{1/2} \theta u^{-\theta}\, d\theta = -\frac{1}{2u^{1/2}\log u} - \frac{1}{u^{1/2}(\log u)^2} + \frac{1}{(\log u)^2}$$

(integration by parts) and $h \ge e$ to find

$$\frac{4}{\pi^2}\int_h^\infty (\log u)^{-2}|\Psi(xu)|^2\, du$$
$$= \frac{4}{\pi^2}\int_h^\infty\int_0^{1/2} \theta u^{-\theta}|\Psi(xu)|^2\, d\theta\, du$$
$$+ \frac{4}{\pi^2}\int_h^\infty \frac{u^{-1/2}}{\log u}\left[\frac{1}{2} + \frac{1}{\log u}\right]|\Psi(xu)|^2\, du$$
$$\le \frac{4}{\pi^2}\int_0^{1/2} \theta W(h,\theta)\, d\theta + \frac{6}{\pi^2}\int_h^\infty u^{-1/2}|\Psi(xu)|^2\, du$$
$$\le \frac{4}{\pi^2}\int_0^{1/2} \frac{K\, d\theta}{\delta^{1/2}h^\theta \log X} + \frac{6}{\pi^2}\frac{K}{\delta^{1/2}\frac{1}{2}h^{1/2}\log X}$$
$$= \frac{4}{\pi^2}\frac{K}{\delta^{1/2}\log X}\left(\frac{1}{\log h} - \frac{1}{h^{1/2}\log h}\right) + \frac{12K}{\pi^2\delta^{1/2}h^{1/2}\log X}.$$

If h is large, the dominant term is the first one, which is less than a constant times $\delta^{-1/2}(\log X)^{-1}(\log h)^{-1}$. Thus with the above restrictions on the parameters plus the condition that h be sufficiently large

$$(5)\qquad \frac{1}{2\pi i}\int_{(1/2)-i\infty}^{(1/2)+i\infty}|I(s)|^2\,ds \le \frac{K_2k^2\log h}{\delta^{1/2}\log X}+\frac{K_3}{\delta^{1/2}\log X\log h}$$
$$+\,K_1k^2X^2(\log X)^2+K_1X^2\frac{(\log X)^2}{(\log h)^2}.$$

Choose X as large as possible, namely, $X=\delta^{-1/15}$. Then the third term is still less than $K_1k^2X^3<K_1k^2\delta^{-1/5}$ which is insignificant compared to the first term (when δ is small and h large) and in the same way the fourth term is insignificant compared to the second. If the first two terms are to have the same order of magnitude, then $k^2(\log h)^2$ must have the order of magnitude 1; this motivates setting $h=e^{\pi/k}$ as in the previous proof. Since h is to become large, this implies $k\longrightarrow 0$. Loosely speaking, the major shortcoming of the proof of Section 11.2 was the fact that in it k did not go to zero; this meant that the subdivision of the interval $\{A\le t\le B\}$ never became very fine. Thus $k\longrightarrow 0$ is a major aspect of Selberg's proof. On the other hand—as is not surprising—it is essential that $k\longrightarrow 0$ very, very slowly. For this reason set $k=(a\log\delta^{-1})^{-1}$, where a is a very small positive constant to be determined later. Finally, for notational convenience set $T=\delta^{-1}$. Then the parameters have been reduced to two, namely, T (large) and a (small) and the others have been related to these two by

$$(6)\qquad \delta=T^{-1},\qquad x=e^{-i\pi/4}e^{i/T},\qquad X=T^{1/15},\qquad k=(a\log T)^{-1},$$

and by $h=e^{\pi/k}$ for the parameter h, which does not appear in the final result. What has been proved (except, of course, for the proof of the lemma) is that when the parameters of I are chosen in this way, then for every $a>0$ the inequality

$$\frac{1}{2\pi i}\int_{(1/2)-i\infty}^{(1/2)+i\infty}|I(s)|^2\,ds\le K_4\frac{k}{\delta^{1/2}\log X}=\frac{K_5T^{1/2}}{a(\log T)^2}$$

holds for all sufficiently large T.

Consider now the estimation of the average value of $J(t)$. Let

$$F(v)=\frac{1}{\pi}\frac{|\xi(\frac{1}{2}+iv)|}{v^2+\frac{1}{4}}|\phi(\tfrac{1}{2}+iv)|^2e^{\pi v/4}e^{-v\delta}$$

so that $J(t)=\int_{t-k}^{t+k}F(v)\,dv$. It is natural to begin by estimating the average magnitude of the positive function $F(v)$. It was shown in the preceding section that

$$\frac{|\xi(\frac{1}{2}+iv)|e^{\pi v/4}}{v^2+\frac{1}{4}}\ge\text{const}\,v^{-1/4}|\zeta(\tfrac{1}{2}+iv)|,$$

so $F(v)\ge\text{const}\,v^{-1/4}|\zeta(\frac{1}{2}+iv)||\phi(\frac{1}{2}+iv)|^2e^{-v\delta}$ and it will suffice to find a lower bound on the average magnitude of $\zeta(\frac{1}{2}+iv)[\phi(\frac{1}{2}+iv)]^2$. This can be done by considering the integral of $\zeta(s)\phi(s)^2$ around the boundary of a rectangle $\{\frac{1}{2} \le \operatorname{Re} s \le 2, A \le \operatorname{Im} s \le B\}$ where $B > A \ge 1$ as follows. The whole integral around the boundary is zero by Cauchy's theorem. By the definition of Lindelöf's μ-function, by $\mu(\frac{1}{2}) \le \frac{1}{4}$, and by Lindelöf's theorem (see Section 9.2) $|\zeta(\sigma + it)|$ is less than a constant times $t^{1/4+\epsilon}$ for all $\sigma + it$ in the half-strip $\{\frac{1}{2} \le \sigma \le 2, t \ge 1\}$. On the other hand $|\phi(s)|$ in this half-strip is $|\sum \beta_n n^{-\sigma-it}| \le \sum |\beta_n| n^{-\sigma} \le \sum_{n \le X} n^{-1/2}$, which by Euler–Maclaurin is about $\int_1^X u^{-1/2}\,du \le 2X^{1/2}$ and which is therefore less than a constant times $X^{1/2}$. Therefore the integral of $\zeta(s)\phi(s)^2$ over the side $\operatorname{Im} s = A$ of the rectangle has modulus at most a constant times $A^{1/4+\epsilon}X$ and the integral over the side $\operatorname{Im} s = B$ modulus at most the same constant times $B^{1/4+\epsilon}X$. On the side $\operatorname{Re} s = 2$ of the rectangle, the integrand can be written in the form

$$\zeta(s)\phi(s)^2 = 1 + \sum_{n=2}^{\infty} \frac{a_n}{n^s},$$

where $|a_n|$ is less than the coefficient of n^{-s} in the expansion of $\zeta(s)^3$ (multiplication of absolutely convergent series); so

$$\int_{2+iA}^{2+iB} \zeta(s)\phi(s)^2\,ds = i(B - A) + \sum_{n=2}^{\infty} a_n \int_{2+iA}^{2+iB} \frac{ds}{n^s},$$

and the integral over this side differs from $i(B - A)$ by at most

$$\sum_{n=2}^{\infty} |a_n| \frac{2}{n^2 \log n} \le \frac{2}{\log 2} \sum_{n=1}^{\infty} \frac{|a_n|}{n^2} \le \frac{2}{\log 2}\zeta(2)^3 \le \text{const.}$$

Therefore the integral over the side $\operatorname{Re} s = \frac{1}{2}$ differs from $i(B - A)$ by a quantity whose modulus is at most

$$\text{const} + \text{const}\, A^{1/4+\epsilon} + \text{const}\, B^{1/4+\epsilon}$$

which shows that

$$\left| \int_A^B \zeta(\tfrac{1}{2} + iv)\phi(\tfrac{1}{2} + iv)^2\,dv \right| \ge B - A - \text{const}\, B^{1/4+\epsilon}.$$

Therefore

$$\begin{aligned}
\int_A^B F(v)\,dv &\ge \text{const}\, B^{-1/4}e^{-B\delta} \int_A^B |\zeta(\tfrac{1}{2} + iv)||\phi(\tfrac{1}{2} + iv)|^2\,dv \\
&\ge \text{const}\, B^{-1/4}e^{-B\delta} \left| \int_A^B \zeta(\tfrac{1}{2} + iv)\phi(\tfrac{1}{2} + iv)^2\,dv \right| \\
&\ge \text{const}\, B^{-1/4}e^{-B\delta}(B - A) - \text{const}\, B^{\epsilon},
\end{aligned}$$

$$\begin{aligned}
\int_A^B J(t)\,dt &= \int_A^B \int_{t-k}^{t+k} F(v)\,dv\,dt \ge \int_{A+k}^{B-k} \int_{v-k}^{v+k} F(v)\,dt\,dv \\
&= \int_{A+k}^{B-k} 2kF(v)\,dv \\
&\ge \text{const}\, 2kB^{-1/4}e^{-B\delta}(B - A - 2k) - \text{const}\, 2kB^{\epsilon}.
\end{aligned}$$

As before, define A and B by $(B + k)\delta = 1$, $B - A = \frac{1}{2}\delta^{-1}$. In other words, set

(7) $$B = T - k, \qquad A = \tfrac{1}{2}T - k.$$

Then since $k \to 0$, it follows easily from the above that

(8) $$\int_A^B J(t)\,dt \geq \text{const } 2kT^{3/4} = \frac{K_6 T^{3/4}}{a \log T}$$

for all sufficiently large T when a is fixed and the other parameters are determined as in (6) and (7). This estimate is less exact than the corresponding estimate in Section 11.2 because that estimate gave a lower bound for $\int_S J(t)\,dt$ where S is a *subset* of the interval $\{A \leq t \leq B\}$. However, this estimate can be made to serve a similar purpose by combining it with the following estimate of $\int_{-\infty}^{\infty} J(t)^2\,dt$.

Parseval's equation applied to the transform equation (3) gives

$$\frac{1}{2\pi i}\int_{1/2-i\infty}^{1/2+i\infty} \frac{4|\xi(s)|^2}{|s|^2|s-1|^2}|\phi(s)|^2|\phi(1-s)|^2|x^{s-1}|^2\,ds$$
$$= \int_0^\infty \left|\Psi(xu) - \frac{\phi(1)\phi(0)}{xu}\right|^2 du.$$

When the formula for $F(v)$ and the familiar symmetry of the integral on the right are used, this becomes

$$2\pi\int_{-\infty}^{\infty} F(v)^2\,dv = 2\int_1^\infty \left|\Psi(xu) - \frac{\phi(1)\phi(0)}{xu}\right|^2 du$$
$$\leq 4\int_1^\infty |\Psi(xu)|^2\,du + 4|\phi(1)\phi(0)|^2\int_1^\infty u^{-2}\,du.$$

The second integral on the right is $4|\phi(1)\phi(0)|^2$ which is less than a constant times $X^2(\log X)^2$ which is less than $T^{1/5}$. The first integral on the right is $W(1, 0)$, but since the lemma does not apply when $\theta = 0$, an estimate of $W(1, 0)$ has yet to be made. Now

$$\int_1^\infty |\Psi(xu)|^2\,du = \int_1^{T^2} |\Psi(xu)|^2\,du + \int_{T^2}^\infty |\Psi(xu)|^2\,du$$
$$\leq e^2\int_1^{T^2} |\Psi(xu)|^2 e^{-\log u/\log T}\,du + \int_{T^2}^\infty |\Psi(xu)|^2\,du$$
$$\leq e^2 W\left(1, \frac{1}{\log T}\right) + \int_{T^2}^\infty |\Psi(xu)|^2\,du$$
$$\leq e^2 15KT^{1/2} + \int_{T^2}^\infty |\Psi(xu)|^2\,du$$

by the lemma, so it will suffice to show that $\int_{T^2}^\infty |\Psi(xu)|^2\,du$ is less than $T^{1/2}$ in order to find an upper estimate of $\int F^2\,dv$. Expand $|\Psi(xu)|^2$ as a sextuple sum

$$\sum_{m,n,\mu,\nu,\kappa,\lambda} \frac{\beta_\kappa\beta_\lambda\beta_\mu\beta_\nu}{\lambda\nu} e^{-\pi m^2\kappa^2u^2x^2/\lambda^2} e^{-\pi n^2\mu^2u^2\bar{x}^2/\nu^2}.$$

For fixed m, n the sum over μ, ν, κ, λ consists of at most X^4 terms the largest of which has modulus at most $\exp[-\pi m^2 X^{-2} u^2 \operatorname{Re} x^2 - \pi n^2 X^{-2} u^2 \operatorname{Re} x^2] = \exp[-\pi(m^2 + n^2)X^{-2}u^2 \sin 2\delta]$, so $|\Psi(xu)|^2$ is at most X^4 times the double sum of this exponential over all m, n. But this is less than

$$X^4 \int_0^\infty \int_0^\infty e^{-\pi(x^2+y^2)X^{-2}u^2 \sin 2\delta}\, dx\, dy$$

$$= X^4 \int_0^{\pi/2} \int_0^\infty e^{-\pi r^2 X^{-2} u^2 \sin 2\delta} r\, dr\, d\theta$$

$$= \frac{\pi}{2} X^4 \cdot \frac{1}{2} \int_0^\infty e^{-\pi v X^{-2} u^2 \sin 2\delta}\, dv \le \text{const}\, \frac{X^2}{u^2 \delta},$$

so its integral du from T^2 to ∞ is at most a constant times $(X^2/\delta)T^{-2} = X^2/T \le T^{-13/15}$. Thus $\int_{-\infty}^\infty F(v)^2\, dv$ is less than a constant times $T^{1/2}$, from which it follows that

$$\int_{-\infty}^\infty J(t)^2\, dt = \int_{-\infty}^\infty \left[\int_{t-k}^{t+k} F(v)\, dv\right]^2 dt$$

$$\le \int_{-\infty}^\infty \left(\int_{t-k}^{t+k} 1^2\, dv\right)\left(\int_{t-k}^{t+k} F(v)^2\, dv\right) dt$$

$$= 2k \int_{-\infty}^\infty \int_{t-k}^{t+k} F(v)^2\, dv\, dt$$

$$= 2k \int_{-\infty}^\infty \int_{v-k}^{v+k} F(v)^2\, dt\, dv$$

$$= 4k^2 \int_{-\infty}^\infty F(v)^2\, dv$$

$$\le \text{const}\, k^2 T^{1/2} = \frac{K_7 T^{1/2}}{a^2(\log T)^2}.$$

Now let ν be the number of zeros of $\xi(\frac{1}{2} + it)$ in the interval $\{0 \le t \le B + k\}$. Let the entire real axis be divided into intervals of length k, and for each of the ν zeros strike out the interval which contains it and the two intervals which adjoin this one. Let S be the subset of $\{A \le t \le B\}$ consisting of points which do not lie in stricken intervals and let $\tilde{S}$ be the subset of $\{A \le t \le B\}$ consisting of those which do. Then the total length of the intervals of $\tilde{S}$ is at most $3\nu k$ and $|I(\frac{1}{2} + it)| = J(t)$ on S so

$$\int_S |I(\tfrac{1}{2} + it)|\, dt = \int_S J(t)\, dt = \int_A^B J(t)\, dt - \int_{\tilde{S}} J(t)\, dt$$

$$\ge \frac{K_6 T^{3/4}}{a \log T} - \left(\int_{\tilde{S}} 1^2\, dt\right)^{1/2} \left(\int_{\tilde{S}} J(t)^2\, dt\right)^{1/2}$$

$$\ge \frac{K_6 T^{3/4}}{a \log T} - (3\nu k)^{1/2} \left[\int_{-\infty}^\infty J(t)^2\, dt\right]^{1/2}$$

$$\ge \frac{K_6 T^{3/4}}{a \log T} - (3\nu k)^{1/2} \frac{K_7^{1/2} T^{1/4}}{a \log T}.$$

On the other hand

$$\begin{aligned}\int_S |I(\tfrac{1}{2}+it)|\,dt &\le \Big(\int_S 1^2\Big)^{1/2}\Big(\int_S |I(\tfrac{1}{2}+it)|^2\,dt\Big)^{1/2}\\ &\le \Big(\int_A^B dt\Big)^{1/2}\Big(\frac{1}{i}\int_{1/2-i\infty}^{1/2+i\infty} |I(s)|^2\,ds\Big)^{1/2}\\ &\le \text{const}\ T^{1/2}\frac{T^{1/4}}{a^{1/2}\log T}\\ &= \frac{K_8T^{3/4}}{a^{1/2}\log T}.\end{aligned}$$

Thus

$$(3\nu k)^{1/2} \ge K_7^{-1/2}(K_6T^{1/2} - a^{1/2}K_8T^{1/2}).$$

For fixed a sufficiently small the right side is a positive constant times $T^{1/2}$; hence

$$3\nu k \ge K_9T, \qquad \nu \ge \frac{K_9T}{3k} = \frac{K_9a}{3}T\log T,$$

and Selberg's theorem is proved.

11.4 PROOF OF A LEMMA

The core of Selberg's proof—and the only part of the proof which makes use of the special choice of the function $\phi(s)$—is the estimate of the integral $W(z, \theta)$ stated in the lemma of the preceding section. This section is devoted to the proof of this estimate.

Recall the following definitions from Section 11.3. There δ is a small positive number; $x = e^{-i\pi/4}e^{i\delta}$; X is a large positive number; α_n is the coefficient of n^{-s} in the Dirichlet series expansion of $\zeta(s)^{-1/2}$; β_n is zero for $n \ge X$ and $(\log X)^{-1}\log(X/n)\alpha_n$ if $n \le X$; $\Psi(u)$ is defined by

$$\Psi(u) = 2\sum_{n=1}^{\infty}\sum_{\mu=1}^{\infty}\sum_{\nu=1}^{\infty} e^{-\pi n^2\mu^2u^2/\nu^2}\frac{\beta_\mu\beta_\nu}{\nu};$$

and $W(z, \theta)$ is defined by

$$W(z,\theta) = \int_z^{\infty} u^{-\theta}|\Psi(xu)|^2\,du,$$

z, θ being positive real numbers. The lemma to be proved states that there exist positive constants K, δ_0 such that $W(z, \theta) \le K(\delta^{1/2}\theta z^{\theta}\log X)^{-1}$ whenever the parameters lie in the ranges

$$0 < \delta \le \delta_0, \qquad 0 < \theta \le \tfrac{1}{2}, \qquad 1 \le X \le \delta^{-1/15}, \qquad 1 \le z \le \delta^{-1/15}.$$

Substitution of the definition of Ψ into the definition of W gives an expression of W as the integral of a sextuple sum of terms of the form

$$4u^{-\theta}\frac{\beta_\kappa\beta_\lambda\beta_\mu\beta_\nu}{\lambda\nu}e^{-\pi m^2\kappa^2u^2x^2/\lambda^2}e^{-\pi n^2\mu^2u^2\bar{x}^2/\nu^2}$$

$$= 4u^{-\theta}\frac{\beta_\kappa\beta_\lambda\beta_\mu\beta_\nu}{\lambda\nu}$$

$$\times \exp\left[-\pi u^2\left(\frac{m^2\kappa^2}{\lambda^2}+\frac{n^2\mu^2}{\nu^2}\right)\operatorname{Re}x^2 - i\pi u^2\left(\frac{m^2\kappa^2}{\lambda^2}-\frac{n^2\mu^2}{\nu^2}\right)\operatorname{Im}x^2\right].$$

Let Σ_1 denote the integral of the sum of the terms in which the imaginary part of the exponential is zero, that is, those terms in which $(m\kappa/\lambda) = (n\mu/\nu)$, and let Σ_2 denote the integral of the sum of the remaining terms. The difficult part of the proof is the estimation of Σ_1, after which it is comparatively easy to show that Σ_2 is much smaller than the estimate which is obtained for Σ_1.

For each fixed quadruple $(\kappa, \lambda, \mu, \nu)$ there is an infinite sequence of terms of Σ_1, namely, the terms corresponding to (m, n) where $m/n = (\lambda\mu/\kappa\nu)$. Let a/b be the expression of $(\lambda\mu/\kappa\nu)$ in lowest terms; then the pairs (m, n) are (a, b), $(2a, 2b)$, $(3a, 3b)$, . . . and therefore

$$\Sigma_1 = \int_z^\infty 4u^{-\theta}\sum_{\kappa\lambda\mu\nu}\frac{\beta_\kappa\beta_\lambda\beta_\mu\beta_\nu}{\lambda\nu}\sum_{r=1}^\infty \exp\left[-2\pi u^2\frac{r^2a^2\kappa^2}{\lambda^2}\sin 2\delta\right]du,$$

where a depends on $(\kappa, \lambda, \mu, \nu)$ as above. This expression can be made more symmetrical by defining q to be the greatest common divisor of $\lambda\mu$ and $\kappa\nu$ (the factor canceled when $\lambda\mu/\kappa\nu$ is reduced to lowest terms) so that $a = \lambda\mu/q$, $b = \kappa\nu/q$, and the above becomes

$$\Sigma_1 = 4\int_z^\infty u^{-\theta}\sum_{\kappa\lambda\mu\nu}\frac{\beta_\kappa\beta_\lambda\beta_\mu\beta_\nu}{\lambda\nu}\sum_{r=1}^\infty \exp[-2\pi u^2r^2\mu^2\kappa^2q^{-2}\sin 2\delta]\,du.$$

The sum over $\kappa\lambda\mu\nu$ is finite because β_j is zero for $j \geq X$. Therefore this sum can be taken out from under the integral sign, and Σ_1 becomes a finite sum of terms of the form

$$\text{const}\int_z^\infty u^{-\theta}\sum_{r=1}^\infty e^{-u^2r^2\eta^2}\,du$$

where $\eta = \kappa\mu q^{-1}(2\pi\sin 2\delta)^{1/2}$. Such a term can be estimated as follows:

$$\begin{aligned}\int_z^\infty u^{-\theta}\sum_{r=1}^\infty e^{-u^2r^2\eta^2}\,du &= \sum_{r=1}^\infty\int_z^\infty u^{1-\theta}e^{-u^2r^2\eta^2}\,d\log u\\ &= \sum_{r=1}^\infty\int_{zr\eta}^\infty\left(\frac{y}{r\eta}\right)^{1-\theta}e^{-y^2}\,d\log y\\ &= \eta^{\theta-1}\sum_{r=1}^\infty r^{\theta-1}\int_{zr\eta}^\infty y^{-\theta}e^{-y^2}\,dy\\ &= \eta^{\theta-1}\int_{z\eta}^\infty\Big(\sum_{1\leq r\leq y/z\eta} r^{\theta-1}\Big)y^{-\theta}e^{-y^2}\,dy.\end{aligned}$$

The sum in the integrand can be estimated by Euler–Maclaurin summation

$$\sum_{1 \le r \le N} r^{\theta-1} = \int_1^N v^{\theta-1}\,dv + \frac{1}{2}[1^{\theta-1} + N^{\theta-1}] + \int_1^N \bar{B}_1(v)(\theta-1)v^{\theta-2}\,dv$$
$$= \frac{N^\theta}{\theta} - \frac{1}{\theta} + \frac{1}{2} + \frac{1}{2}N^{\theta-1} + \int_1^\infty \bar{B}_1(v)(\theta-1)v^{\theta-2}\,dv$$
$$- \int_N^\infty \bar{B}_1(v)(\theta-1)v^{\theta-2}\,dv.$$

Let N be the smallest integer greater than $y/z\eta$. Then subtracting $N^{\theta-1}$ from both sides gives

$$\sum_{1 \le r \le y/z\eta} r^{\theta-1} = \frac{N^\theta}{\theta} - \frac{1}{\theta} + \frac{1}{2} - \frac{1}{2}N^{\theta-1} + \int_1^\infty \bar{B}_1(v)(\theta-1)v^{\theta-2}\,dv$$
$$- \int_N^\infty \bar{B}_1(v)(\theta-1)v^{\theta-2}\,dv.$$

Since $\theta - 1 < 0$, this differs from

$$\frac{1}{\theta}\left(\frac{y}{z\eta}\right)^\theta - \frac{1}{\theta} + \frac{1}{2} + \int_1^\infty \bar{B}_1(v)(\theta-1)v^{\theta-2}\,dv$$

by at most

$$\left|\frac{N^\theta}{\theta} - \frac{1}{\theta}\left(\frac{y}{z\eta}\right)^\theta - \frac{1}{2}N^{\theta-1} - (\theta-1)\int_N^\infty \bar{B}_1(v)v^{\theta-2}\,dv\right|$$
$$\le \int_{y/z\eta}^N u^{\theta-1}\,du + \frac{1}{2}\left(\frac{y}{z\eta}\right)^{\theta-1}$$
$$+ |\theta-1|\int_0^{1/2}\left|u - \frac{1}{2}\right|(N+u)^{\theta-2}\,du$$
$$\le \frac{3}{2}\left(\frac{y}{z\eta}\right)^{\theta-1} + \frac{1}{2}\cdot\frac{1}{2}N^{\theta-2}$$
$$< \frac{3}{2}\left(\frac{y}{z\eta}\right)^{\theta-1} + \frac{1}{4}N^{\theta-1} < 2\left(\frac{y}{z\eta}\right)^{\theta-1}.$$

Substitution of this estimate into the original integral shows that

$$\text{(1)} \qquad \int_z^\infty u^{-\theta}\sum_{r=1}^\infty e^{-u^2r^2\eta^2}\,du$$

differs from

$$\text{(2)} \quad \eta^{\theta-1}\int_{z\eta}^\infty\left[\frac{1}{\theta}\left(\frac{y}{z\eta}\right)^\theta - \frac{1}{\theta} + \frac{1}{2} + \int_1^\infty \bar{B}_1(v)(\theta-1)v^{\theta-2}\,dv\right]y^{-\theta}e^{-y^2}\,dy$$

by at most

$$2\eta^{\theta-1}(z\eta)^{1-\theta}\int_{z\eta}^\infty y^{\theta-1}y^{-\theta}e^{-y^2}\,dy$$
$$= 2z^{1-\theta}\int_{z\eta}^1 y^{-1}e^{-y^2}\,dy + 2z^{1-\theta}\int_1^\infty y^{-1/2}e^{-y^2}\,dy$$
$$\le \text{const}\; z^{1-\theta}\,|\log z\eta| + \text{const}\; z^{1-\theta}.$$

If the integral (2) is extended to the interval $\{0 \leq y < \infty\}$, it becomes

$$\eta^{-1}z^{-\theta}\frac{1}{\theta}\int_0^\infty e^{-y^2}\,dy + \eta^{\theta-1}\Big(-\frac{1}{\theta}+\frac{1}{2}+\int_1^\infty \bar{B}_1(v)(\theta-1)v^{\theta-2}\,dv\Big) \times \Big(\int_0^\infty y^{-\theta}e^{-y^2}\,dy\Big) = \frac{\pi^{1/2}}{2\eta\theta z^\theta}+\frac{A(\theta)}{\theta}\eta^{\theta-1},$$

where $A(\theta)$ is a bounded function of $\{0 \leq \theta \leq \frac{1}{2}\}$, and this differs from the original integral (2) by at most

$$\eta^{-1}z^{-\theta}\frac{1}{\theta}\int_0^{z\eta} e^{-y^2}\,dy + \eta^{\theta-1}\Big(-\frac{1}{\theta}+\frac{1}{2}+\int_1^\infty \bar{B}_1(v)(\theta-1)v^{\theta-2}\,dv\Big) \times \Big(\int_0^{z\eta} y^{-\theta}e^{-y^2}\,dy\Big)$$
$$\leq \frac{z^{1-\theta}}{\theta}+\eta^{\theta-1}\,\text{const}\,\frac{A(\theta)}{\theta}\int_0^{z\eta} y^{-\theta}\,dy \leq \text{const}\,\frac{z^{1-\theta}}{\theta}$$

because $A(\theta)$ is bounded. Therefore the original integral (1) differs from

$$\frac{\pi^{1/2}}{2\eta\theta z^\theta}+\frac{A(\theta)}{\theta}\eta^{\theta-1}$$

by at most

$$\text{const}\,z^{1-\theta}\,|\log z\eta| + \text{const}\,z^{1-\theta}/\theta.$$

Finally, using this estimate of (1) in Σ_1 shows that Σ_1 differs from

$$(3)\qquad 4\sum_{\kappa\lambda\mu\nu}\frac{\beta_\kappa\beta_\lambda\beta_\mu\beta_\nu}{\lambda\nu} \times\left[\frac{\pi^{1/2}q}{2\kappa\mu(2\pi\sin 2\delta)^{1/2}\theta z^\theta}+\frac{A(\theta)}{\theta}(2\pi\sin 2\delta)^{(\theta-1)/2}\Big(\frac{q}{\kappa\mu}\Big)^{1-\theta}\right]$$

by at most a constant times

$$(4)\qquad \sum_{\kappa\lambda\mu\nu}\frac{|\beta_\kappa\beta_\lambda\beta_\mu\beta_\nu|}{\lambda\nu}\left\{\frac{z^{1-\theta}}{\theta}+z^{1-\theta}\left|\log\left[\frac{z\kappa\mu}{q}(2\pi\sin 2\delta)^{1/2}\right]\right|\right\}.$$

In the nonzero terms κ, λ, μ, ν are all at most X, so $\kappa\mu/q$ lies between X^{-2} and X^2. On the other hand $\sin 2\delta$ is of the order of magnitude of δ; so, since X and z are at most $\delta^{-1/15}$, the logarithm in the error estimate (4) is at most a constant times $\log(1/\delta)$ in absolute value, and therefore grows rather slowly as $\delta \to 0$. Specifically

$$\frac{z\kappa\mu}{q}(2\pi\sin 2\delta)^{1/2} \geq \frac{1\cdot 1\cdot 1}{X^2}(2\pi)^{1/2}\delta^{1/2} \geq \text{const}\,\delta^{2/15}\delta^{1/2} \geq \delta,$$

$$\frac{z\kappa\mu}{q}(2\pi\sin 2\delta)^{1/2} \leq \frac{\delta^{-1/15}X^2}{1}(2\pi)^{1/2}(2\delta)^{1/2} \leq \text{const}\,\delta^{-1/5}\delta^{1/2} \leq 1,$$

$$\log\delta \leq \log\frac{z\kappa\mu}{q}(2\pi\sin 2\delta)^{1/2} \leq 0,$$

for all sufficiently small δ. Since

$$\sum_{\kappa\lambda\mu\nu} \frac{|\beta_\kappa\beta_\lambda\beta_\mu\beta_\nu|}{\lambda\nu} \le \sum_{\kappa\lambda\mu\nu} \frac{1}{\nu\lambda} = \Big(\sum_\kappa 1\Big)\Big(\sum_\lambda \frac{1}{\lambda}\Big)\Big(\sum_\mu 1\Big)\Big(\sum_\nu \frac{1}{\nu}\Big) \le X^4,$$

this shows that the error estimate (4) is at most a constant times

$$\begin{aligned} X^4 \frac{z^{1-\theta}}{\theta} \log\Big(\frac{1}{\delta}\Big) &\le \Big(\frac{1}{\delta}\Big)^{4/15}\Big(\frac{1}{\delta}\Big)^{1/15} \frac{1}{\theta z^\theta} \log\Big(\frac{1}{\delta}\Big) \\ &\le \frac{\delta^{-1/3}\delta^{1/2} \log(1/\delta) \log X}{\delta^{1/2}\theta z^\theta \log X} \\ &\le \frac{\delta^{1/6} \cdot \frac{1}{15}[\log(1/\delta)]^2}{\delta^{1/2}\theta z^\theta \log X} \end{aligned}$$

for all sufficiently small δ. Since this is smaller than the desired estimate $K(\delta^{1/2}\theta z^\theta \log X)^{-1}$ for small δ, it can be neglected and the desired estimate of Σ_1 is equivalent to the statement that the approximating quantity (3) is less than a constant times $(\delta^{1/2}\theta z^\theta \log X)^{-1}$ for all sufficiently small δ.

Consider now the approximating quantity (3). With the new notation

$$S(\sigma) = \sum_{\kappa\lambda\mu\nu} \Big(\frac{q}{\kappa\mu}\Big)^\sigma \frac{\beta_\kappa\beta_\lambda\beta_\mu\beta_\nu}{\lambda\nu}$$

(where for a given quadruple κ, λ, μ, ν the integer q is defined to be the greatest common divisor of $\kappa\nu$ and $\lambda\mu$), this quantity can be written

$$\text{(3')} \qquad \frac{\sqrt{2}}{\theta z^\theta (\sin 2\delta)^{1/2}} S(1) + 4\frac{A(\theta)}{\theta}(2\pi \sin 2\delta)^{(\theta-1)/2} S(1-\theta).$$

Since $A(\theta)$ is bounded and since $\sin 2\delta$ is essentially equal to 2δ, in order to estimate this quantity it will suffice to estimate the quadruple sum $S(\sigma)$. This will be done by first reducing the estimation of $S(\sigma)$ to the estimation of a double sum, by then reducing this to the estimation of a single sum, and by finally estimating the single sum.

The first step in the estimation of $S(\sigma)$ makes use of a generalization of Euler's famous identity

$$\text{(5)} \qquad n = \sum_{m|n} \phi(m)$$

where ϕ is the Euler ϕ-function. For present purposes† the most convenient definition of ϕ is

$$\phi(n) = n \sum_{m|n} \frac{\mu(m)}{m} = n \prod_{p|n} \Big(1 - \frac{1}{p}\Big).$$

In order to prove that the function ϕ so defined has property (5), let $Q(x)$

†It is easily shown that $\phi(n)$ is equal to the number of integers between 0 and n relatively prime to n, and this property is the usual definition of ϕ.

$= \sum_{n<x} n$ and let $\Phi(x) = \sum_{n<x} \phi(n)$ with the usual adjustment for integral values of x. Then the definition of ϕ is tantamount to

$$\Phi(x) = \sum_{m=1}^{\infty} \mu(m) Q\left(\frac{x}{m}\right)$$

[at any integer $x = n$ both sides jump by $\phi(n) = \sum_{m|n} \mu(m)n/m$] which by Möbius inversion is equivalent to

$$Q(x) = \sum_{m=1}^{\infty} \Phi\left(\frac{x}{m}\right)$$

which implies the desired conclusion (5) [because at $x = n$ the left side jumps by n and the right side by $\sum_{m|n} \phi(m)$]. In exactly the same way it follows that the function $\phi_a(n)$ defined by

$$\phi_a(n) = n^{1+a} \sum_{m|n} \frac{\mu(m)}{m^{1+a}} = n^{1+a} \prod_{p|n} \left(1 - \frac{1}{p^{1+a}}\right)$$

satisfies

$$n^{1+a} = \sum_{m|n} \phi_a(m).$$

This identity and the fact that q is the greatest common divisor of $\kappa\nu$ and $\lambda\mu$ imply that $S(\sigma)$ can be rewritten in the form

(6)
$$\begin{aligned} S(\sigma) &= \sum_{\kappa\lambda\mu\nu} \left[\sum_{\rho|\kappa\nu, \rho|\lambda\mu} \phi_{\sigma-1}(\rho) \right] \frac{\beta_\kappa \beta_\lambda \beta_\mu \beta_\nu}{\kappa^\sigma \mu^\sigma \lambda \nu} \\ &= \sum_{\rho} \phi_{\sigma-1}(\rho) \sum_{\rho|\kappa\nu, \rho|\lambda\mu} \frac{\beta_\kappa \beta_\lambda \beta_\mu \beta_\nu}{\kappa^\sigma \mu^\sigma \lambda \nu} \\ &= \sum_{\rho} \phi_{\sigma-1}(\rho) \left[\sum_{\rho|\kappa\nu} \frac{\beta_\kappa \beta_\nu}{\kappa^\sigma \nu} \right]^2, \end{aligned}$$

where for given ρ the inner sum is over all pairs of positive integers κ, ν such that ρ divides $\kappa\nu$. Since $\phi_{\sigma-1}(\rho)$ is easy to estimate, this reduces the estimation of $S(\sigma)$ to the estimation of a double sum.

Now let ρ be given. For each pair κ, ν with $\rho | \kappa\nu$ let d denote the smallest factor of κ such that κ/d is relatively prime to ρ, and let d_1 denote the smallest factor of ν such that ν/d_1 is relatively prime to ρ. In other words, let d and d_1, respectively, be the product of all prime factors of κ and ν, respectively, which divide ρ. Set $\kappa' = \kappa/d$, $\nu' = \nu/d_1$. Since $\rho | \kappa\nu$ implies $\rho | dd_1$ and since, given d, d_1, the range of κ', ν' is all integers relatively prime to ρ,

$$\sum_{\rho|\kappa\nu} \frac{\beta_\kappa \beta_\nu}{\kappa^\sigma \nu} = \sum_{\rho|dd_1} \left(\sum_{\kappa'} \frac{\beta_{d\kappa'}}{(d\kappa')^\sigma} \right) \left(\sum_{\nu'} \frac{\beta_{d_1\nu'}}{d_1 \nu'} \right),$$

where d, d_1 range over positive integers all of whose prime factors divide ρ, and κ', ν' range over positive integers none of whose prime factors divide ρ. Now since d, κ' are relatively prime, formula (2) of Section 11.3 for α_ν gives

$\alpha_{d\kappa'} = \alpha_d\alpha_{\kappa'}$; hence

$$\sum_{\kappa'} \frac{\beta_{d\kappa'}}{(d\kappa')^\sigma} = \sum_{d\kappa' \le X} \frac{\alpha_d\alpha_{\kappa'}}{\log X} \log\left(\frac{X}{d\kappa'}\right)(d\kappa')^{-\sigma}$$
$$= \frac{\alpha_d}{d^\sigma \log X} \sum_{\kappa' \le X/d} \frac{\alpha_{\kappa'}}{(\kappa')^\sigma} \log \frac{X}{d\kappa'},$$

and therefore

(7) $$\sum_{\rho|\kappa\nu} \frac{\beta_\kappa \beta_\nu}{\kappa^\sigma \nu} = \frac{1}{(\log X)^2} \sum_{\rho|dd_1} \frac{\alpha_d\alpha_{d_1}}{d^\sigma d_1} \sum_{\kappa' \le X/d} \frac{\alpha_{\kappa'}}{(\kappa')^\sigma}$$
$$\times \log \frac{X}{d\kappa'} \sum_{\nu' \le X/d_1} \frac{\alpha_{\nu'}}{\nu'} \log \frac{X}{d_1\nu'},$$

where ρ is given, d and d_1 range over positive integers whose prime factors all divide ρ, and κ' and ν' range over positive integers relatively prime to ρ.

The single sum

(8) $$\sum_{\kappa' \le Y} \frac{\alpha_{\kappa'}}{(\kappa')^\sigma} \log\left(\frac{Y}{\kappa'}\right),$$

where κ' ranges over positive integers relatively prime to ρ, can be estimated as follows. For $\sigma > 1$

$$\sum_{\kappa'} \frac{\alpha_{\kappa'}}{(\kappa')^\sigma} = \prod_{p \nmid \rho} \left(1 - \frac{1}{p^\sigma}\right)^{1/2} = \frac{1}{[\zeta(\sigma)]^{1/2}} \prod_{p|\rho} \left(1 - \frac{1}{p^\sigma}\right)^{-1\ 2}$$

virtually by the definition of $\alpha_{\kappa'}$. This is not applicable in the range $\{\frac{1}{2} \le \sigma \le 1\}$ which is under consideration ($\sigma = 1 - \theta$ or $\sigma = 1$), but it shows that in terms of Fourier analysis the problem of estimating (8) can be restated: Let the operator whose transform is $\zeta(s)^{-1/2} \prod_{p|\rho} (1 - p^{-s})^{-1/2}$ in $\operatorname{Re} s > 1$, namely,

$$f(x) \mapsto \sum_{\kappa'} \alpha_{\kappa'} f(\kappa' x),$$

be modified to

$$f(x) \mapsto \sum_{\kappa' \le Y} \alpha_{\kappa'} \log \frac{Y}{\kappa'} f(\kappa' x)$$

(for a large parameter Y) so as to be finite and hence to have a transform defined throughout the s-plane. Estimate the value of the transform of the modified operator at points of the line segment $\{\frac{1}{2} \le \sigma \le 1, t = 0\}$ of the s-plane. This can be done as follows.

Integration by parts in the basic formula of Section 3.3 gives

$$\frac{1}{2\pi i} \int_{a-i\infty}^{a+i\infty} x^s \frac{ds}{s^2} = \frac{1}{2\pi i} \int_{a-i\infty}^{a+i\infty} (\log x) x^s \frac{ds}{s} = \begin{cases} 0, & 0 < x \le 1, \\ \log x, & 1 \le x < \infty. \end{cases}$$

Thus termwise integration gives, for any series $\sum b_n$,

$$\frac{1}{2\pi i} \int_{a-i\infty}^{a+i\infty} \left\{\sum_{n=1}^{\infty} b_n n^{-s-\sigma}\right\} Y^s \frac{ds}{s^2} = \sum_{n \le Y} b_n\left(\log \frac{Y}{n}\right) n^{-\sigma}$$

provided a is such that $\sum b_n n^{-a-\sigma}$ converges absolutely. This formula expresses the transform of the modified operator in terms of the transform of the unmodified one and in the case at hand shows that

$$\sum_{\kappa' \leq Y} \frac{\alpha_{\kappa'}}{(\kappa')^\sigma} \log \frac{Y}{\kappa'} = \frac{1}{2\pi i} \int_{a-i\infty}^{a+i\infty} \frac{1}{[\zeta(s+\sigma)]^{1/2}} \prod_{p|\rho} \left(1 - \frac{1}{p^{s+\sigma}}\right)^{-1/2} Y^s \frac{ds}{s^2} \tag{9}$$

for any a such that $a + \sigma > 1$, that is, any a such that $a > 1 - \sigma$. If $\sigma \neq 1$, then the singularity of the integrand at $s = 1 - \sigma$ is not serious and the line of integration can be moved over to $a = 1 - \sigma$. To prove a precise statement to this effect, it is necessary to have an estimate of $|\zeta(s+\sigma)^{-1/2}|$ in the half-plane $\operatorname{Re} s \geq 1 - \sigma$ or, what is the same, an estimate of $|\zeta(s)|^{-1}$ in the half-plane $\operatorname{Re} s \geq 1$. A simple estimate of this sort is

$$\left|\frac{1}{\zeta(s)}\right| \leq A|s-1| \qquad (\operatorname{Re} s \geq 1) \tag{10}$$

for some positive constant A. This estimate is proved below. Using it, it is easy to prove that the integral (9) converges for $a = 1 - \sigma$, that the equation still holds, and that the value of the left side is therefore at most

$$\begin{aligned}
&\frac{1}{2\pi} \int_{-\infty}^{\infty} A^{1/2} |t|^{1/2} \prod_{p|\rho} \left(1 - \frac{1}{p}\right)^{-1/2} Y^{1-\sigma} \frac{dt}{[(1-\sigma)^2 + t^2]} \\
&\quad = \left[\prod_{p|\rho} \left(1 + \frac{1}{p}\right)\left(1 - \frac{1}{p^2}\right)^{-1}\right]^{1/2} Y^{1-\sigma} \frac{A^{1/2}}{\pi} \\
&\qquad \times \int_0^{\infty} \frac{t^{3/2}}{t^2 + (1-\sigma)^2}\, d \log t \\
&\quad \leq \prod_{p|\rho} \left(1 + \frac{1}{p}\right)^{1/2} \left[\prod_{p} \left(1 - \frac{1}{p^2}\right)^{-1}\right]^{1/2} Y^{1-\sigma} \frac{A^{1/2}}{\pi} \\
&\qquad \times \int_0^{\infty} \frac{[(1-\sigma)u]^{3/2}}{(1-\sigma)^2(u^2+1)}\, d \log u \\
&\quad = \text{const} \prod_{p|\rho} \left(1 + \frac{1}{p}\right)^{1/2} Y^{1-\sigma}(1-\sigma)^{-1/2}
\end{aligned}$$

in absolute value when $\sigma < 1$. If $\sigma = 1$, then the line of integration cannot be moved over to $a = 1 - \sigma$ because of the factor s^2 in the denominator of the integrand. Instead, take as the path of integration the line from $-i\infty$ to $-iB$, the semicircle in $\operatorname{Re} s \geq 0$ from $-iB$ to $+iB$, and the line from iB to $i\infty$, where B is a constant to be determined later. Using Cauchy's theorem and the estimate (10), it is easily seen that formula (9) holds with this new path of integration in place of the line $\operatorname{Re} s = a$. Along the semicircle $dz/z = i\, d\theta$, and the integral along this portion has modulus at most

$$\frac{1}{2\pi}(AB)^{1/2} \prod_{p|\rho} \left(1 - \frac{1}{p}\right)^{-1/2} Y^B \int_{-\pi}^{\pi} \frac{d\theta}{B}$$

[using the assumption $Y > 1$ which is justified by the fact that the quantity (8) to be estimated is zero if $Y \leq 1$] which is less than a constant times

$$B^{-1/2} \prod_{p|\rho} \left(1 + \frac{1}{p}\right)^{1/2} e^{B \log Y}.$$

The integral over the portion of the path of integration along the imaginary axis has modulus at most

$$\frac{2}{2\pi} \int_B^\infty A^{1/2} t^{1/2} \prod_{p|\rho} \left(1 - \frac{1}{p}\right)^{-1/2} \frac{dt}{t^2} \leq \text{const} \prod_{p|\rho} \left(1 + \frac{1}{p}\right)^{1/2} B^{-1/2}.$$

These two estimates are of the same order of magnitude when $B \log Y$ is of the order of magnitude of 1. Therefore assume $Y > 1$ and set $B = (\log Y)^{-1}$. It follows that when $\sigma = 1$ the quantity (9) in question is at most a constant times $(\log Y)^{1/2} \prod_{p|\rho} (1 + p^{-1})^{1/2}$. This estimate extends from $\sigma = 1$ to $1 - \frac{1}{2}(\log Y)^{-1} \leq \sigma \leq 1$ if the path of integration is taken to be the line $\operatorname{Re} s = 1 - \sigma$ with a detour around the right side of the circle $|s| = (\log Y)^{-1}$ where it intercepts the line. The result is again that the quantity (9) to be estimated has modulus at most a constant times $(\log Y)^{1/2} \prod_{p|\rho} (1 + p^{-1})^{1/2}$. But for $\frac{1}{2} \leq \sigma \leq 1 - \frac{1}{2}(\log Y)^{-1}$ the previous estimate shows it has modulus at most a constant times $\prod_{p|\rho} (1 + p^{-1})^{1/2} Y^{1-\sigma} (\log Y)^{1/2}$. Therefore in any case (even $Y \leq 1$) there is a constant K' such that

$$\left| \sum_{\kappa' \leq Y} \frac{\alpha_{\kappa'}}{(\kappa')^\sigma} \log \frac{Y}{\kappa'} \right| \leq K' \prod_{p|\rho} \left(1 + \frac{1}{p}\right)^{1/2} Y^{1-\sigma} |\log Y|^{1/2}. \tag{11}$$

This completes the estimate of the single sum except for the proof of the needed estimate (10) of $|\zeta(s)|^{-1}$. Since this estimate implies $\zeta(1 + it) \neq 0$, its proof cannot be expected to be entirely elementary. The trigonometric inequality $3 + 4 \cos \theta + \cos 2\theta \geq 0$ of Section 5.2 combined with the formula $\log \zeta(s) = \int_0^\infty x^{-s} \, dJ(x)$ of Section 1.11 gives

$$\begin{aligned} &\operatorname{Re}\{3 \log \zeta(\sigma) + 4 \log \zeta(\sigma + it) + \log \zeta(\sigma + 2it)\} \\ &\quad = \int_0^\infty x^{-\sigma} \operatorname{Re}\{3 + 4x^{-it} + x^{-2it}\} \, dJ(x) \geq 0, \end{aligned}$$

$$\begin{aligned} 4 \log |\zeta(\sigma + it)| &\geq -3 \log |\zeta(\sigma)| - \log |\zeta(\sigma + 2it)| \\ |\zeta(\sigma + it)| &\geq |\zeta(\sigma)|^{-3/4} |\zeta(\sigma + 2it)|^{-1/4}, \end{aligned}$$

for all $\sigma > 1$. Since $(s - 1)\zeta(s)$ is bounded on the interval $1 \leq s \leq 2$ and since $|\zeta(\sigma + 2it)|$ is less than a constant times $\log(2t)$ for $\sigma \geq 1$, $2t \geq 1$ (see Section 9.2), this shows that there is a positive constant K such that

$$|\zeta(\sigma + it)| \geq K(\sigma - 1)^{3/4} (\log t)^{-1/4}$$

for all $t \geq 1$, $1 < \sigma \leq 2$. This will be used to find a lower estimate of

$|\zeta(1+it)|$ by combining it with the fundamental theorem of calculus $\zeta(\sigma+it)-\zeta(1+it)=\int_1^\sigma \zeta'(u+it)\,du$ and the following estimate of $\zeta'(s)$ by Euler–Maclaurin summation:

$$\begin{aligned}
-\zeta'(s) &= \sum_{n=1}^{\infty} \frac{\log n}{n^s} \\
&= \sum_{n=1}^{N-1} \frac{\log n}{n^s} + \int_N^\infty \frac{\log u}{u^s}\,du + \frac{1}{2}\frac{\log N}{N^s} \\
&\quad + \int_N^\infty \bar{B}_1(u)\left(\frac{1}{u^{s+1}} - \frac{s\log u}{u^{s+1}}\right) du \\
&= \sum_{n=1}^{N-1} \frac{\log n}{n^s} + \frac{\log N}{(s-1)N^{s-1}} + \frac{1}{(s-1)^2N^{s-1}} + \frac{1}{2}\frac{\log N}{N^s} \\
&\quad + \int_N^\infty \bar{B}_1(u)\frac{1-s\log u}{u^{s+1}}\,du
\end{aligned}$$

holds at first for $\operatorname{Re} s > 1$ but then by analytic continuation holds for the entire halfplane $\operatorname{Re} s > 0$ where the integral on the right converges. With $s=\sigma+it$, $N=[t]$ this gives

$$\begin{aligned}
|\zeta'(\sigma+it)| \le \sum_{n=1}^{N-1} \frac{\log t}{n} &+ \frac{\log t}{[(\sigma-1)^2+t^2]^{1/2}} + \frac{1}{(\sigma-1)^2+t^2} \\
&+ \frac{\log t}{2} + \frac{1}{2} + |\sigma+it|\left|\int_N^\infty \bar{B}_1(u)\frac{\log u}{u^{s+1}}\,du\right|
\end{aligned}$$

for $\sigma \ge 1$; hence $|\zeta'(\sigma+it)|$ is less than a constant times $(\log t)^2$ for $t \ge 2$, which gives

$$\begin{aligned}
|\zeta(1+it)| &\ge |\zeta(\sigma+it)| - \int_1^\sigma |\zeta'(u+it)|\,du \\
&\ge K_1\frac{(\sigma-1)^{3/4}}{(\log t)^{1/4}} - K_2(\sigma-1)(\log t)^2
\end{aligned}$$

for $1<\sigma<2$, $t\ge 2$. These two terms are of the same order of magnitude when $(\sigma-1)^{1/4}(\log t)^{9/4}$ is of the order of magnitude of 1, so choose σ by $(\sigma-1)=c(\log t)^{-9}$ to find

$$|\zeta(1+it)| \ge (K_1c^{3/4}-K_2c)(\log t)^{-7}$$

which gives $|\zeta(1+it)| \ge K_3(\log t)^{-7}$ when c is sufficiently small. Therefore $|\zeta(1+it)^{-1}t^{-1}| \le K_3^{-1}(\log t)^7 t^{-1} \le K_4$ for $t \ge 2$, so the function $[\zeta(s)(s-1)]^{-1}$ is bounded on the line segment $\{\operatorname{Re} s = 1, \operatorname{Im} s \ge 2\}$. Since it is also bounded on the rectangle $\{1 \le \operatorname{Re} s \le 2, |\operatorname{Im} s| \le 2\}$ and on the halfplane $\operatorname{Re} s \ge 2$, it follows from Lindelöf's theorem (Section 9.2) that it is bounded on the strip $\{1 \le \operatorname{Re} s \le 2\}$; hence it is bounded on the halfplane $\operatorname{Re} s \ge 1$ and the estimate (10) follows.

The estimate (11) of the single sum can be used to find that the double sum (7) is at most

$$\frac{1}{(\log X)^2} \sum_{\rho|dd_1, d<X, d_1<X} \frac{|\alpha_d||\alpha_{d_1}|}{d^\sigma d_1} K' \prod_{p|\rho}\left(1+\frac{1}{p}\right)^{1/2}\left(\frac{X}{d}\right)^{1-\sigma}\left(\log\frac{X}{d}\right)^{1/2}$$
$$\times K' \prod_{p|\rho}\left(1+\frac{1}{p}\right)^{1/2}\left(\log\frac{X}{d_1}\right)^{1/2}$$

which, since $[\log(X/d)\log(X/d_1)]^{1/2} \le \log X$, is at most a constant times

$$\frac{1}{\log X}\prod_{p|\rho}\left(1+\frac{1}{p}\right)X^{1-\sigma}\sum_{\rho|dd_1, d<X, d_1<X}\frac{|\alpha_d||\alpha_{d_1}|}{dd_1},$$

where ρ and X are given and where d, d_1 range over integers all of whose prime factors divide ρ. Let D represent numbers all of whose prime factors divide ρ. Then

$$\sum_{\rho|dd_1, d<X, d_1<X}\frac{|\alpha_d||\alpha_{d_1}|}{dd_1} = \sum_D \sum_{\substack{dd_1=\rho D\\ d<X, d_1<X}}\frac{|\alpha_d||\alpha_{d_1}|}{\rho D} \tag{12}$$

and for any fixed D the terms of the sum corresponding to D are at most

$$\frac{1}{\rho D}\sum_{dd_1=\rho D}|\alpha_d||\alpha_{d_1}|,$$

where $dd_1 = \rho D$ ranges over all possible factorizations of ρD. Now since

$$\left|\binom{+\frac{1}{2}}{n}\right| \le (-1)^n\binom{-\frac{1}{2}}{n},$$

the terms of the expansion of $(1-p^{-s})^{-1/2}$ as a Dirichlet series, which are all positive, dominate in absolute value the corresponding terms of the expansion of $(1-p^{-s})^{1/2}$. Consequently the terms of the expansion of $\prod_p(1-p^{-s})^{-1/2} = \zeta(s)^{1/2}$ dominate in absolute value the corresponding terms of the expansion of $\prod_p(1-p^{-s})^{1/2} = \zeta(s)^{-1/2}$. Since these latter terms have absolute value $|\alpha_n|/n^s$, squaring both sides shows that $\sum_{jk=n}|\alpha_j||\alpha_k|$ is less than the coefficient of n^{-s} in $[\zeta(s)^{1/2}]^2 = \zeta(s)$ which is one. Therefore (12) is at most

$$\rho^{-1}\sum_D D^{-1} = \rho^{-1}\prod_{p|\rho}\left(1-\frac{1}{p}\right)^{-1} = \rho^{-1}\prod_{p|\rho}\left(1-\frac{1}{p^2}\right)^{-1}\left(1+\frac{1}{p}\right)$$
$$\le \rho^{-1}\zeta(2)\prod_{p|\rho}\left(1+\frac{1}{p}\right)$$

and the double sum (7) to be estimated is at most a constant times

$$\frac{1}{\log X}\prod_{p|\rho}\left(1+\frac{1}{p}\right)^2 X^{1-\sigma}\frac{1}{\rho}.$$

Therefore the quadruple sum $S(\sigma)$ has, by (6), modulus at most a constant times

$$\sum_{\rho \le X^2} |\phi_{\sigma-1}(\rho)| \left[\prod_{p|\rho} \left(1 + \frac{1}{p}\right)^2 \frac{X^{1-\sigma}}{\rho \log X}\right]^2$$

because $\rho > X^2$ and $\rho | \kappa\nu$ imply $\beta_\kappa \beta_\nu = 0$. Now $|\phi_{\sigma-1}(\rho)| = \rho^\sigma \prod_{p|\rho} (1 - p^{-\sigma}) < \rho^\sigma$ and $\prod_{p|\rho}[1 + (1/p)]^4$ is less than a constant times $\prod_{p|\rho}(1 + p^{-1/2})$. [To prove the latter fact, it suffices to note that $(1 + x)^4 < 1 + x^{1/2}$ for all sufficiently small x, so the quotient $(1 + p^{-1})^4/(1 + p^{-1/2})$ is less than one for all but a finite number of primes p.] Therefore $|S(\sigma)|$ is at most a constant times

$$\begin{aligned} \frac{X^{2-2\sigma}}{(\log X)^2} \sum_{\rho \le X^2} \rho^{\sigma-2} \prod_{p|\rho}\left(1 + \frac{1}{p^{1/2}}\right) &\le \frac{X^{2-2\sigma}}{(\log X)^2} \sum_{\rho \le X^2} \rho^{\sigma-2} \sum_{n|\rho} n^{-1/2} \\ &= \frac{X^{2-2\sigma}}{(\log X)^2} \sum\sum_{mn \le X^2} (mn)^{\sigma-2} n^{-1/2} \\ &\le \frac{X^{2-2\sigma}}{(\log X)^2} \sum_{n=1}^{\infty} n^{-3/2} \sum_{m \le X^2} m^{-1} \end{aligned}$$

which is at most a constant times $X^{2-2\sigma}(\log X)^{-1}$. Finally, this estimate of $S(\sigma)$ shows that (3′) has modulus at most a constant times

$$\frac{1}{\theta z^\theta (\sin 2\delta)^{1/2}} \frac{1}{\log X} + \frac{(\sin 2\delta)^{\theta/2}}{\theta(\sin 2\delta)^{1/2}} \frac{X^{2\theta}}{\log X}.$$

Since $(\sin 2\delta)^{\theta/2}$ is less than a constant times $\delta^{\theta/2}$, while $X^{2\theta} \le \delta^{-2\theta/15}$, and $z^\theta \le \delta^{-\theta/15}$, the second term is less than or equal to a constant times the first. Therefore (3) and hence Σ_1 are less than a constant times $(\delta^{1/2}\theta z^\theta \log X)^{-1}$ for δ sufficiently small. This completes the estimate of Σ_1.

It remains to show that Σ_2 is insignificant compared to $(\delta^{1/2}\theta z^\theta \log X)^{-1}$. Now Σ_2 is the integral of a sum of terms of the form

$$4u^{-\theta} \frac{\beta_\kappa \beta_\lambda \beta_\mu \beta_\nu}{\lambda\nu} \exp(-Pu^2 - iQu^2),$$

where

$$P = \pi\left(\frac{m^2\kappa^2}{\lambda^2} + \frac{n^2\mu^2}{\nu^2}\right) \sin 2\delta > 0,$$

$$Q = \pi\left(\frac{m^2\kappa^2}{\lambda^2} - \frac{n^2\mu^2}{\nu^2}\right) \cos 2\delta \ne 0.$$

Every term with $Q > 0$ is the complex conjugate of a term with $Q < 0$, so when the finite sum over $\kappa\lambda\mu\nu$ is taken outside the integral sign, Σ_2 is twice the real part of

$$\sum_{\kappa\lambda\mu\nu} \frac{4\beta_\kappa \beta_\lambda \beta_\mu \beta_\nu}{\lambda\nu} \int_z^\infty \sum_{m=1}^{\infty} \sum_{n\mu/\nu < m\kappa/\lambda} u^{-\theta} e^{-Pu^2 - iQu^2}\, du,$$

where P, Q are defined in terms of the sextuple κ, λ, μ, ν, m, n as above. The sum over m, n can be integrated termwise because it is dominated by the integral of $\sum e^{-P}$ so $|\Sigma_2|$ is at most a constant times

$$\sum_{\kappa\lambda\mu\nu mn, Q>0} \left| \int_z^\infty u^{-\theta} e^{-Pu^2 - iQu^2}\, du \right|.$$

A typical term of this sextuple sum can be estimated by writing

$$\begin{aligned}\int_z^\infty & u^{1-\theta} e^{-Pu^2 - iQu^2}\, d\log u \\ &= \int_{Qz^2}^\infty e^{-Pv/Q} e^{-iv} (v/Q)^{(1-\theta)/2} \tfrac{1}{2}\, d\log v \\ &= Q^{(\theta-1)/2} \tfrac{1}{2} \int_{Qz^2}^\infty e^{-Pv/Q} (\cos v - i \sin v) v^{-(\theta+1)/2}\, dv.\end{aligned}$$

By the lemma of Section 9.7 the real and imaginary parts of this integral have absolute value at most

$$Q^{(\theta-1)/2} e^{-Pz^2} (Qz^2)^{-(\theta+1)/2} = Q^{-1} e^{-Pz^2} z^{-1-\theta} \le Q^{-1} e^{-P} z^{-1-\theta}$$

so Σ_2 has modulus at most a constant times

$$\frac{1}{z^{1+\theta}} \sum_{\kappa\lambda\mu\nu mn, Q>0} e^{-P} Q^{-1}. \tag{13}$$

For fixed κ, λ, μ, ν, m the sum over n is at most $z^{-1-\theta}$ times

$$\begin{aligned}\frac{1}{\pi \cos 2\delta} & \sum_{n\mu/\nu < m\kappa/\lambda} e^{-\pi m^2 \kappa^2 \lambda^{-2} \sin 2\delta} \left(\frac{m\kappa}{\lambda} - \frac{n\mu}{\nu}\right)^{-1} \left(\frac{m\kappa}{\lambda} + \frac{n\mu}{\nu}\right)^{-1} \\ &\le \frac{e^{-\pi m^2 X^{-2} \sin 2\delta}}{(\pi \cos 2\delta)(m\kappa/\lambda)} \sum_{n < m\kappa\nu/\lambda\mu} \left(\frac{m\kappa\nu - n\lambda\mu}{\lambda\nu}\right)^{-1} \\ &\le \frac{\lambda^2 \nu e^{-\pi m^2 X^{-2} \sin 2\delta}}{m\kappa\pi \cos 2\delta} \sum_{n < m\kappa\nu/\lambda\mu} \frac{1}{m\kappa\nu - n\lambda\mu}.\end{aligned}$$

The sum here is a sum of reciprocal positive integers spaced at intervals of $\mu\lambda$; the largest term of the sum is at most one and the other terms can be estimated using the inequality

$$\frac{1}{k} \le \frac{1}{\mu\lambda} \int_{k-\mu\lambda}^k \frac{dv}{v} \qquad (k > \mu\lambda)$$

to find that the above sum is at most

$$\begin{aligned}\frac{\lambda^2 \nu e^{-\pi m^2 X^{-2} \sin 2\delta}}{m\kappa\pi \cos 2\delta} & \left(1 + \frac{1}{\mu\lambda} \int_1^{m\kappa\nu - \lambda\mu} \frac{dv}{v}\right) \\ &\le \frac{X^3 e^{-\pi m^2 X^{-2} \sin 2\delta}}{m\pi \cos 2\delta} [1 + \log(m\kappa\nu)] \\ &\le \frac{X^3 e^{-\pi m^2 X^{-2} \sin 2\delta}}{m\pi \cos 2\delta} [1 + \log(mX^2)].\end{aligned}$$

This is independent of $\kappa, \lambda, \mu, \nu$ so summation over these variables multiplies the estimate by X^4 at most and the sextuple sum (13) is at most

$$\frac{X^7}{z^{1+\theta}\pi \cos 2\delta} \sum_{m=1}^{\infty} e^{-\pi m^2 X^{-2} \sin 2\delta} \left(\frac{1}{m} + \frac{\log(mX^2)}{m} \right).$$

For $m \geq e$ the function $m^{-1}[1 + \log(mX^2)]$ is less than $2m^{-1} \log(mX^2)$ and this function decreases, so this is at most

$$\frac{X^7}{z^{1+\theta}\pi \cos 2\delta} \left[\sum_{m=1}^{3} e^{-\pi m^2 X^{-2} \sin 2\delta} \left(\frac{1}{m} + \frac{\log(mX^2)}{m} \right) + 2 \int_3^\infty e^{-\pi u^2 X^{-2} \sin 2\delta} \frac{\log(uX^2)}{u}\, du \right].$$

The first term in brackets is at most $3[1 + \log(3X^2)] \leq \text{const} + \text{const} \log X \leq \text{const} \log(1/\delta)$. If the integral in brackets is split at the point where the exponent is $-\pi$, that is, where† $u = X(\sin 2\delta)^{-1/2}$, it is seen to be at most 2 times

$$\int_3^{X(\sin 2\delta)^{-1/2}} \log(uX^2)\, d \log(uX^2) + \int_1^\infty e^{-\pi v^2} \log \frac{vX^3}{(\sin 2\delta)^{1/2}}\, d \log v$$

$$\leq \frac{1}{2} \left[\log\left(\frac{X^3}{(\sin 2\delta)^{1/2}} \right) \right]^2 + \log \frac{X^3}{(\sin 2\delta)^{1/2}} \int_1^\infty e^{-\pi v^2}\, d \log v + \int_1^\infty e^{-\pi v^2} \log v\, d \log v.$$

Since $X^3(\sin 2\delta)^{-1/2} \leq \delta^{-3/15}\, \delta^{-1/2} \leq \delta^{-1}$, this is less than a constant times $[\log(1/\delta)]^2$ as $\delta \downarrow 0$. Therefore Σ_2 is less than a constant times

$$\frac{X^7}{z^{1+\theta}} \left(\log \frac{1}{\delta} \right)^2.$$

Since this quantity multiplied by $\delta^{1/2}\theta z^\theta \log X$ is at most $\frac{1}{2} z^{-1} \delta^{-7/15} \delta^{1/2} \cdot [\log(1/\delta)]^2 (1/15)[\log(1/\delta)]$ and because this quantity approaches zero as $\delta \downarrow 0$, this completes the proof of the lemma.

Note added in second printing: The theorems of this chapter have now been superseded by Levinson's theorem that *a third of the zeros lie on the line*, that is, $(NT) > \frac{1}{3}(T/2\pi)\log(T/2\pi)$ for all sufficiently large T. See his article "More than one third of zeros of Riemann's zeta-function are on $\sigma = \frac{1}{2}$," *Advances in Mathematics*, **13** (1974) pp. 383–436.

†Note that this point lies to the right of $u = 3$ when δ is sufficiently small.

Chapter 12

Miscellany

12.1 THE RIEMANN HYPOTHESIS AND THE GROWTH OF $M(x)$

Let dM be the Stieltjes measure such that the formula

$$\frac{1}{\zeta(s)} = \sum_{n=1}^{\infty} \frac{\mu(n)}{n^s} \qquad (\mathrm{Re}\, s > 1) \tag{1}$$

[(1) of Section 5.6] takes the form

$$\frac{1}{\zeta(s)} = \int_0^{\infty} x^{-s}\, dM(x) \qquad (\mathrm{Re}\, s > 1).$$

Then $M(x) = \int_0^x dM$ is a step function which is zero at $x = 0$, which is constant except at positive integers, and which has a jump of $\mu(n)$ at n. As usual, the value of M at a jump is by definition $\frac{1}{2}[M(n - \varepsilon) + M(n + \varepsilon)] = \sum_{j=1}^{n-1} \mu(j) + \frac{1}{2}\mu(n)$. Integration by parts gives for $\mathrm{Re}\, s > 1$

$$\begin{aligned}\frac{1}{\zeta(s)} &= \int_0^{\infty} d[x^{-s}M(x)] - \int_0^{\infty} M(x)\, d(x^{-s})\\ &= \lim_{X\to\infty}\left[X^{-s}M(X) + s\int_0^X M(x)x^{-s-1}\, dx\right]\\ &= s\int_0^{\infty} M(x)x^{-s-1}\, dx\end{aligned}$$

because the obvious inequality $|M(x)| \le x$ implies that $x^{-s}M(x) \to 0$ as $x \to \infty$ and that $\int_0^{\infty} M(x)x^{-s-1}\, dx$ converges, both provided $\mathrm{Re}\, s > 1$. Now if $M(x)$ grows less rapidly than x^a for some $a > 0$, then this integral for $1/\zeta(s)$ converges for all s in the halfplane $\{\mathrm{Re}(a - s) < 0\} = \{\mathrm{Re}\, s > a\}$, and therefore, by analytic continuation, the function $1/\zeta(s)$ is analytic in this halfplane. Since $1/\zeta(s)$ has poles on the line $\mathrm{Re}\, s = \frac{1}{2}$, this shows that $M(x)$ *does not* *grow less rapidly than x^a for any $a < \frac{1}{2}$.* Moreover, it shows that *in order to prove the Riemann hypothesis, it would suffice to prove that $M(x)$ grows less rapidly than $x^{(1/2)+\varepsilon}$ for all $\varepsilon > 0$.* Littlewood in his 1912 note [L12] on the three circles theorem proved that this sufficient condition for the Riemann hypothesis is also necessary; that is, he proved the following theorem.

Theorem The Riemann hypothesis is equivalent to the statement that for every $\varepsilon > 0$ the function $M(x)x^{-(1/2)-\varepsilon}$ approaches zero as $x \longrightarrow \infty$.

Proof It was shown above that the second statement implies the Riemann hypothesis. Assume now that the Riemann hypothesis is true. Then Backlund's proof in Section 9.4 shows [using the Riemann hypothesis to conclude that $F(s) = \zeta(s)$] that for every $\varepsilon > 0$, $\delta > 0$, and $\sigma_0 > 1$ there is a T_0 such that $|\log \zeta(\sigma + it)| < \delta \log t$ whenever $t \geq T_0$ and $\frac{1}{2} + \varepsilon \leq \sigma \leq \sigma_0$. Since $|\log \zeta(s)|$ is bounded on the halfplane $\{\text{Re } s \geq \sigma_0\}$, this implies that on the quarterplane $\{s = \sigma + it\colon \sigma = \frac{1}{2} + \varepsilon, t \geq T_0\}$ there is a constant K such that $|1/\zeta(s)| \leq Kt^\delta$. This is the essential step of the proof. Littlewood omits the remainder of the proof, stating merely that it follows from known theorems. One way of completing the proof is as follows.

The estimates (2) and (3) of Section 3.3 show that the error in the approximation

$$M(x) = \sum_{n<x} \mu(n) \sim \frac{1}{2\pi i} \int_{2-iT}^{2+iT} \left[\sum_{n=1}^{\infty} \mu(n)\left(\frac{x}{n}\right)^s\right] \frac{ds}{s}$$
$$= \frac{1}{2\pi i} \int_{2-iT}^{2+iT} \frac{x^s}{\zeta(s)} \frac{ds}{s},$$

for x not an integer is at most

$$\sum_{n<x} \frac{|\mu(n)|\,(x/n)^2}{\pi T \log(x/n)} + \sum_{n>x} \frac{|\mu(n)|\,(x/n)^2}{\pi T \log(n/x)}$$
$$\leq \frac{x^2}{\pi T}\left[\sum_{n<x} \frac{1}{n^2 \log(x/n)} + \sum_{n>x} \frac{1}{n^2 \log(n/x)}\right].$$

The first sum in brackets is at most

$$\sum_{n \leq x/2} \frac{1}{n^2 \log(x/n)} + \sum_{x/2<n<x} \frac{1}{n^2 \log\{1 + [(x-n)/n]\}}$$

assuming, of course, that x is not an integer. Since $\log(1 + y) \geq \frac{1}{2}y$ for $0 \leq y \leq 1$ this is at most†

$$\frac{1}{\log 2} \sum_{n \leq x/2} \frac{1}{n^2} + \sum_{x/2<n<x} \frac{1}{n^2} \cdot \frac{2n}{x-n} \leq \frac{\zeta(2)}{\log 2} + \frac{4}{x} \sum_{0<x-n<x/2} \frac{1}{x-n}$$
$$\leq \frac{\zeta(2)}{\log 2} + \frac{4}{x} \cdot \frac{1}{x-[x]} + \frac{4}{x} \sum_{0<j<x/2} \frac{1}{j}.$$

†As usual, $[x]$ denotes the largest integer less than x.

This shows that it is bounded for large x provided $x - [x]$ is not too small, say, for example, if x is a half integer (half an odd integer). The second sum in brackets can be estimated in a similar way to arrive at the conclusion that for half-integer values of x the error in the approximation

$$M(x) \sim \frac{1}{2\pi i} \int_{2-iT}^{2+iT} \frac{x^s}{\zeta(s)} \frac{ds}{s}$$

is less than a constant times x^2/T as $x \to \infty$. But by Cauchy's theorem (and the Riemann hypothesis) the integral on the right is equal to

$$\frac{1}{2\pi i} \int_{2-iT}^{(1/2)+\varepsilon-iT} \frac{x^s}{\zeta(s)} \frac{ds}{s} + \frac{1}{2\pi i} \int_{(1/2)+\varepsilon-iT}^{(1/2)+\varepsilon+iT} \frac{x^s}{\zeta(s)} \frac{ds}{s} + \frac{1}{2\pi i} \int_{(1/2)+\varepsilon+iT}^{2+iT} \frac{x^s}{\zeta(s)} \frac{ds}{s}.$$

The estimate $|1/\zeta(s)| \le Kt^\delta$ shows that the first integral and the third integral are each less than a constant times $x^2 \cdot KT^{\delta-1}$ while the middle integral is at most

$$\left| \frac{1}{2\pi i} \int_{(1/2)+\varepsilon-iT_0}^{(1/2)+\varepsilon+iT_0} \frac{x^s}{\zeta(s)} \frac{ds}{s} \right| + \frac{2}{2\pi} \int_{T_0}^{T} \frac{x^{(1/2)+\varepsilon} K t^\delta}{t} dt$$
$$\le x^{(1/2)+\varepsilon} \text{ const} + \frac{x^{(1/2)+\varepsilon} K T^\delta}{\pi\delta}.$$

Setting $T = x^2$ then shows that $M(x)$ is less than a constant times $x^{(1/2)+\varepsilon+2\delta}$ for all large half integers x. Since $M(x)$ changes by at most ± 1 between half integers, the same is true for all values of x and, since $\varepsilon > 0$ and $\delta > 0$ were arbitrary, it follows that the Riemann hypothesis implies $M(x)$ grows less rapidly than $x^{1/2+\varepsilon}$ for all $\varepsilon > 0$ as desired.

Corollary If the Riemann hypothesis is true, then the series (1) converges throughout the halfplane $\{\operatorname{Re} s > \frac{1}{2}\}$ to the function $1/\zeta(s)$.

Proof

$$\sum_{n<x} \mu(n)/n^s = \int_0^x u^{-s}\, dM(u) = x^{-s}M(x) + s \int_0^x M(u) u^{-s-1}\, du.$$

The theorem shows that if the Riemann hypothesis is true and if $\operatorname{Re} s > \frac{1}{2}$, then the limit of this expression as $x \to \infty$ exists and is equal to $s \int_0^\infty M(u) u^{-s-1}\, du$. This limit is $1/\zeta(s)$ for $\operatorname{Re} s > 1$ and hence by analytic continuation for $\operatorname{Re} s > \frac{1}{2}$ as well.

Stieltjes wrote to Hermite in 1885 that he had succeeded in proving the even stronger statement that $M(x) = O(x^{1/2})$—that is, $M(x)/x^{1/2}$ remains bounded as $x \to \infty$—and he observed that this implies the Riemann hypothesis. This letter [S6] from Stieltjes to Hermite is what lies behind Hadamard's startling statement in his paper [H2] that he is publishing his proof that the zeta function has no zeros on the line $\{\text{Re } s = 1\}$ only because Stieltjes' proof that it has no zeros in the halfplane $\{\text{Re } s > \frac{1}{2}\}$ has not yet been published and is probably much more difficult!

In retrospect it seems very unlikely that Stieltjes had actually proved the Riemann hypothesis. Although in his time this was quite new territory—Stieltjes was among the first to penetrate the mysteries of Riemann's paper—enough work has now been done on the Riemann hypothesis to justify extreme skepticism about any supposed proof, and one must be very skeptical indeed in view of the fact that Stieltjes himself was unable in later years to reconstruct his proof. All he says about it is that it was very difficult, that it was based on arithmetic arguments concerning $\mu(n)$, and that he put it aside hoping to find a simpler proof of the Riemann hypothesis based on the theory of the zeta function rather than on arithmetic. Moreover, even *assuming* the Riemann hypothesis, Stieltjes' stronger claim $M(x) = O(x^{1/2})$ has never been proved. All in all, except to remember that a first-rate mathematician once believed that the most fruitful approach to the Riemann hypothesis was through a study of the growth of $M(x)$ as $x \to \infty$, the incident is probably best forgotten.

12.2 THE RIEMANN HYPOTHESIS AND FAREY SERIES

It was shown in the preceding section that the Riemann hypothesis is equivalent to the arithmetic statement "$M(x) = o(x^{(1/2)+\varepsilon})$ for all $\varepsilon > 0$." Similarly, it was shown in Section 5.5 that the Riemann hypothesis is equivalent to the arithmetic statement "$\psi(x) - x = o(x^{(1/2)+\varepsilon})$ for all $\varepsilon > 0$." A third arithmetic statement equivalent to the Riemann hypothesis was found by Franel and Landau [F1] in the 1920s. Theirs deals with *Farey series.*

For a given real number $x > 1$ consider the rational numbers which, when expressed in lowest terms, have the denominators less than x. (For the sake of convenience assume x is not an integer.) The *Farey series*† corresponding to x is a complete set of representatives modulo 1 of these rational numbers, namely, the positive rationals less than or equal to 1 which can be expressed with denominators less than x. For example, for $x = 7\frac{1}{2}$ the Farey series is

$$\tfrac{1}{7}, \tfrac{1}{6}, \tfrac{1}{5}, \tfrac{1}{4}, \tfrac{2}{7}, \tfrac{1}{3}, \tfrac{2}{5}, \tfrac{3}{7}, \tfrac{1}{2}, \tfrac{4}{7}, \tfrac{3}{5}, \tfrac{2}{3}, \tfrac{5}{7}, \tfrac{3}{4}, \tfrac{4}{5}, \tfrac{5}{6}, \tfrac{6}{7}, 1.$$

†On the history of the name see Hardy and Wright [H7]. Note that the Farey series is not a series at all but a finite sequence.

A fascinating property of this sequence—and in fact the property which first attracted attention to it—is that if p/q, r/s are successive terms of the sequence, then $qr - ps = 1$. This property will not, however, play a role here. Let $A(x)$ denote the number of terms in the Farey series corresponding to x, for example, $A(7\frac{1}{2}) = 18$. Since the $A(x)$ terms of the Farey series are unequally spaced through the interval from 0 to 1, they will in general differ from the the equally spaced points $1/A(x), 2/A(x), \ldots, A(x)/A(x) = 1$. For $\nu = 1, 2, \ldots, A(x)$ let δ_ν denote the amount by which the νth term of the Farey series differs from $\nu/A(x)$; for example, when $x = 7\frac{1}{2}$, $\delta_5 = (2/7) - (5/18) = 1/126$. The theorem of Franel and Landau is that *the Riemann hypothesis is equivalent to the statement that* $|\delta_1| + |\delta_2| + \cdots + |\delta_{A(x)}| = o(x^{(1/2)+\varepsilon})$ *for all* $\varepsilon > 0$ as $x \longrightarrow \infty$.

The connection between Farey series and the zeta function, so surprising at first, can be deduced from the following formula. Let f be a real-valued function defined on the interval [0, 1]—or, perhaps better, let f be a *periodic* real-valued function of a real variable with period one—and let $r_1, r_2, \ldots, r_{A(x)} = 1$ denote the terms of the Farey series corresponding to x. Then†

$$\sum_{\nu=1}^{A(x)} f(r_\nu) = \sum_{k=1}^{\infty} \sum_{j=1}^{k} f\left(\frac{j}{k}\right) M\left(\frac{x}{k}\right). \tag{1}$$

Thus the rather irregular operation of summing f over the Farey series can be expressed more regularly using the function M and a double sum. Note that the right side of (1) is defined for all x and that it gives a natural extension of the left side to integer values $x = n$, namely, the mean of the value for $n + \varepsilon$ and the value for $n - \varepsilon$ or, what is the same, the sum of f over all positive rationals less than or equal to 1 with denominators less than or equal to n, counting those with denominator exactly n with weight $\frac{1}{2}$.

Formula (1) can be proved as follows. Let $D(x)$ be the function

$$D(x) = \begin{cases} 1 & \text{for } x > 1, \\ \frac{1}{2} & \text{for } x = 1, \\ 0 & \text{for } x < 1. \end{cases}$$

Then the definition of M gives $M(x) = \sum \mu(n)D(x/n)$. But by Möbius inversion this is equivalent to $D(x) = \sum M(x/n)$. Now for any fraction in lowest terms p/q ($0 < p \le q$, p and q relatively prime integers) the term $f(p/q) = f(2p/2q) = f(3p/3q) = \cdots$ occurs on the right side of (1) with the coefficient $M(x/q) + M(x/2q) + M(x/3q) + \cdots = D(x/q)$ which is one if $q < x$ and zero if $q > x$, which is its coefficient on the left side of (1). This completes the proof of (1).

†Note that the sum on the right is finite because all terms with $k > x$ are zero.

Now formula (1) applied to $f(u) = e^{2\pi iu}$ gives

$$\sum_{\nu=1}^{A(x)} e^{2\pi i r_\nu} = \sum_{k=1}^{\infty} \sum_{j=1}^{k} e^{2\pi i j/k} M\left(\frac{x}{k}\right).$$

But since $\sum_{j=1}^{k} e^{2\pi i j/k}$ is a sum of k complex numbers equally spaced around the unit circle, it is zero unless $k = 1$ in which case it is 1. Thus the right side is simply $M(x)$. Hence, with $A(x)$ abbreviated to A,

$$\begin{aligned} M(x) &= \sum_{\nu=1}^{A} e^{2\pi i r_\nu} = \sum_{\nu=1}^{A} e^{2\pi i[(\nu/A)+\delta_\nu]} \\ &= \sum_{\nu=1}^{A} e^{2\pi i\nu/A}(e^{2\pi i\delta_\nu} - 1) + \sum_{\nu=1}^{A} e^{2\pi i\nu/A} \\ |M(x)| &\le \sum_{\nu=1}^{A} |e^{2\pi i\delta_\nu} - 1| + 0 = \sum_{\nu=1}^{A} |e^{\pi i\delta_\nu} - e^{-\pi i\delta_\nu}| \\ &= 2\sum |\sin \pi\delta_\nu| \le 2\pi \sum_{\nu=1}^{A} |\delta_\nu| \end{aligned}$$

which proves one half of the Franel–Landau theorem, namely, the half which states that $\sum |\delta_\nu| = o(x^{(1/2)+\epsilon})$ implies the Riemann hypothesis.

The key step in the proof of the converse half is to apply formula (1) to the function† $\bar{B}_1(u) = u - [u] + \frac{1}{2}$ of Section 6.2. The technique used in Section 6.2 to prove $B_n(2u) = 2^{n-1}[B_n(u) + B_n(u + \frac{1}{2})]$ gives immediately the identity

$$B_n(ku) = k^{n-1}\left[B_n(u) + B_n\left(u + \frac{1}{k}\right) + \cdots + B_n\left(u + \frac{k-1}{k}\right)\right].$$

The same identity applies to $\bar{B}_n$ because by the periodicity of $\bar{B}_n$ it suffices to consider the case $0 \le u < 1/k$, in which all the values of $\bar{B}_n$ in the identity coincide with those of B_n. Thus

$$\bar{B}_1\left(u + \frac{1}{k}\right) + \bar{B}_1\left(u + \frac{2}{k}\right) + \cdots + \bar{B}_1(u + 1) = \bar{B}_1(ku) \tag{2}$$

and

$$\begin{aligned} \sum_{\nu=1}^{A} \bar{B}_1(u + r_\nu) &= \sum_{k=1}^{\infty} \sum_{j=1}^{k} \bar{B}_1\left(u + \frac{j}{k}\right) M\left(\frac{x}{k}\right) \\ &= \sum_{k=1}^{\infty} \bar{B}_1(ku) M\left(\frac{x}{k}\right). \end{aligned}$$

Let G denote this function. The two expressions for G lead to two different ways of evaluating the definite integral

$$I = \int_0^1 [G(u)]^2 \, du$$

†For the sake of neatness one should stipulate $\bar{B}_1(0) = 0$ so that the value at the jump is the middle value. This is not necessary here.

and these lead to the proof of the remaining half of the Franel–Landau theorem.

Consider first the evaluation of I using $G(u) = \sum_{\nu=1}^{A} \bar{B}_1(u + r_\nu)$. This equation shows that G jumps downward by 1 at each Farey fraction and increases like Au between Farey fractions. Moreover, since the r_ν other than $r_A = 1$ are symmetrically distributed around $\frac{1}{2}$, $\sum_{\nu=1}^{A-1} \bar{B}_1(r_\nu) = 0$, so the right-hand limit $G(0^+) = \lim_{u\downarrow 0} G(u) = \lim_{u\downarrow 0} \bar{B}(u+1)$ is $-\frac{1}{2}$. Thus between r_ν and $r_{\nu+1}$ the value of G is given by the formula $G(u) = -\frac{1}{2} + Au - \nu$. Hence

$$I = \sum_{\nu=1}^{A} \int_{r_{\nu-1}}^{r_\nu} \left(-\frac{1}{2} + Au - \nu + 1\right)^2 du$$

$$= \sum_{\nu=1}^{A} \frac{1}{A} \frac{(Au - \nu + \frac{1}{2})^3}{3}\Bigg|_{r_{\nu-1}}^{r_\nu}$$

provided that r_0 is defined to be 0. Since $Ar_\nu = A[r_\nu - (\nu/A) + (\nu/A)] = A\delta_\nu + \nu$, this gives

$$I = \frac{1}{3A} \sum_{\nu=1}^{A} \left[\left(A\delta_\nu + \frac{1}{2}\right)^3 - \left(A\delta_{\nu-1} - \frac{1}{2}\right)^3\right]$$

$$= \frac{1}{3A} \sum_{\nu=1}^{A} \left[\left(A\delta_\nu + \frac{1}{2}\right)^3 - \left(A\delta_\nu - \frac{1}{2}\right)^3\right]$$

(using $A\delta_0 - \frac{1}{2} = -\frac{1}{2} = A\delta_A - \frac{1}{2}$)

$$= \frac{1}{3A} \sum_{\nu=1}^{A} \left[2\cdot 3(A\delta_\nu)^2 \cdot \frac{1}{2} + 2\left(\frac{1}{2}\right)^3\right]$$

$$= A \sum_{\nu=1}^{A} \delta_\nu^2 + \frac{1}{12}$$

as an exact formula for I in terms of the δ_ν.

Now consider the evaluation of I using $G(u) = \sum_{k=1}^{\infty} \bar{B}_1(ku)M(x/k)$. Since the sum is finite, this gives immediately

$$I = \sum_{a=1}^{\infty} \sum_{b=1}^{\infty} M\left(\frac{x}{a}\right) M\left(\frac{x}{b}\right) \int_0^1 \bar{B}_1(au)\bar{B}_1(bu)\, du.$$

The coefficients $I_{ab} = \int_0^1 \bar{B}_1(au)\bar{B}_1(bu)\, du$ of this double series can be evaluated explicitly as follows. If $b = 1$, then it is

$$\int_0^a \bar{B}_1(v)\bar{B}_1\left(\frac{v}{a}\right) a^{-1}\, dv = a^{-1} \sum_{k=0}^{a-1} \int_0^1 \bar{B}_1(k+t)\bar{B}_1\left(\frac{k}{a} + \frac{t}{a}\right) dt$$

$$= a^{-1} \int_0^1 \bar{B}_1(t)\bar{B}_1\left(a \cdot \frac{t}{a}\right) dt$$

[by the periodicity of $\bar{B}_1$ and by (2)]

$$= a^{-1} \int_0^1 \left(t - \frac{1}{2}\right)^2 dt = (12a)^{-1}.$$

If b is relatively prime to a, then the same sequence of steps shows, since $(bk/a) \bmod 1$ for $k = 0, 1, \ldots, a-1$ gives each fraction j/a exactly once, that

$$\begin{aligned} I_{ab} &= a^{-1} \int_0^1 \bar{B}_1(t)\bar{B}_1\left(a \cdot \frac{bt}{a}\right) dt \\ &= a^{-1} \int_0^1 B_1(t)\bar{B}_1(bt)\, dt \\ &= a^{-1} I_{b1} = (12ab)^{-1}. \end{aligned}$$

Finally, if $c = (a, b)$ is the greatest common divisor of a and b, then $a = c\alpha$ and $b = c\beta$, where α and β are relatively prime and

$$\begin{aligned} I_{ab} &= \int_0^1 \bar{B}_1(c\alpha u)\bar{B}_1(c\beta u)\, du \\ &= c^{-1} \int_0^c \bar{B}_1(\alpha t)\bar{B}_1(\beta t)\, dt \\ &= I_{\alpha\beta} = (12\alpha\beta)^{-1} = c^2/12ab. \end{aligned}$$

Thus the final formula is

$$I = \sum_{a=1}^{\infty} \sum_{b=1}^{\infty} M\left(\frac{x}{a}\right) M\left(\frac{x}{b}\right) \frac{c^2}{12ab},$$

where $c = (a, b)$ is the greatest common divisor of a and b.

Now if the Riemann hypothesis is true, then for every $\varepsilon > 0$ there is a C such that $M(x) < Cx^{(1/2)+\varepsilon}$ for all x. Hence the Riemann hypothesis implies

$$\begin{aligned} I &< \sum_{a=1}^{\infty} \sum_{b=1}^{\infty} C^2 \left(\frac{x}{a}\right)^{1/2+\varepsilon} \left(\frac{x}{b}\right)^{1/2+\varepsilon} \frac{c^2}{12ab} \\ &= x^{1+2\varepsilon} \frac{C^2}{12} \sum_{a=1}^{\infty} \sum_{b=1}^{\infty} \frac{c^2}{(\alpha c \beta c)^{(3/2)+\varepsilon}} \\ &< x^{1+2\varepsilon} K \sum_{\alpha=1}^{\infty} \sum_{\beta=1}^{\infty} \sum_{c=1}^{\infty} \frac{1}{\alpha^{3/2}\beta^{3/2}c^{1+2\varepsilon}} \end{aligned}$$

(replacing a sum over relatively prime α, β with a sum over all α, β) which shows that I is less than a constant times $x^{1+2\varepsilon}$ for all $\varepsilon > 0$. The other expression for I shows then that for every $\varepsilon > 0$ the Riemann hypothesis implies

$$A \sum_{\nu=1}^{A} \delta_\nu^2 < Kx^{1+2\varepsilon},$$

where K is a constant depending on ε. But then by the Schwarz inequality

$$\begin{aligned} \sum |\delta_\nu| &= |\sum (\pm 1)\delta_\nu| \le [\sum (\pm 1)^2]^{1/2} [\sum (\delta_\nu)^2]^{1/2} \\ &= (A \sum \delta_\nu^2)^{1/2} < K^{1/2} x^{(1/2)+\varepsilon} \end{aligned}$$

as was to be shown.

12.3 DENJOY'S PROBABILISTIC INTERPRETATION OF THE RIEMANN HYPOTHESIS

One of the things which makes the Riemann hypothesis so difficult is the fact that there is no plausibility argument, no hint of a reason, however unrigorous, why it should be true. This fact gives some importance to Denjoy's probabilistic interpretation of the Riemann hypothesis which, though it is quite absurd when considered carefully, gives a fleeting glimmer of plausibility to the Riemann hypothesis.

Suppose an unbiased coin is flipped a large number of times, say N times. By the de Moivre–Laplace limit theorem the probability that the number of heads deviates by less than $KN^{1/2}$ from the expected number of $\frac{1}{2}N$ is nearly equal to $\int_{-(2K^2/\pi)^{1/2}}^{(2K^2/\pi)^{1/2}} \exp(-\pi x^2)\,dx$ in the sense that the limit of these probabilities as $N \longrightarrow \infty$ is equal to this integral. Thus if the total number of heads is subtracted from the total number of tails, the probability that the resulting number is less than $2KN^{1/2}$ in absolute value is nearly equal to $2\int_0^{(2K^2/\pi)^{1/2}} \exp(-\pi x^2)\,dx$, and therefore the probability that it is less than $N^{(1/2)+\varepsilon}$ for some fixed $\varepsilon > 0$ is nearly $2\int_0^{N^\varepsilon(2\pi)^{1/2}} \exp(-\pi x^2)\,dx$. The fact that this approaches 1 as $N \longrightarrow \infty$ can be regarded as saying that *with probability one the number of heads minus the number of tails grows less rapidly than* $N^{(1/2)+\varepsilon}$.

Consider now a very large square-free integer n, that is, a very large integer n with $\mu(n) \neq 0$. Then $\mu(n) = \pm 1$. It is perhaps plausible to say that $\mu(n)$ is plus or minus one "with equal probability" because n will normally have a large number of factors (the density of primes $1/\log x$ approaches zero) and there seems to be no reason why either an even or an odd number of factors would be more likely. Moreover, by the same principle it is perhaps plausible to say that successive evaluations of $\mu(n) = \pm 1$ are "independent" since knowing the value of $\mu(n)$ for one n would not seem to give any† information about its values for other values of n. But then the evaluation of $M(x)$ would be like flipping a coin once for each square-free integer less than x and subtracting the number of heads from the number of tails. It was shown above that for any given $\varepsilon > 0$ the outcome of this experiment for a large‡ number of flips is, with probability nearly one, less than the number of flips raised to the power $\frac{1}{2} + \varepsilon$ and *a fortiori* less than $x^{(1/2)+\varepsilon}$. Thus these probabi-

†An exception to this statement is that for any prime p, $\mu(pn)$ is either $-\mu(n)$ or zero. However, this principle can only be applied once for any p because $\mu(p^2 n) = 0$ and this "information" really says little more than that μ is determined by a formula and is not, in fact, a random phenomenon.

‡The number of flips goes to infinity as $x \longrightarrow \infty$ because, among other reasons, there are infinitely many primes, hence *a fortiori* infinitely many square-free integers (products of distinct primes).

listic assumptions about the values of $\mu(n)$ lead to the conclusion, ludicrous as it seems, that $M(x) = O(x^{(1/2)+\varepsilon})$ with probability one and hence that the Riemann hypothesis is true with probability one!

12.4 AN INTERESTING FALSE CONJECTURE

Riemann says in his memoir on $\pi(x)$ that "the known approximation $\pi(x) \sim \text{Li}(x)$ is correct only up to terms of the order $x^{1/2}$ and gives a value which is slightly too large." It appears from the context that he means that the *average* value of $\pi(x)$ is less than $\text{Li}(x)$ because he ignores the "periodic" terms $\text{Li}(x^\rho)$ in the formula for $J(x)$, but in the tables in Sections 1.1 and 1.17 it will be noticed that even the *actual* value of $\pi(x)$ is markedly less than $\text{Li}(x)$ in all cases considered. This makes it natural to ask whether it is indeed true that $\pi(x) < \text{Li}(x)$. This conjecture is supported by all the numerical evidence and by Riemann's observation that the next largest term in the formula for $\pi(x)$ is the negative term $-\frac{1}{2}\text{Li}(x^{1/2})$. Nonetheless it has been shown by Littlewood [L13] that this conjecture is *false* and that there exist numbers x for which $\pi(x) > \text{Li}(x)$. In fact Littlewood showed that it is false to such an extent that for very $\varepsilon > 0$ there exist values of x such that $\pi(x) > \text{Li}(x) + x^{(1/2)-\varepsilon}$.

This example shows the danger of basing conjectures on numerical evidence, even such seemingly overwhelming evidence as Lehmer's computations of $\pi(x)$ up to ten million. As a matter of fact (see Lehman [L6]) no actual value of x is known for which $\pi(x) > \text{Li}(x)$, although Littlewood's proof can be used to produce a very large X with the property that some x less than X has this property, which reduces the problem of finding such an x to a finite problem. More importantly, though, this example shows the danger of assuming that relatively small oscillatory terms can be neglected on the assumption that they probably will not reinforce each other enough to overwhelm a larger principal term. In the light of these observations, the evidence for the Riemann hypothesis provided by the computations of Rosser *et al.* and by the empirical verification of Gram's law loses all its force.

12.5 TRANSFORMS WITH ZEROS ON THE LINE

The problem of the Riemann hypothesis motivated a great deal of study of the circumstances under which an invariant operator on R^+ has a transform with zeros on $\text{Re}\, s = \frac{1}{2}$, that is, under which an integral of the form $\int_0^\infty x^{-s}F(x)\,dx$ has all of its zeros on $\text{Re}\, s = \frac{1}{2}$.

A very general theorem on this subject was proved by Polya [P1] in 1918. Stated in the terminology of Chapter 10, Polya's theorem is that *a real self-adjoint operator of the form* $f(x) \mapsto \int_{1/a}^{a} f(ux)F(u)\,du$ [where $F(u)$ is real and satisfies $u^{-1}F(u^{-1}) = F(u)$] *which has the property that*† $u^{1/2}F(u)$ *is nondecreasing on the interval* [1, *a*] *has the property that the zeros of its transform all lie on the line* Re $s = \frac{1}{2}$.

Simple and elegant though this theorem is, it gives little promise of leading to a proof of the Riemann hypothesis because the function $H(x)$ which occurs in the formula $2\xi(s) = \int_0^\infty u^{-s}H(u)\,du$ very definitely does not have the property that $u^{1/2}H(u)$ is nondecreasing and, in fact, as obviously must be the case if a positive function $H(u)$ is to have a transform $\int_0^\infty u^{-s}H(u)\,du$ defined for all s, the decrease of $H(u)$ for large u is very strong.

In 1927 Polya published [P2] a very different sort of theorem on the same general subject. This theorem states that *if* ϕ *is a polynomial which has all its roots on the imaginary axis, or if* ϕ *is an entire function which can be written in a suitable way as a limit of such polynomials, then if* $\int_0^\infty u^{-s}\,F(u)\,du$ *has all its zeros on* Re $s = \frac{1}{2}$, *so does* $\int_0^\infty u^{-s}F(u)\phi(\log u)\,du$. Here the conditions on F can be quite weak; it will suffice to consider the case $F(u) = o(\exp(-|\log u|^{2+\delta}))$ in which, in particular, $F(u)$ goes to zero much more rapidly than any power of u as $u \to 0$ or $u \to \infty$. Polya also proved that, conversely, if ϕ is an entire function of genus 0 or 1 which preserves in this way the property of a transform's having zeros on the line Re $s = \frac{1}{2}$, then ϕ must be a polynomial with purely imaginary roots or a limit of such polynomials.

The idea of the proof of this theorem is roughly as follows. If $P(t)$ is a polynomial with distinct real roots, then so is $rP(t) - P'(t)$ for any real number r; if $r = 0$, this follows from the fact that there is a zero of $P'(t)$ between any two consecutive zeros of $P(t)$ [thus accounting for all the zeros of $P'(t)$], and if $r \neq 0$, it follows from the fact that $rP(t) - P'(t)$ changes sign on each of the intervals (two of them half infinite) into which the real line is divided by the zeros of $P'(t)$ [thus accounting for all the zeros of $rP(t) - P'(t)$]. The change of variable $s = \frac{1}{2} + it$ then shows that if $P(s)$ is a polynomial with distinct roots all of which lie on Re $s = \frac{1}{2}$, then the same is true of $-irP(s) - P'(s)$ for any real r. In other words, the operator $-d/ds - ir$ preserves the property of a polynomial's having distinct roots all of which lie on Re $s = \frac{1}{2}$. Thus if $\int_0^\infty uF(u)^{-s}\,du$ has all its roots on Re $s = \frac{1}{2}$, and if it is a nice entire function which can be written in a suitable way as a limit of polynomials,

†The unnatural-seeming factor $u^{1/2}$ can be eliminated by renormalizing so that $\int_0^\infty x^{-s}F(x)\,dx$ is written $\int_0^\infty x^{(1/2)-s}[x^{1/2}F(x)]\,d\log x = \int_0^\infty x^{-z}\tilde{F}(x)\,d\log x$. Then the self-adjointness condition is simply $\tilde{F}(x) = \tilde{F}(x^{-1})$, Polya's condition is that $\tilde{F}$ be nondecreasing on [1, *a*], and the conclusion of the theorem is that the zeros lie on Im $z = 0$.

then it is reasonable to expect that

$$-ir \int_0^\infty u^{-s} F(u)\, du - \frac{d}{ds} \int_0^\infty u^{-s} F(u)\, du$$
$$= \int_0^\infty u^{-s} (\log u - ir) F(u)\, du$$

has the same property. Iterating this statement then gives Polya's theorem for any polynomial ϕ with imaginary roots and hence, on passage to the limit, for any suitable limit of such polynomials. For the actual proof see Polya [P2].

In order to apply Polya's theorem to obtain integrals of the form $\int_0^\infty u^{-s} F(u)\, du$ with zeros on the line $\operatorname{Re} s = \frac{1}{2}$, it is necessary to begin with such an integral. From the theory of Bessel functions it was known that $\int_0^\infty u^{-s} u^{-1/2} \exp(-\pi u^2 - \pi u^{-2})\, du$ is such an integral. Polya proved this directly, without reference to the theory of Bessel functions, as follows. For fixed s let

$$w(a) = \int_0^\infty u^{-s} u^{-1/2} e^{-au^2 - au^{-2}}\, du.$$

This can be regarded as a deformation of the given integral $w(\pi)$ to $0 = w(\infty)$. It satisfies a second-order linear differential equation, as can be seen by applying $(a\, d/da)^2$ to find

$$\left(a \frac{d}{da}\right)^2 w(a) = \int_0^\infty u^{-s-(1/2)} \left(a \frac{d}{da}\right)(-au^2 - au^{-2}) e^{-au^2 - au^{-2}}\, du$$
$$= \int_0^\infty u^{-s-(1/2)} [(-au^2 - au^{-2}) + (au^2 + au^{-2})^2] e^{-au^2 - au^{-2}}\, du.$$

The second part of this integrand is similar to $(u\, d/du)^2$ applied to $\exp(-au^2 - au^{-2})$ which is

$$\left(u \frac{d}{du}\right)(-2au^2 + 2au^{-2}) e^{-au^2 - au^{-2}}$$
$$= [(-4au^2 - 4au^{-2}) + 4(-au^2 + au^{-2})^2] e^{-au^2 - au^{-2}}$$
$$= 4[(-au^2 - au^{-2}) + (au^2 + au^{-2})^2 - 4a^2] e^{-au^2 - au^{-2}}.$$

Hence

$$\int_0^\infty u^{-s-1/2} \left[\left(u \frac{d}{du}\right)^2 e^{-au^2 - au^{-2}}\right] du = 4\left(a \frac{d}{da}\right)^2 w(a) - 16a^2 w(a).$$

Integration by parts on the left then gives, since the adjoint of $u\, d/du$ is $-(d/du)u$ which carries $u^{-s-(1/2)}$ to $(s - \frac{1}{2}) u^{-s-(1/2)}$,

$$[(s - \tfrac{1}{2})^2 + 16a^2] w(a) = 4\left(a \frac{d}{da}\right)^2 w(a)$$

which is the desired differential equation satisfied by $w(a)$. Let $W(a) =$

$a\,d/da\; w(a)$. Then

$$a\frac{d}{da}[W\bar{w}] = \tfrac{1}{4}[(s-\tfrac{1}{2})^2 + 16a^2]w(a)\bar{w}(a) + W\bar{W}.$$

Divide by a and integrate both sides from π to ∞ to find

$$-W(\pi)\bar{w}(\pi) = \int_\pi^\infty \frac{1}{a}\left\{\left[\frac{1}{4}\left(s-\frac{1}{2}\right)^2 + 4a^2\right]|w|^2 + |W|^2\right\} da$$

$$-\operatorname{Im} W(\pi)\bar{w}(\pi) = \frac{1}{4}2xy\int_\pi^\infty \frac{1}{a}|w|^2\, da,$$

where $s - \frac{1}{2} = x + iy$. If $w(\pi) = 0$, then, because the integral cannot be zero, either x or y must be zero. But if $y = 0$, then u^{-s} is positive real; hence $w(\pi) \neq 0$ directly from its definition. Thus $w(\pi) = 0$ implies $x = 0$, Re $s = \frac{1}{2}$, as was to be shown.

Thus $\int_0^\infty u^{-s}\phi(\log u)u^{-1/2}\exp(-\pi u^2 - \pi u^{-2})\, du$ has its zeros on the line Re $s = \frac{1}{2}$ whenever ϕ is as above. Although there seems to be no way to use this fact to prove the Riemann hypothesis, Polya used it to prove that a certain "approximation" to $\xi(s)$ does have its zeros on the line Re $s = \frac{1}{2}$. In the formula

$$2\xi(s) = \int_0^\infty u^{-s}\frac{d}{du}u^2\frac{d}{du}\sum_1^\infty e^{-\pi n^2u^2}\, du$$

the term with $n = 1$ predominates for u large. This term is

$$\int_0^\infty u^{-s}(4\pi^2u^4 - 6\pi u^2)e^{-\pi u^2}\, du.$$

If this is replaced by

$$2\xi^{**}(s) = \int_0^\infty u^{-s}[4\pi^2(u^4 + u^{-5}) - 6\pi(u^2 + u^{-3})]e^{-\pi u^2 - \pi u^{-2}}\, du,$$

the approximation is still good for large u, and since the integral is now the transform of a self-adjoint operator, the approximation must also be good for u near 0. Thus ξ^{**} is in some sense "like" ξ. However, ξ^{**} *does* have its zeros on the line Re $s = \frac{1}{2}$, a fact which follows from the above theorems once it is shown that ϕ defined by

$$\phi(\log u) = 4\pi^2(u^{9/2} + u^{-9/2}) - 6\pi(u^{5/2} + u^{-5/2}),$$

that is,

$$\phi(z) = 8\pi^2\cosh(9z/2) - 12\pi\cosh(5z/2)$$

can be written as a suitable limit of polynomials with imaginary zeros. By making appeal to the theory of entire functions, the proof of this statement can be reduced to the statement that ϕ itself has all its zeros on the imaginary axis. This can be done as follows.

Consider the function $P(y) = 8\pi^2 y^9 + 8\pi^2 y^{-9} - 12\pi y^5 - 12\pi y^{-5}$. This function has precisely 18 nonzero roots in the complex y-plane, and, since $\phi(z) = P(e^{z/2})$, in order to prove that the zeros of ϕ are all pure imaginary it will suffice to prove that these 18 roots all lie on the circle $|y| = 1$. Now on the unit circle $\bar{y} = y^{-1}$, so $P(y) = 2\,\mathrm{Re}\{Q(y)\}$ where $Q(y) = 8\pi^2 y^9 - 12\pi y^5$. Since $8\pi^2 > 12\pi$, all 9 roots of Q lie inside the unit circle. Therefore the integral of the logarithmic derivative of Q around the unit circle is $18\pi i$, that is, $Q(y) = re^{i\theta}$, where θ increases by 18π as y goes once around the unit circle. But since $P(y) = \mathrm{Re}\,Q(y) = r\cos\theta$, this implies $P(y)$ has 18 zeros on the circle $|y| = 1$ and accounts for all the zeros of P.

12.6 ALTERNATIVE PROOF OF THE INTEGRAL FORMULA

An interesting alternative proof of the Riemann–Siegel integral formula

(1) $$\frac{2\xi(s)}{s(s-1)} = F(s) + \overline{F(1-\bar{s})},$$

(2) $$F(s) = \pi^{-s/2}\Pi\left(\frac{s}{2} - 1\right)\int_{0\searrow 1} \frac{e^{-i\pi x^2}x^{-s}\,dx}{e^{i\pi x} - e^{-i\pi x}}$$

(see Section 7.9) was given by Kuzmin [K3] in 1934. Kuzmin's proof is altogether different from the proof given in Section 7.9, and it shows an interesting connection between formula (1)—which can be regarded as Riemann's third proof of the functional equation of ξ—and Riemann's second proof of the functional equation (see Section 1.7). What follows is a simplified proof of (1) based on Kuzmin's. It depends on the functional equation $G(x) = (1/x)G(1/x)$ (see Section 10.4) but not on the definite integral formula (5) of Section 7.4 which is the basis of Riemann's proof of (1).

Let $G(x) = \sum_{n=-\infty}^{\infty} \exp(-\pi n^2 x^2)$ as before. Then the formula

(3) $$\frac{2\xi(s)}{s(s-1)} = \int_0^\infty u^{s-1}[G(u) - 1]\,du \qquad (\mathrm{Re}\, s > 1),$$

which is easily proved by using absolute convergence to justify interchange of summation and integration,

$$\begin{aligned}
\int_0^\infty u^{s-1}\left[2\sum_{n=1}^\infty e^{-\pi n^2 u^2}\right]du &= 2\sum_{n=1}^\infty \int_0^\infty e^{-\pi n^2 u^2}\, u^s\, d\log u \\
&= \sum_{n=1}^\infty \int_0^\infty e^{-v}\left(\frac{v}{\pi n^2}\right)^{s/2} d\log v \\
&= \pi^{-s/2}\sum_{n=1}^\infty \frac{1}{n^s}\int_0^\infty e^{-v}\, v^{(s/2)-1}\,dv \\
&= \pi^{-s/2}\Pi\left(\frac{s}{2} - 1\right)\zeta(s) \\
&= \frac{\pi^{-s/2}\Pi(s/2)(s-1)\zeta(s)}{(s/2)(s-1)} = \frac{2\xi(s)}{s(s-1)},
\end{aligned}$$

is essentially the formula (1) of Section 1.7 on which Riemann bases his second proof of the functional equation. He breaks the integral at $u = 1$ and obtains what amounts to

$$\begin{aligned}\frac{2\xi(s)}{s(s-1)} &= \int_0^1 u^{s-1}[G(u) - 1]\,du + \int_1^\infty u^{s-1}[G(u) - 1]\,du \\ &= \int_1^\infty v^{-s}\left[\frac{1}{v}G\left(\frac{1}{v}\right) - \frac{1}{v}\right]dv + \int_1^\infty u^{s-1}[G(u) - 1]\,du \\ &= \int_1^\infty (v^{-s} + v^{s-1})[G(v) - 1]\,dv + \int_1^\infty v^{-s}\left[1 - \frac{1}{v}\right]dv \\ &= \int_1^\infty (v^{-s} + v^{s-1})[G(v) - 1]\,dv + \frac{1}{s-1} - \frac{1}{s}\end{aligned}$$

at first for $\operatorname{Re} s > 1$ but then by analytic continuation for all s. If the integral (3) is broken at $u = b$ instead of $u = 1$, the same sequence of steps gives

$$\begin{aligned}\frac{2\xi(s)}{s(s-1)} &= \int_{b^{-1}}^\infty v^{-s}\left[\frac{1}{v}G\left(\frac{1}{v}\right) - \frac{1}{v}\right]dv + \int_b^\infty u^{s-1}[G(u) - 1]\,du \\ &= \int_{b^{-1}}^\infty v^{-s}[G(v) - 1]\,dv + \frac{b^{s-1}}{s-1} - \frac{b^s}{s} \\ &\quad + \int_b^\infty u^{s-1}[G(u) - 1]\,du.\end{aligned}$$

Let $F_b(s)$ be the function which is defined by

$$F_b(s) = \int_b^\infty u^{s-1}G(u)\,du$$

for $\operatorname{Re} s < 0$ and therefore by

$$F_b(s) = \int_b^\infty u^{s-1}[G(u) - 1]\,du - \frac{b^s}{s}$$

for all $s \neq 0$. Then the above formula is simply

$$\text{(4)} \qquad \frac{2\xi(s)}{s(s-1)} = F_b(s) + F_{b^{-1}}(1-s)$$

and Riemann's second proof of the functional equation is simply the case $b = b^{-1} = 1$ of this formula. But $F_b(s)$ is defined not only for positive real b but for all values of b in the wedge $\{|\operatorname{Im}\log b| < \pi/4\}$ where G is defined; for example, the integral from b to ∞ in the definition of $F_b(s)$ can be taken to be the integral over the half-line $\{b + t \colon t \text{ positive real}\}$ parallel to the real axis. The complex conjugate of $F_b(s)$ is $F_{\bar b}(\bar s)$, so formula (4) has the same form as (1) whenever $b^{-1} = \bar b$; that is,

$$\text{(5)} \qquad \frac{2\xi(s)}{s(s-1)} = F_b(s) + \overline{F_b(1-\bar s)} \qquad (|b| = 1)$$

whenever b lies on the unit circle between $(-i)^{1/2}$ and $i^{1/2}$. Kuzmin proves the

Riemann–Siegel formula (1) by proving it is the limiting case of this formula (5) as $b \longrightarrow (-i)^{1/2}$; that is,

$$\lim_{b\to(-i)^{1/2}} F_b(s) = F(s) \qquad (s \neq 0)$$

when b approaches $(-i)^{1/2}$ along the unit circle. This clearly suffices to prove (1).

Kuzmin had already† studied formula (5) in 1930, prior to the publication of the Riemann–Siegel formula, and he had already shown that the limiting case could be written in the form

$$\text{(6)} \qquad \lim_{b\to(-i)^{1/2}} F_b(s) = \int_{(-i)^{1/2}}^{\infty} u^{s-1}[G(u) - 1]\,du - \frac{(-i)^{s/2}}{s}$$
$$= 2\int_{(-i)^{1/2}}^{\infty} u^{s-1} \sum_{n=1}^{\infty} e^{-\pi n^2 u^2}\,du - \frac{(-i)^{s/2}}{s}$$
$$= 2\sum_{n=1}^{\infty} \int_{(-i)^{1/2}}^{\infty} u^{s-1} e^{-\pi n^2 u^2}\,du - \frac{(-i)^{s/2}}{s}$$

for $s \neq 0$. These manipulations will be justified below. If Re $s < 0$, then the final formula can be written more simply as

$$\text{(7)} \qquad \lim_{b\to(-i)^{1/2}} F_b(s) = \sum_{n=-\infty}^{\infty} \int_{(-i)^{1/2}}^{\infty} u^{s-1} e^{-\pi n^2 u^2}\,du.$$

On the other hand, $F(s)$ can also be written as a sum over all integers by using the elementary‡ formula

$$\frac{1}{e^{i\pi x} - e^{-i\pi x}} = \frac{1}{2\pi i} \sum_{-\infty}^{\infty} \frac{(-1)^n x}{x^2 - n^2}$$

†It is interesting to note that this work [K2] of Kuzmin's, which preceded the publication of the Riemann–Siegel formula, was motivated by the wish to be able to compute $\zeta(\frac{1}{2} + it)$ for large t, as was Riemann's. With this and with the Hardy–Littlewood approximate functional equation, one has the feeling that after 70 years other mathematicians were getting up to where Riemann had been. However, it still seems rather doubtful that the Riemann–Siegel asymptotic formula would have been found to this day had it not been found by Riemann.

‡One way to prove this formula is to expand $f(t) = e^{2\pi ixt}$ as a Fourier series on the interval $\{-\frac{1}{2} \leq t \leq \frac{1}{2}\}$ to find $f(t) = \sum_{-\infty}^{\infty} a_n e^{2\pi int}$, where

$$a_n = \frac{(-1)^n(e^{i\pi x} - e^{-i\pi x})}{2\pi i(x - n)}$$

from which, with $t = 0$,

$$\frac{1}{e^{i\pi x} - e^{-i\pi x}} = \sum_{n=-\infty}^{\infty} \frac{(-1)^n}{2\pi i(x - n)} = \sum_{n=-\infty}^{\infty} \frac{(-1)^n x}{2\pi i(x^2 - n^2)}$$

as desired; this holds for real nonintegral x and hence for all nonintegral x by analytic continuation. Another way to prove it is to note that the two sides have the same poles and to then use a method like that of Chapter 2.

in the definition of $F(s)$ to find

$$(8) \qquad F(s) = \sum_{-\infty}^{\infty} \pi^{-s/2}\Pi\left(\frac{s}{2}-1\right)\frac{(-1)^n}{2\pi i}\int_{0\searrow 1} \frac{e^{-i\pi x^2}x^{1-s}\,dx}{x^2-n^2}$$

provided termwise integration is valid. Thus in order to deduce the Riemann–Siegel formula from his 1930 formula, Kuzmin had only to justify the termwise integration (8) and to prove

$$(9) \qquad \pi^{-s/2}\Pi\left(\frac{s}{2}-1\right)\frac{(-1)^n}{2\pi i}\int_{0\searrow 1} \frac{e^{-i\pi x^2}x^{1-s}\,dx}{x^2-n^2} = \int_{(-i)^{1/2}}^{\infty} u^{s-1}e^{-\pi n^2u^2}\,du$$

for Re $s < 0$, since this proves $\lim F_b(s) = F(s)$ for Re $s < 0$ and hence by analytic continuation for all s.

Consider the function

$$(10) \qquad \int_{i\infty}^{-i\infty} \frac{e^{\alpha x^2}x^{1-s}\,dx}{x^2-n^2}$$

for Re $s < 0$. This integral converges not only for positive real α but for all α in the halfplane Re $\alpha > 0$. If the line of integration is tilted slightly away from the imaginary axis and toward the line of slope -1 through the origin, then the halfplane of convergence of the integral is rotated slightly and comes to include the negative imaginary α-axis. Thus the function (10) can be continued analytically to have a value at $\alpha = -i\pi$. But then the line of integration can be moved to $0 \searrow 1$ without changing the value of the integral when $\alpha = -i\pi$. In other words, *the integral*

$$\int_{0\searrow 1} \frac{e^{-i\pi x^2}x^{1-s}\,dx}{x^2-n^2}$$

can be evaluated by finding the analytic continuation of the function (10) *to the negative imaginary α-axis and setting $\alpha = i\pi$* (s being fixed with Re $s < 0$). But a formula for the function (10) which "remains valid for α in the slit plane" can be found simply by the manipulations

$$\begin{aligned}
\int_{i\infty}^{-i\infty} \frac{e^{\alpha x^2}x^{1-s}\,dx}{x^2-n^2} &= \int_{i\infty}^{-i\infty} \frac{e^{\alpha x^2}x^{-s}\,d\log x}{1-x^{-2}n^2} \\
&= \int_0^{\infty} \frac{e^{-\alpha v^2}(-iv)^{-s}\,d\log v}{1+v^{-2}n^2} - \int_0^{\infty} \frac{e^{-\alpha v^2}(iv)^{-s}\,d\log v}{1+v^{-2}n^2} \\
&= [i^s + (-i)^s]\int_0^{\infty} \frac{e^{-\alpha v^2}v^{-s}\,d\log v}{1+v^{-2}n^2} \\
&= 2i\sin\frac{s\pi}{2}e^{\alpha n^2}\int_0^{\infty} \frac{e^{-\alpha(v^2+n^2)}v^{1-s}\,dv}{v^2+n^2} \\
&= 2i\sin\frac{s\pi}{2}e^{\alpha n^2}\int_0^{\infty}\left[\int_{\alpha}^{\infty} e^{-w(v^2+n^2)}\,dw\right]v^{1-s}\,dv \\
&= 2i\sin\frac{s\pi}{2}e^{\alpha n^2}\int_{\alpha}^{\infty} e^{-wn^2}\left[\int_0^{\infty} e^{-wv^2}v^{2-s}\,d\log v\right]dw
\end{aligned}$$

$$= 2i \sin \frac{s\pi}{2} e^{\alpha n^2} \int_\alpha^\infty e^{-wn^2} \int_0^\infty e^{-u}\left(\frac{u}{w}\right)^{(2-s)/2} \frac{1}{2}\, d \log u\, dw$$

$$= 2i \sin \frac{s\pi}{2} e^{\alpha n^2} \int_\alpha^\infty e^{-wn^2} w^{(s/2)-1} \Pi\left(-\frac{s}{2}\right)\frac{1}{2}\, dw$$

$$= 2i\Pi\left(-\frac{s}{2}\right)\left(\sin \frac{s\pi}{2}\right) e^{\alpha n^2} \int_{(\alpha/\pi)^{1/2}}^\infty e^{-\pi z^2 n^2}(\pi z^2)^{s/2}\, d \log z$$

$$= 2i\pi^{s/2}\Pi\left(-\frac{s}{2}\right)\left(\sin \frac{s\pi}{2}\right) e^{\alpha n^2} \int_{(\alpha/\pi)^{1/2}}^\infty e^{-\pi z^2 n^2} z^{s-1}\, dz$$

at first for α real and positive (say) but then by analytic continuation for all α in the slit plane. Thus with $\alpha = -i\pi$

$$\pi^{-s/2}\Pi\left(\frac{s}{2} - 1\right)\frac{(-1)^n}{2\pi i} \int_{0\searrow 1} \frac{e^{-i\pi x^2} x^{1-s}\, dx}{x^2 - n^2}$$

$$= \Pi\left(\frac{s}{2} - 1\right)\Pi\left(-\frac{s}{2}\right)\left(\sin \frac{s\pi}{2}\right)\frac{(-1)^n e^{-i\pi n^2}}{\pi} \int_{(-i)^{1/2}}^\infty e^{-\pi n^2 u^2} u^{s-1}\, du$$

$$= \int_{(-i)^{1/2}}^\infty e^{-\pi n^2 u^2} u^{s-1}\, du$$

for Re $s < 0$ as was to be shown. (The only complications for Re $s \geq 0$ occur in the terms with $n = 0$.)

This reduces the proof of the Riemann–Siegel formula (1) to Kuzmin's formula (7) and the termwise integration (8). The termwise integration (8) is easily justified by noting that for any x on the path of integration $0 \searrow 1$ the point x/n lies between $0 \searrow 1$ and the line of slope 1 through the origin ($n \neq 0$). Hence x^2/n^2 is bounded away from 1, say $|(x/n)^2 - 1| \geq K$, and therefore $|x^2 - n^2|^{-1} \leq K^{-1}n^{-2}$, from which it follows that the integrand converges uniformly and can therefore be integrated termwise on finite intervals. Toward the ends of the line of integration, the integrand is dominated by a constant times $\exp(-\pi |x|^2)|x|^{1-s}$ and can therefore be integrated termwise by the Lebesgue dominated convergence theorem. Similarly elementary arguments suffice to prove that the termwise integration

$$F_b(s) = \sum_{n=-\infty}^{\infty} \int_b^\infty u^{s-1}e^{-\pi n^2 u^2}\, du \qquad (\text{Re } s < 0)$$

is valid for all b *inside* the wedge $\{\text{Im} \log b < \pi/4\}$ where G is defined, and it suffices to prove that the limit as $b \to (-i)^{1/2}$ can be taken termwise. But integration by parts

$$\int u^{s-1}e^{-\pi n^2 u^2}\, du = -\frac{1}{2\pi n^2}u^{s-2}e^{-\pi n^2 u^2} + \left(\frac{s}{2} - 1\right)\frac{1}{\pi n^2}\int u^{s-3}e^{-\pi n^2 u^2}\, du$$

shows that if Re $s < 0$ and if b is inside or on the wedge and outside or on the unit circle, then

$$\left|\int_b^\infty u^{s-1}e^{-\pi n^2 u^2}\, du\right| \leq \frac{1}{n^2}\left[\frac{1}{2\pi} + \frac{|(s/2) - 1|}{\pi \cdot 2}\right],$$

and therefore the series for $F_b(s)$ converges uniformly (for fixed s) for b in this region. Therefore the limit as $b \longrightarrow (-i)^{1/2}$ can be taken termwise and the proof is complete.

12.7 TAUBERIAN THEOREMS

Perhaps the simplest formulation of the idea of the prime number theorem is the approximation $d\psi(x) \sim dx$. Since $d\psi(x) = (\log x)\, dJ(x)$ (see Section 3.1), this is equivalent to Riemann's approximation $dJ(x) \sim dx/\log x$ (see Section 1.18). The theory of *Tauberian theorems* gives a natural interpretation of the approximate formula $d\psi(x) \sim dx$ and shows a direct heuristic connection between it and the simple pole of $\zeta(s)$ at $s = 1$.

Of course the statement $d\psi(x) \sim dx$ makes no sense at all except as a statement about the *average* density of the point measure $d\psi(x)$. The theory of Tauberian theorems deals precisely with the notion of "average" and its various interpretations. Let the Abel average of a sequence $s_1, s_2, s_3, \ldots$ be defined to be

$$\lim_{r \uparrow 1} \frac{s_1 r + s_2 r^2 + s_3 r^3 + \cdots}{r + r^2 + r^3 + \cdots} = L \tag{1}$$

when this limit exists (and when, in particular, the infinite series in the numerator is convergent for all $r < 1$). In other words, the Abel average is found by taking the weighted average of the sequence $\{s_n\}$, counting the nth term with weight r^n, and then letting $r \uparrow 1$. Since for fixed r the weights r^n approach zero rather rapidly as $n \longrightarrow \infty$, the sum $\sum s_n r^n$ will converge unless the s_n grow rapidly in absolute value; on the other hand, for fixed n the weight r^n of the nth term approaches 1 as $r \uparrow 1$, and so for any fixed N the terms beyond the Nth eventually far outweigh the terms up to the Nth once r is near enough to 1. *Abel's theorem*† states that if the sequence $\{s_n\}$ converges to a limit L, then the Abel average exists and is equal to L. *Tauber's theorem* [T1] states that if $\{s_n\}$ is slowly changing in the sense that $|s_{n+1} - s_n| = o(1/n)$, then the converse is true. More precisely, Tauber's theorem says that if for every $\varepsilon > 0$ there is an N such that $|s_{n+1} - s_n| < \varepsilon/n$ whenever $n \geq N$ and if the limit (1) exists [it is easily shown that the condition on $\{s_n\}$ implies that the

†Abel's theorem is, however, more frequently stated for the series $a_1 = s_1, a_2 = s_2 - s_1, a_3 = s_3 - s_2, \ldots, a_n = s_n - s_{n-1}, \ldots$ of which the s_n are the partial sums (see, for example, [E1]). If the series converges, the s_n are bounded, so $\sum s_n r^n$ converges for $r < 1$ and by multiplication of power series $(\sum s_n r^n)(1 - r) = s_1 r + (s_2 - s_1) r^2 + \cdots = \sum a_n r^n$. Thus the statement (1) is identical to $\lim_{r \uparrow 1} \sum a_n r^n = L$, which is the statement that the series $\sum a_n$ is Abel summable. The usual method of proof of Abel's theorem is to put it in the form (1) by partial summation.

numerator of (1) converges for $r < 1$], then in fact the sequence must be convergent to L.

Both Abel's theorem and Tauber's theorem are "Tauberian theorems" in the modern sense, provided that the statement

$$\lim_{n\to\infty} s_n = L \tag{2}$$

is thought of as one possible interpretation of the statement that the "average" of $\{s_n\}$ is L. Then Abel's theorem states that if $\{s_n\}$ has the average L in the sense of (2), it has the average L in the sense of (1), and Tauber's theorem states that if it has the average L in the sense of (1) *and* if $|s_{n+1} - s_n| = o(1/n)$, then it has the average L in the sense of (2). In general, a "Tauberian theorem" is a theorem like these which permits a conclusion about one kind of average, given information about another kind of average.

An important step forward in the theory of Tauberian theorems, and perhaps the real beginning of the theory as such, was Littlewood's discovery in 1910 that the condition in Tauber's theorem can be very significantly weakened to $|s_{n+1} - s_n| = O(1/n)$. At about the same time Hardy proved the analog of Tauber's theorem—with Littlewood's modification—for the "average" defined by

$$\lim_{N\to\infty} \frac{s_1 + s_2 + s_3 + \cdots + s_N}{1 + 1 + 1 + \cdots + 1} = L. \tag{3}$$

That is, Hardy showed that if $\{s_n\}$ has the average L in the sense of (3) and if there is a K such that $|s_{n+1} - s_n| < K/n$ for all n, then $\{s_n\}$ has the average L in the sense of (2). An average in the sense of (3) is called a *Cesaro average*. In 1914 Hardy and Littlewood [H4] in collaboration proved that for positive sequences an Abel average implies a Cesaro average; that is, if $s_n \geq 0$ and if (1), then (3).

Now let $d\phi(x)$ be the point measure which is s_n at n and zero elsewhere. Then the three types of average (1), (2), and (3) can be restated in terms of $d\phi(x)$ as

$$\lim_{r\uparrow 1} \frac{\int_0^\infty r^x \, d\phi(x)}{\int_0^\infty r^x \, d([x])} = L, \tag{1'}$$

where $d([x])$ is the point measure which is 1 at integers and zero elsewhere

$$\lim_{A\to\infty} \int_A^{A+1} d\phi(x) = L \tag{2'}$$

[if A is an integer $A = n$, this integral is by definition $\frac{1}{2}(s_n + s_{n+1})$] and

$$\lim_{A\to\infty} \frac{\int_0^A d\phi(x)}{\int_0^A d([x])} = L. \tag{3'}$$

The second type of average will not be needed in what follows, and the first and third can be rewritten somewhat more simply as

$$\lim_{r\uparrow 1} \frac{\int_0^\infty r^x \, d\phi(x)}{\int_0^\infty r^x \, dx} = L, \qquad \lim_{A\to\infty} \frac{\int_0^A d\phi(x)}{\int_0^A dx} = L,$$

respectively [because $\int_0^\infty r^x \, d([x]) \sim \int_0^\infty r^x \, dx$ and $\int_0^A d([x]) \sim \int_0^A dx$]. It is natural to take these two statements as the definition of what it means to say that $d\phi(x) \sim L \, dx$ as an Abel average or a Cesaro average, respectively.

With this terminology the prime number theorem $\psi(x) \sim x$ is equivalent to the statement that $d\psi(x) \sim dx$ as a Cesaro average. Now since $d\psi(x) \geq 0$, the Hardy–Littlewood theorem cited above implies that in order to prove the prime number theorem, it would suffice to prove that $d\psi(x) \sim dx$ as an Abel average. Hardy and Littlewood were able to prove $d\psi(x) \sim dx$ as an Abel average more simply than it is possible to prove the prime number theorem directly and thereby were able to give a simpler proof of the prime number theorem or, more exactly, a proof of the prime number theorem in which a significant amount of the work is done by a Tauberian theorem. However, it is natural to hope to give a proof in which *all* of the work is done by a Tauberian theorem because there is a sense of "average" in which it is trivial to prove $d\psi(x) \sim dx$, namely, the sense of

$$\text{(4)} \qquad \lim_{s\downarrow 1} \frac{\int_1^\infty x^{-s} \, d\psi(x)}{\int_1^\infty x^{-s} \, dx} = 1.$$

Since $\int_1^\infty x^{-s} \, dx = (s-1)^{-1}$, this amounts to saying $\lim_{s\downarrow 1}(s-1)[-\zeta(s)/\zeta'(s)] = 1$, which can be proved by taking the logarithmic derivative of the analytic function $(s-1)\zeta(s)$, multiplying by $(s-1)$, and letting $s \to 1$. In short, to say that $d\psi(x) \sim dx$ in the sense of (4) amounts to saying that $-\zeta'(s)/\zeta(s)$ has a pole like $(s-1)^{-1}$ at $s = 1$ or, what is the same, that $\zeta(s)$ has a simple pole at $s = 1$. Thus the study of the prime number theorem suggests that one study conditions on measures $d\phi(x)$ under which one can assert that $d\phi(x) \sim dx$ in the sense of

$$\text{(4}'\text{)} \qquad \lim_{s\downarrow 1} \frac{\int_1^\infty x^{-s} \, d\phi(x)}{\int_1^\infty x^{-s} \, dx} = 1$$

implies $d\phi(x) \sim dx$ in the sense of

$$\text{(3}''\text{)} \qquad \lim_{A\to\infty} \frac{\int_0^A d\phi(x)}{\int_0^A dx} = 1.$$

The attempt to prove the prime number theorem in this way stimulated a great deal of study of Tauberian theorems in the 1920s and early 1930s, culminating in Wiener's general Tauberian theorem [W3]. Although Wiener's theory was immensely successful in revealing the true nature of Tauberian theorems, its conclusions with respect to the prime number theorem were largely negative in that it showed that to justify the implication (4′) ⇒ (3″), it is essential to study the Fourier transform $\int_0^\infty x^{-s}\,d\phi(x)$ on the line $\operatorname{Re} s = 1$ and hence that this approach to the prime number theorem leads to essentially the same ideas and techniques as those used by Hadamard and de la Vallée Poussin of Fourier inversion of $-\zeta'(s)/\zeta(s)$ and use of $\zeta(1 + it) \neq 0$. However, the general theory did give a concise theorem concerning the implication (4′) ⇒ (3″).

Ikehara's Theorem If the measure $d\phi(x)$ is positive, then the implication (4′) ⇒ (3″) is valid provided the function

$$g(s) = \frac{\int_1^\infty x^{-s}\,d\phi(x)}{\int_1^\infty x^{-s}\,dx}$$

has, in addition to the property that $g(s)$ is defined for $s > 1$ and $\lim_{s\downarrow 1} g(s)$ exists, the property that the function $[g(s) - g(1)]/(s - 1)$ has a continuous extension from the open halfplane $\operatorname{Re} s > 1$ (where it is necessarily defined and analytic) to the closed halfplane $\operatorname{Re} s \geq 1$. [Here $g(1)$ is written for $\lim_{s\downarrow 1} g(s)$.]

Ikehara's original proof [I1] of this theorem was a deduction from Wiener's general Tauberian theorem, but Bochner [B6] and others have given direct proofs independent of the general theory. Since in the case $d\phi(x) = d\psi(x)$ the function $g(s)$ is $(s - 1)[-\zeta'(s)/\zeta(s)]$ which is analytic in the entire plane except for poles at the zeros of $\zeta(s)$, the proof of the prime number theorem amounts to the proof that $\zeta(1 + it) \neq 0$ and to the proof that Ikehara's theorem is true in the particular case $d\phi(x) = d\psi(x)$.

12.8 CHEBYSHEV'S IDENTITY

Chebyshev's work on the distribution of primes consists of just two papers which occupy a total of only about 40 pages in his collected works (available in French [C4] as well as Russian [C5]). These two papers are very clearly written and are well worth reading.

The first of them is a study of the approximation $\pi(x) \sim \int_2^x (dx/\log x)$. It is based on an analysis of the function $\zeta(s) - (s - 1)^{-1}$ for *real* s as $s \downarrow 1$, in the course of which Chebyshev succeeds in proving that if there is a best value for A in the approximation

$$\pi(x) \sim \frac{x}{\log x - A},$$

that value is $A = 1$ and that, more generally, no other approximation to $\pi(x)$

of the same form as

$$\pi(x) \sim \frac{x}{\log x} + \frac{x}{(\log x)^2} + \frac{2x}{(\log x)^3} + \cdots + \frac{n!x}{(\log x)^{n+1}} \qquad [\sim \mathrm{Li}(x)]$$

can be a better approximation than this one (see Section 5.4). Thus the basic idea of the paper involves the relationship between the approximation $\pi(x) \sim \mathrm{Li}(x)$ and the pole of the zeta function, a relationship which was much more thoroughly exploited by Riemann and de Vallée Poussin and which was well incorporated into the mainstream of the study of the prime number theorem.

The second paper, on the other hand, is based on a very different idea, one which was until rather recently relegated to the status of a curiosity, showing what sorts of results can be obtained by "elementary" methods, that is, by methods which do not use the theory of Fourier analysis or functions of a complex variable. In the late 1940s, however, Selberg and Erdös showed that this idea of Chebyshev's can be taken further and that from it one can deduce the prime number theorem itself by entirely "elementary" arguments which do not appeal to Fourier analysis or functions of a complex variable (see Section 12.10). Consequently, there has been a great renewal of interest in it. Briefly, the idea is as follows.

Let T be the step function which for positive nonintegral values of x is $\sum_{n<x} \log n$ and which for integral values of x is, as usual, the middle value $T(n) = \frac{1}{2}[T(n+\varepsilon) + T(n-\varepsilon)]$. The value of $T(x)$ for x not an integer can also be described as the logarithm of $[x]$ factorial where $[x]$ is the integer part of x, that is, $T(x) = \log \Pi([x])$ (x not an integer). The identity on which Chebyshev's proof is based is

$$T(x) = \psi(x) + \psi(x/2) + \psi(x/3) + \psi(x/4) + \cdots. \tag{1}$$

This formula can be proved as follows.

Since $\psi(x/n)$ is a step function which jumps only when x is a multiple of n, both sides of (1) are step functions which jump only at integer values of x. Since, moreover, both sides are 0 at $x = 0$ and both assume the middle value at jumps, in order to prove they are equal, it suffices to prove that their jumps are equal at each integer. But at $x = n$ the left side jumps by $\log n$ and the right side jumps by $\sum \Lambda(d)$ where d runs over all divisors of n and where $\Lambda(d)$ is defined as in Section 3.2. Now if $n = p_1^{\alpha_1} p_2^{\alpha_2} \cdots p_k^{\alpha_k}$ is the prime factorization of n, then obviously the divisors of n include precisely α_1 powers of p_1 (namely, $p_1, p_1^2, p_1^3, \ldots, p_1^{\alpha_1}$), α_2 powers of p_2, etc., and no other prime powers. Hence $\sum \Lambda(d) = \alpha_1 \log p_1 + \alpha_2 \log p_2 + \cdots + \alpha_k \log p_k = \log n$, which proves (1).

In terms of Fourier analysis Chebyshev's identity† is the inverse transform

†Credit for the discovery of the identity is shared by de Polignac and Chebyshev (see Landau [L3]), but Chebyshev made better use of it.

of the identity

$$-\zeta'(s) = [-\zeta'(s)/\zeta(s)]\zeta(s)$$

because the left side $-\zeta'(s) = \sum (\log n)n^{-s}$ is the transform of the operator $f(x) \mapsto \int_0^\infty f(ux)\, dT(u)$, whereas the right side is the transform of the composition of the operators $f(x) \mapsto \int_0^\infty f(ux)\, d\psi(u)$ and $f(x) \mapsto \sum_1^\infty f(nx)$, which can be written $f(x) \mapsto \sum_n \int_0^\infty f(nux)\, d\psi(u) = \int_0^\infty f(vx)\, d[\sum \psi(v/n)]$. This suggests $dT(u) = d[\sum \psi(u/n)]$ and hence (1). It is not difficult to make this into a proof of (1), but the elementary proof above, which is essentially the one given by Chebyshev, is to be preferred.

Now Möbius inversion applied to Chebyshev's identity (1) gives

$$\psi(x) = \sum_{n=1}^{\infty} \mu(n) T\left(\frac{x}{n}\right). \tag{2}$$

On the other hand, a good approximation to $T(x)$ can be obtained using Stirling's formula (Euler–Maclaurin summation of $\log n$), and hence this formula should give some information about $\psi(x)$, perhaps even the prime number theorem $\psi(x) \sim x$. The difficulty is of course the irregularity of the coefficients $\mu(n)$ which prevents any straightforward analysis of formula (2). Chebyshev circumvents this difficulty by replacing the right side of (2) by

$$T(x) - T\left(\frac{x}{2}\right) - T\left(\frac{x}{3}\right) - T\left(\frac{x}{5}\right) + T\left(\frac{x}{30}\right) \tag{3}$$

and observing that when (1) is substituted in this expression the resulting series in ψ

$$\psi(x) - \psi\left(\frac{x}{6}\right) + \psi\left(\frac{x}{7}\right) - \psi\left(\frac{x}{10}\right) + \psi\left(\frac{x}{11}\right) - \psi\left(\frac{x}{12}\right) + \cdots$$

has the remarkable property that it *alternates*. (More specifically, the series in ψ is $\sum A_n \psi(x/n)$, where A_n depends only on the congruence class of n mod 30 and where, by explicit computation

$$A_n = 1, 0, 0, 0, 0, -1, 1, 0, 0, -1, 1, -1, 1, 0, -1,$$
$$0, 1, -1, 1, -1, 0, 0, 1, -1, 0, 0, 0, 0, 1, -1$$

for $n = 1, 2, 3, \ldots, 30$, respectively. Chebyshev does not say how he discovered this particular fact.) Thus (3) is less than $\psi(x)$ but greater than $\psi(x) - \psi(x/6)$, and this, together with Stirling's formula for $T(x)$, gives Chebyshev his estimates of $\psi(x)$.

Specifically, the weak form $T(x) = x \log x - x + O(\log x)$ of Stirling's formula gives easily

$$T(x) - T\left(\frac{x}{2}\right) - T\left(\frac{x}{3}\right) - T\left(\frac{x}{5}\right) + T\left(\frac{x}{30}\right) = Ax + O(\log x)$$

where A is the constant $A = \frac{1}{2}\log 2 + \frac{1}{3}\log 3 + \frac{1}{5}\log 5 - \frac{1}{30}\log 30 = 0.921\ldots$. Thus

$$\psi(x) > Ax + O(\log x), \qquad \psi(x) - \psi\left(\frac{x}{6}\right) < Ax + O(\log x).$$

If the second inequality is iterated

$$\psi\left(\frac{x}{6}\right) - \psi\left(\frac{x}{6^2}\right) < A\frac{x}{6} + O(\log x),$$
$$\psi\left(\frac{x}{6^2}\right) - \psi\left(\frac{x}{6^3}\right) < A\frac{x}{6^2} + O(\log x),$$
$$\vdots$$

only $\log x/\log 6$ steps are required to reach $\psi(x/6^n) = 0$. Adding these then gives

$$\begin{aligned}\psi(x) &< Ax\left(1 + \frac{1}{6} + \frac{1}{6^2} + \cdots + \frac{1}{6^n}\right) + O((\log x)^2)\\ &< \frac{6}{5}Ax + O((\log x)^2)\end{aligned}$$

and shows that in the limit as $x \longrightarrow \infty$, the quotient $\psi(x)/x$ lies between $A = 0.921\ldots$ and $6A/5 = 1.105\ldots$. In particular $\psi(x) = O(x)$, a fact which will be needed in the following sections.

12.9 SELBERG'S INEQUALITY

Chebyshev's formula $\sum \psi(x/n) = T(x)$ taken together with Stirling's formula $T(x) = (x + \frac{1}{2})\log x - x + O(1)$ lends credence to the prime number theorem $\psi(x) \sim x$ because if $\psi(x/n)$ is replaced by x/n, then the sum $\sum \psi(x/n)$ is replaced by $x \sum n^{-1} \sim x \log x \sim T(x)$. More specifically, choose as an approximation to $\psi(x)$ a function of the form

$$g(x) = \begin{cases} 0, & x \leq a, \\ x - a, & x \geq a, \end{cases}$$

where a is a positive constant. Then for large x

$$\sum_{n=1}^{\infty} g\left(\frac{x}{n}\right) = \sum_{x/n \geq a}\left[\left(\frac{x}{n}\right) - a\right] = x \sum_{n \leq x/a} \frac{1}{n} - a\left[\frac{x}{a}\right].$$

Now by Euler–Maclaurin summation [in its simplest version (3) of Section 6.2]

$$\begin{aligned}
\sum_{n\le y}\frac{1}{n} &= \sum_{n\le[y]}\frac{1}{n} = \int_1^{[y]}\frac{dy}{y} + \frac{1}{2}\left(1+\frac{1}{[y]}\right) - \int_1^{[y]}\frac{\bar{B}_1(u)}{u^2}\,du \\
&= \log[y] + \frac{1}{2} - \int_1^{\infty}\frac{\bar{B}_1(u)}{u^2}\,du + \frac{1}{2[y]} + \int_{[y]}^{\infty}\frac{\bar{B}_1(u)}{u^2}\,du \\
&= \log y + \text{const} + \frac{1}{2y}\left(\frac{y-[y]}{[y]}+1\right) + \log\left(1-\frac{y-[y]}{y}\right) \\
&\quad + O\left(\frac{1}{[y]^2}\right) \\
&= \log y + \gamma + O\left(\frac{1}{y}\right),
\end{aligned}$$

where the constant γ is by definition (see Section 3.8) Euler's constant. This gives

$$\begin{aligned}
\sum_{n=1}^{\infty} g\left(\frac{x}{n}\right) &= x\left[\log\frac{x}{a} + \gamma + O\left(\frac{a}{x}\right)\right] - x - O(a) \\
&= x\log x + (\gamma - \log a)x - x + O(1)
\end{aligned}$$

and shows, therefore, that if a is chosen to be $a = e^{\gamma}$, then

$$\sum_{n=1}^{\infty} g\left(\frac{x}{n}\right) = x\log x - x + O(1) = T(x) + O(\log x) = \sum_{n=1}^{\infty}\psi\left(\frac{x}{n}\right) + O(\log x).$$

Thus setting

$$r(x) = \sum_{n=1}^{\infty}\psi\left(\frac{x}{n}\right) - \sum_{n=1}^{\infty} g\left(\frac{x}{n}\right),$$

Chebyshev's identity and Stirling's formula give $r(x) = O(\log x)$. On the other hand, by Möbius inversion $\psi(x) - g(x) = \sum \mu(n)r(x/n)$, so the prime number theorem is the statement that $\sum \mu(n)r(x/n) = o(x)$. This leads to the question of whether estimates of the growth of $\sum \mu(n)r(x/n)$ can be deduced from estimates of the growth of $r(x)$.

It is very difficult to obtain sharp estimates of the growth of $\sum \mu(n)r(x/n)$ because the real reason for its slow growth involves *cancellation* between terms, so that the distribution of the signs $\mu(n) = \pm 1$ and the rate of change of r are crucial. As was shown in the preceding section, Chebyshev dealt with this difficulty by replacing the actual Möbius inverse $\sum \mu(n)r(x/n)$ by an approximate Möbius inverse $r(x) - r(x/2) - r(x/3) - r(x/5) + r(x/30)$. The first step in the elementary proof of the prime number theorem is to replace $\sum \mu(n)r(x/n)$ by the approximate Möbius inverse suggested by Selberg's proof in Section 11.3, namely, to replace it by the expression

(1) $$\sum_{n=1}^{\infty}\mu(n)\left(1-\frac{\log n}{\log x}\right)r\left(\frac{x}{n}\right).$$

(Note that $r(x/n) = 0$ for $n \geq x$ so the weights $[1 - (\log n/\log x)]$ are positive in the nonzero terms.) This leads to an estimate of $\psi(x)$ known as Selberg's inequality which, as will be shown in the next section, is a major step toward the prime number theorem.

The first step in the derivation of Selberg's inequality is to note that the expression (1) grows less rapidly than x as $x \to \infty$ and that in fact it is $O(x/\log x)$. This follows easily from $r(y) < K \log y$ ($y \geq 1$, K a constant independent of y) because this shows that the absolute value of (1) is at most

$$\sum_{n<x} \frac{\log(x/n)}{\log x} K \log\left(\frac{x}{n}\right) = \frac{K}{\log x} \sum_{n<x} \left(\log \frac{x}{n}\right)^2.$$

This sum can be estimated using Euler–Maclaurin summation, but the result is only that it is $O(x/\log x)$, a result which can be obtained much more easily by using $\log y < K' y^\varepsilon$ ($y \geq 1$, K' a constant depending on ε) to find

$$\begin{aligned}\sum_{n<x} \left(\log \frac{x}{n}\right)^2 &< K' \sum_{n<x} \left(\frac{x}{n}\right)^{2\varepsilon} \\ &= K' x^{2\varepsilon} \sum_{n<x} n^{-2\varepsilon} < K' x^{2\varepsilon}\left[1 + \int_1^x u^{-2\varepsilon}\, du\right] \\ &< K' x^{2\varepsilon}\left(1 + \frac{x^{-2\varepsilon+1}}{-2\varepsilon+1}\right) < K'' x\end{aligned}$$

which gives the desired result that (1) is $O(x/\log x)$.

The second step in the derivation of Selberg's inequality is to give a precise sense to the idea that the operation (1) is an "approximate Möbius inversion." One can in fact give an explicit expression for

$$\sum_{n=1}^{\infty} \mu(n)\left(1 - \frac{\log n}{\log x}\right) F\left(\frac{x}{n}\right) \tag{2}$$

when F is a function of the form $F(x) = \sum_{n=1}^{\infty} f(x/n)$ with f a function which is identically zero near zero. This explicit expression can be derived as follows. The sum (2) is equal to

$$\begin{aligned}&\frac{1}{\log x} \sum_{n=1}^{\infty} \mu(n) \log\left(\frac{x}{n}\right) \sum_{m=1}^{\infty} f\left(\frac{x}{mn}\right) \\ &= \frac{1}{\log x} \sum_{n=1}^{\infty} \sum_{m=1}^{\infty} \mu(n)\left[\log\left(\frac{x}{mn}\right) + \log m\right] f\left(\frac{x}{mn}\right) \\ &= \frac{1}{\log x} \sum_{n=1}^{\infty} \sum_{m=1}^{\infty} \mu(n)\left[\log\left(\frac{x}{mn}\right) f\left(\frac{x}{mn}\right)\right] \\ &\quad + \frac{1}{\log x} \sum_{n=1}^{\infty} \sum_{m=1}^{\infty} \mu(n)(\log m) f\left(\frac{x}{mn}\right).\end{aligned}$$

Now by ordinary Möbius inversion the first of these two sums is simply $(\log x)^{-1} (\log x) f(x) = f(x)$, so the second sum gives the amount by which (2) is only an "approximate" Möbius inverse. Note that it is $(\log x)^{-1}$ times a composition of the operators $f(x) \mapsto \sum_{m=1}^{\infty} (\log m) f(x/m)$ and $f(x) \mapsto \sum_{n=1}^{\infty} \mu(n) f(x/n)$, or, what is the same, a composition of $f(x) \mapsto \int_0^\infty f(x/u)\, dT(u)$ and $f(x) \mapsto \int_0^\infty f(x/u)\, dM(u)$. Since these operators have transforms $-\zeta'(-s)$ and $1/\zeta(-s)$, respectively, their composition has the transform $-\zeta'(-s)/\zeta(-s)$ which is the transform of $f(x) \mapsto \int_0^\infty f(x/u)\, d\psi(u)$. This leads to the conjecture that the second sum above is $(\log x)^{-1} \int_0^\infty f(x/u)\, d\psi(u)$, a conjecture which is easily verified (without appeal to Fourier analysis) by using Chebyshev's identity and Möbius inversion to write, for x not an integer,

$$\psi(x) = \sum_{n=1}^{\infty} \mu(n) T\left(\frac{x}{n}\right) = \sum_{n=1}^{\infty} \sum_{m<x/n} \mu(n) \log m = \sum\sum_{mn<x} \mu(n) \log m$$
$$= \sum_{k<x} \sum_{mn=k} \mu(n) \log m,$$

so that

$$\text{(3)} \qquad \int_0^\infty f\left(\frac{x}{u}\right) d\psi(u) = \sum_{k=1}^{\infty} f\left(\frac{x}{k}\right)\left[\sum_{mn=k} \mu(n) \log m\right]$$
$$= \sum_{m=1}^{\infty} \sum_{n=1}^{\infty} \mu(n)(\log m) f\left(\frac{x}{mn}\right).$$

Thus the final formula is

$$\sum_{n=1}^{\infty} \mu(n)\left(1 - \frac{\log n}{\log x}\right) F\left(\frac{x}{n}\right) = f(x) + \frac{1}{\log x} \int_0^\infty f\left(\frac{x}{u}\right) d\psi(u),$$

where f is a function which is identically zero near zero and where $F(x) = \sum_{m=1}^{\infty} f(x/m)$.

Applying this formula in the case $F = r$ and using the fact that (1) is $O(x/\log x)$ gives

$$\text{(4)} \qquad [\psi(x) - g(x)] + \frac{1}{\log x} \int_0^\infty \left[\psi\left(\frac{x}{u}\right) - g\left(\frac{x}{u}\right)\right] d\psi(u)$$
$$= O(x/\log x)$$

which, in essence, is Selberg's inequality. To obtain the inequality in the form stated by Selberg [S3], it is necessary first to estimate the integral

$$\int_0^\infty g\left(\frac{x}{u}\right) d\psi(u) = -\int_0^\infty \psi(u)\, dg\left(\frac{x}{u}\right) = \int_0^\infty \psi\left(\frac{x}{v}\right) dg(v) = \int_a^\infty \psi\left(\frac{x}{v}\right) dv$$
$$= \sum_1^\infty \psi\left(\frac{x}{n}\right) - \sum_1^\infty \psi\left(\frac{x}{n}\right) + \int_1^\infty \psi\left(\frac{x}{v}\right) dv - \int_1^a \psi\left(\frac{x}{v}\right) dv$$
$$= T(x) - \frac{1}{2}\psi(x) - \int_1^\infty \bar{B}_1(v)\, d\psi\left(\frac{x}{v}\right) + O(\psi(x))$$
$$= x \log x + O(x) + O\left[-\int_1^\infty d\psi\left(\frac{x}{v}\right)\right] = x \log x + O(x)$$

using Chebyshev's theorem $\psi(x) = O(x)$. (This calculation assumes that x is not an integer—so that the discontinuities of $\bar{B}_1(v)$ never coincide with those of $\psi(x/v)$—and it assumes that x is not an integral multiple of a—so that $v = a$ is not a discontinuity of $\psi(x/v)$. This excludes only a discrete set of values of x, and since $\int_0^\infty g(x/u)\, d\psi(u)$ is an increasing function of x, the final estimate is obviously valid for these values of x as well.) This shows that Selberg's inequality (4) can also be written in the form

$$\begin{aligned}\psi(x) \log x &+ \int_0^\infty \psi\left(\frac{x}{u}\right) d\psi(u) \\ &= g(x) \log x + \int_0^\infty g\left(\frac{x}{u}\right) d\psi(u) + O(x) \\ &= x \log x + O(\log x) + x \log x + O(x)\end{aligned}$$

and hence finally

$$\psi(x) \log x + \int_0^\infty \psi\left(\frac{x}{u}\right) d\psi(u) = 2x \log x + O(x). \tag{5}$$

This is almost Selberg's statement of it except that Selberg [S3] deals with θ rather than ψ (see Section 4.4 for the definition of θ) and his inequality is

$$\theta(x) \log x + \int_0^\infty \theta\left(\frac{x}{u}\right) d\theta(u) = 2x \log x + O(x).$$

[Note that the integral is simply the finite sum $\sum_{p<x} \theta(x/p) \log p$.] The proof of the inequality in this form is somewhat longer than the proof of (5) and will not be needed in what follows.

12.10 ELEMENTARY PROOF OF THE PRIME NUMBER THEOREM

The deduction of the prime number theorem from Selberg's inequality, although it is "elementary" in the technical sense that it does not use Fourier analysis or complex variables, is by no means simple or straightforward. Selberg's original proof depended on a weakened version of the prime number theorem which Erdös had previously proved by elementary methods, but Selberg never published this proof in full, preferring to give a complete proof *ab initio* and also preferring to eliminate the appeal to the notion of "lim sup" which the original proof contained. Since 1949 many variations, extensions, and refinements of the elementary proof have been given, but none of them seems very straightforward or natural, nor does any of them give much insight into the theorem.

The proof which follows is a combination of Wirsing's proof [W5] and the proof given by Levinson in his expository paper [L10]. Following Wirsing, it is based on the consideration of approximations not to $\psi(x)$ but to the function $\int_0^x u^{-1}\, d\psi(u)$. This function has the advantage that its discontinuities are small for large x $[\Lambda(n)/n \leq (\log n)/n < \varepsilon]$ whereas the discontinuities of $\psi(x)$ are large. As in Section 5.6, this function $\int_0^x u^{-1}\,d\psi(u)$ will be denoted $P(x)$. It was shown in Section 5.6—but not by elementary methods—that $P(x) = \log x - \gamma + \eta(x)$, where γ is Euler's constant and where the remainder $\eta(x)$ goes to zero faster than $(\log x)^{-n}$ for any n. This and the form of the approximation g to ψ in the preceding section suggest as an approximation to $P(x)$

$$G(x) = \begin{cases} \log x - \gamma, & x \geq e^{\gamma}, \\ 0, & x \leq e^{\gamma}. \end{cases}$$

As in Section 5.6, let $\eta(x) = P(x) - G(x)$ be the error in this approximation. In order to prove the prime number theorem it will suffice to prove that $\eta(x) \longrightarrow 0$ as $x \longrightarrow \infty$ because then

$$\begin{aligned} \psi(x) &= \int_0^x d\psi(u) = \int_0^x u\,dP(u) \\ &= \int_0^x u\,dG(u) + \int_0^x u\,d\eta(u) \\ &= \int_{e^{\gamma}}^x u\,d\log u + \int_0^x d[u\eta(u)] - \int_0^x \eta(u)\,du \\ &= x - e^{\gamma} + x\eta(x) - \int_0^x \eta(u)\,du \\ &= x + x\left\{-\frac{e^{\gamma}}{x} + \eta(x) - \text{average of } \eta \text{ on } [0, x]\right\} \\ &= x + o(x). \end{aligned}$$

Thus the goal is to prove by elementary methods that $\eta(x) \longrightarrow 0$.

Note first that η is bounded. This follows easily from the estimate $\int_0^\infty g(x/u)\,d\psi(u) = x \log x + O(x)$ at the end of the preceding section which gives

$$\begin{aligned} P(x) - \int_0^x \frac{1}{u}\,d\psi(u) &= \frac{1}{x}\int_0^x \frac{x}{u}\,d\psi(u) \\ &= \frac{1}{x}\int_0^x g\left(\frac{x}{u}\right) d\psi(u) + \frac{1}{x}\int_0^x \left[\frac{x}{u} - g\left(\frac{x}{u}\right)\right] d\psi(u) \\ &= \frac{1}{x}\int_0^\infty g\left(\frac{x}{u}\right) d\psi(u) + \int_0^{x/a} a\,d\psi(u) + \frac{1}{x}\int_{x/a}^x \frac{x}{u}\,d\psi(u) \\ &= \log x + O(1) + \frac{a\psi(x/a)}{x} + O\left(\frac{a[\psi(x) - \psi(x/a)]}{x}\right) \\ &= \log x + O(1) = G(x) + O(1), \end{aligned}$$

where, as before, $a = e^{\gamma}$. Thus $\eta(x) = O(1)$ as was to be shown.

Chebyshev's identity $\sum \psi(x/n) = T(x) = \sum_{n<x} \log n$ ($x \neq$ integer) implies an analogous identity for P which can be derived as follows:

$$xP(x) = \int_0^x \frac{x}{u}\, d\psi(u) = \int_\infty^1 v\, d\psi\left(\frac{x}{v}\right),$$

$$\sum_{n=1}^\infty \frac{x}{n} P\left(\frac{x}{n}\right) = \sum_{n=1}^\infty \int_\infty^1 v\, d\psi\left(\frac{x}{nv}\right) = \int_\infty^1 v\, dT\left(\frac{x}{v}\right)$$

$$= \int_0^x \frac{x}{u}\, dT(u) = x \sum_{n<x} \frac{\log n}{n},$$

$$\sum_{n=1}^\infty \frac{1}{n} P\left(\frac{x}{n}\right) = \sum_{n<x} \frac{\log n}{n}.$$

The form of this identity suggests that one consider $\sum (1/n)\eta(x/n)$. This gives the following estimate analogous to the estimate $r(x) = O(\log x)$ of the last section:

$$\sum_{n=1}^\infty \frac{1}{n}\eta\left(\frac{x}{n}\right) = \sum_{n=1}^\infty \frac{1}{n} G\left(\frac{x}{n}\right) - \sum_{n=1}^\infty \frac{1}{n} P\left(\frac{x}{n}\right)$$

$$= \sum_{x/n>a} \frac{1}{n}\left(\log \frac{x}{n} - \gamma\right) - \sum_{n<x} \frac{\log n}{n}$$

$$= (\log x - \gamma) \sum_{n<x/a} \frac{1}{n} - \sum_{n<x/a} \frac{\log n}{n} - \sum_{n<x} \frac{\log n}{n}.$$

Now by Euler–Maclaurin summation

$$\sum_{n=1}^N \frac{\log n}{n} = \int_1^N \frac{\log u}{u}\, du + \frac{1}{2}\left[\frac{\log N}{N} + 0\right] + \int_1^N \bar{B}_1(u) \frac{1 - \log u}{u^2}\, du$$

$$= \frac{1}{2}(\log u)^2 \Big|_1^N + \int_1^\infty \frac{\bar{B}_1(u)(1 - \log u)}{u^2}\, du$$

$$- \int_N^\infty \frac{\bar{B}_1(u)(1 - \log u)}{u^2}\, du + \frac{\log N}{2N}$$

$$= \frac{1}{2}(\log N)^2 + \text{const} + O\left(\frac{\log N}{N}\right)$$

and this together with $\sum_{n<x} 1/n = \log x + \gamma + O(1/x)$ gives

$$\sum \frac{1}{n}\eta\left(\frac{x}{n}\right) = \log (x - \gamma)\left[\log \frac{x}{a} + \gamma + O\left(\frac{a}{x}\right)\right] - \frac{1}{2}\left(\log \frac{x}{a}\right)^2$$

$$- \text{const} - O\left(\frac{\log(x/a)}{x/a}\right) - \frac{1}{2}(\log x)^2 - \text{const} - O\left(\frac{\log x}{x}\right)$$

$$= (\log x - \gamma)\left[\log x + O\left(\frac{1}{x}\right)\right] - \frac{1}{2}(\log x - \gamma)^2$$

$$- \frac{1}{2}(\log x)^2 + \text{const} + O\left(\frac{\log x}{x}\right)$$

$$= \text{const} + O\left(\frac{\log x}{x}\right).$$

Let $s(x)$ denote this function $\sum (1/n)\eta(x/n)$. Then Möbius inversion gives $\eta(x) = \sum [\mu(n)/n]s(x/n)$, and the sort of approximate Möbius inversion of the preceding section [see in particular formula (3)] gives

$$\begin{aligned}
&\sum \frac{\mu(n)}{n}\left(1 - \frac{\log n}{\log x}\right)s\left(\frac{x}{n}\right) \\
&= \frac{1}{\log x}\sum_{n=1}^{\infty}\sum_{m=1}^{\infty} \mu(n)\left(\log \frac{x}{n}\right)\frac{1}{mn}\eta\left(\frac{x}{mn}\right) \\
&= \frac{1}{x \log x}\sum\sum \mu(n)\left[\log \frac{x}{mn} + \log m\right]\frac{x}{mn}\eta\left(\frac{x}{mn}\right) \\
&= \frac{1}{x \log x}(\log x)x\eta(x) + \frac{1}{x \log x}\int_0^{\infty} \frac{x}{u}\eta\left(\frac{x}{u}\right) d\psi(u) \\
&= \eta(x) + \frac{1}{\log x}\int_0^{\infty} \eta\left(\frac{x}{u}\right) dP(u).
\end{aligned}$$

Using the estimate $s(x) = \text{const} + O[(\log x)/x]$, it is possible to show that this function of x is $O(1/\log x)$. In fact, since

$$\begin{aligned}
&\frac{1}{\log x}\sum_{n<x} \frac{1}{n}\left(\log \frac{x}{n}\right)\frac{\log(x/n)}{x/n} = \frac{1}{x \log x}\sum_{n<x} \left(\log \frac{x}{n}\right)^2 \\
&= O\left(\frac{1}{\log x}\right)
\end{aligned}$$

(see the estimate of $\sum [\log (x/n)]^2$ in the preceding section), the proof of this reduces immediately to the proof that

$$\sum \frac{\mu(n)}{n} \log \frac{x}{n} = O(1). \tag{1}$$

This can be accomplished as follows.

Let $D(x)$ again represent the function which is 1 for $x > 1$, $\frac{1}{2}$ for $x = 1$, and 0 for $x < 1$. Then $\sum D(x/n)$ is simply the greatest integer function (x not an integer) and Möbius inversion gives

$$D(x) = \sum_{n=1}^{\infty} \mu(n)\left[\frac{x}{n}\right]$$

which for large x is

$$1 = \sum_{n<x} \mu(n)\frac{x}{n} - \sum_{n<x} \mu(n)\left(\frac{x}{n} - \left[\frac{x}{n}\right]\right) = x\sum_{n<x} \frac{\mu(n)}{n} + O(x)$$

so that division by x gives $\sum_{n<x} \mu(n)/n = O(1)$. Then Möbius inversion of the estimate

$$\sum_{n=1}^{\infty} \frac{1}{n}D\left(\frac{x}{n}\right) = \sum_{n<x} \frac{1}{n} = \begin{cases} \log x + \gamma + O\left(\dfrac{1}{x}\right) & \text{for } x \geq 1, \\ 0 & \text{for } x < 1, \end{cases}$$

gives

$$D(x) = \sum_{n<x} \frac{\mu(n)}{n}\left[\log\left(\frac{x}{n}\right) + \gamma + O\left(\frac{n}{x}\right)\right],$$

$$1 = \sum_{n<x} \frac{\mu(n)}{n} \log \frac{x}{n} + \gamma \cdot O(1) + O(1),$$

from which (1) follows.

In summary, then, it has been shown that

$$(2) \qquad \eta(x) + \frac{1}{\log x}\int_0^\infty \eta\left(\frac{x}{u}\right) dP(u) = O\left(\frac{1}{\log x}\right).$$

This is the analog of Selberg's inequality [in the form (4) of Section 12.9] for the error in the approximation $P \sim G$ instead of the error in the approximation $\psi \sim g$. The objective is to use it to prove that $\eta(x) \longrightarrow 0$ as $x \longrightarrow \infty$.

The first step in the proof is to *iterate* Selberg's inequality, which in the present case can be carried out as follows:

$$\int_0^\infty \eta\left(\frac{x}{u}\right) dP(u) = -\eta(x) \log x + O(1),$$

$$\begin{aligned}
\int_0^\infty \int_0^\infty \eta\left(\frac{x}{uv}\right) dP(u)\, dP(v) &= \int_0^{x+\varepsilon} \int_0^\infty \eta\left(\frac{x}{uv}\right) dP(u)\, dP(v) \\
&= -\int_0^{x+\varepsilon} \eta\left(\frac{x}{v}\right) \log\left(\frac{x}{v}\right) dP(v) \\
&\quad + \int_0^{x+\varepsilon} O(1)\, dP(v) \\
&= -\log x \int_0^\infty \eta\left(\frac{x}{v}\right) dP(v) \\
&\quad + \int_0^\infty \eta\left(\frac{x}{v}\right) \log v\, dP(v) + O(P(x)) \\
&= -\log x[-\eta(x) \log x + O(1)] \\
&\quad + \int_0^\infty \eta\left(\frac{x}{v}\right) \log v\, dP(v) + O(G(x)) \\
&= \eta(x)(\log x)^2 + \int_0^\infty \eta\left(\frac{x}{v}\right) \log v\, dP(v) \\
&\quad + O(\log x).
\end{aligned}$$

The double integral on the left is in fact a finite sum, so it can be rearranged and written as $\int_0^\infty \eta(x/w)\, dN(w)$ where $N(w) = \int_0^\infty P(w/u)\, dP(u)$. Thus

$$\int_0^\infty \eta\left(\frac{x}{w}\right) dN(w) - \int_0^\infty \eta\left(\frac{x}{v}\right) \log v\, dP(v) = \eta(x)(\log x)^2 + O(\log x).$$

Since N and P are both increasing functions, taking absolute values gives

$$\begin{aligned}
(3) \qquad (\log x)^2 |\eta(x)| &\le \int_0^\infty \left|\eta\left(\frac{x}{w}\right)\right| dN(w) \\
&\quad + \int_0^\infty \left|\eta\left(\frac{x}{w}\right)\right| \log w\, dP(w) + O(\log x).
\end{aligned}$$

Now $dP(w) = w^{-1}\,d\psi(w)$ is roughly like $w^{-1}\,dw = d\log w$, from which it follows that $dN(w)$ and $\log w\, dP(w)$ are both roughly $\log w\, d\log w$; so the right side is roughly $2\int_0^\infty |\eta(x/w)| \log w\, d\log w$. More precisely,

$$\begin{aligned}
&\int_0^x [dN(w) + \log w\, dP(w)] \\
&\quad = N(x) + \int_0^x \log w\, dP(w) \\
&\quad = \int_0^\infty \left\{ G\left(\frac{x}{w}\right) + \eta\left(\frac{x}{w}\right) \right\} dP(w) + \int_0^x \log w\, dP(w) \\
&\quad = \int_0^\infty \eta\left(\frac{x}{w}\right) dP(w) + \int_0^{x/a} \log \frac{x}{aw}\, dP(w) + \int_0^x \log w\, dP(w) \\
&\quad = -\eta(x)\log x + O(1) + \int_0^x \log\frac{x}{a}\, dP(w) - \int_{x/a}^x \log\frac{x}{aw}\, dP(w) \\
&\quad = -\eta(x)\log\frac{x}{a} + O(1) + \left(\log\frac{x}{a}\right) P(x) + O\left(\int_{x/a}^x \log a\, dP(w)\right) \\
&\quad = G(x)\log\frac{x}{a} + O(1) + O\left(P(x) - P\left(\frac{x}{a}\right)\right) \\
&\quad = \{G(x)\}^2 + O(1) + O\left(\log x - \gamma + \eta(x) - \log\frac{x}{a} + \gamma - \eta\left(\frac{x}{a}\right)\right) \\
&\quad = \{G(x)\}^2 + O(1)
\end{aligned}$$

for all $x > a$. Moreover, the measures $dN(w)$, $\log w\, dP(w)$, and $d\{G(w)^2\}$ are all identically zero for $x \le 1$, so integration by parts shows that the right side of (3) differs from $2\int_0^\infty |\eta(x/w)|\, G(w)\, dG(w)$ by at most $O(\log x)$ plus

$$\begin{aligned}
&\int_1^\infty \left|\eta\left(\frac{x}{w}\right)\right| [dN(w) + \log w\, dP(w) - 2G(w)\, dG(w)] \\
&\qquad = -\int_1^\infty \{O(1)\}\, d\left[\left|\eta\left(\frac{x}{w}\right)\right|\right] = O\left(-\int_x^0 d[|\eta(v)|]\right) \\
&\qquad \le O\left(\int_0^x dG(v) + \int_0^x dP(v)\right) = O(2G(x) + \eta(x)) = O(\log x).
\end{aligned}$$

On the other hand $2\int_0^\infty |\eta(x/w)|\, G(w)\, dG(w)$ can be rewritten in the form

$$\begin{aligned}
&2\int_0^\infty \left|\eta\left(\frac{x}{w}\right)\right| G(w)\, dG(w) \\
&\quad = 2\int_a^\infty \left|\eta\left(\frac{x}{w}\right)\right| \log\frac{w}{a}\, d\log w = 2\int_{x/a}^0 |\eta(v)| \log\frac{x}{va}\, d\log\left(\frac{1}{v}\right) \\
&\quad = 2\int_0^{x/a} |\eta(v)| \left\{\int_v^{x/a} d\log u\right\} d\log v = 2\int_1^{x/a}\int_v^{x/a} |\eta(v)|\, d\log u\, d\log v \\
&\quad = 2\int_1^{x/a}\int_1^u |\eta(v)|\, d\log v\, d\log u;
\end{aligned}$$

so (3) can be rewritten as

(4) $$(\log x)^2 |\eta(x)| \le 2\int_1^{x/a} \int_1^u |\eta(v)|\, d\log v\, d\log u + O(\log x).$$

Now the integral

(5) $$\frac{2}{[\log(x/a)]^2} \int_1^{x/a} \int_1^u |\eta(v)|\, d\log v\, d\log u$$
$$= \frac{2}{[\log(x/a)]^2} \int_1^{x/a} \log u \left(\frac{1}{\log u} \int_1^u |\eta(v)|\, d\log v\right) d\log u$$

can be regarded as the result of applying to $|\eta(v)|$ two averaging processes one after the other, the first being the average $f(x) \mapsto (\log x)^{-1} \int_1^x f(v)\, d\log v$ and the second being the average $f(x) \mapsto [\log(x/a)]^{-2} \int_1^{x/a} f(u)\, d\{(\log u)^2\}$. In particular, since $|\eta|$ is bounded the integral (5) is bounded; so multiplication of (4) by $(\log x)^{-2} = [\log(x/a)]^{-2}[1 + O(1/\log x)]$ gives finally

$$|\eta(x)| \le \frac{2}{\left(\log \frac{x}{a}\right)^2} \int_1^{x/a} \log u \left(\frac{1}{\log u} \int_1^u |\eta(v)|\, d\log v\right) d\log u + O\left(\frac{1}{\log x}\right).$$

(6)

This inequality is the keystone of the proof. It states that $|\eta|$ is dominated, in the limit as $x \to \infty$, by an average of an average of $|\eta|$. Since averaging normally reduces functions unless they are constant, this indicates that $|\eta(x)|$ must be nearly constant in the limit as $x \to \infty$. Since $\eta(x)$ changes rather gradually, this indicates that $\eta(x)$ must also be nearly constant in the limit as $x \to \infty$, that is, $\lim_{x\to\infty} \eta(x)$ exists. However, because of the analogy with Chebyshev's elementary proof that $[\pi(x) - \mathrm{Li}(x)]/\mathrm{Li}(x)$ can have no limit other than zero, one would expect to be able to prove by elementary means that $\lim_{x\to\infty} \eta(x)$, given that it exists, must be zero and thus to complete the proof.

The actual proof will require two more estimates, an estimate of the rate of change of η and an estimate which proves that η can approach no limit other than zero.

Wirsing gives the estimate

(7) $$|\eta(x) - \eta(y)| \le \log(x/y) + O(1/\log x)$$

for $x > y > 0$. Since P increases, $\eta(x) - \eta(y) = [G(x) - G(y)] - [P(x) - P(y)] \le G(x) - G(y) = \int_y^x dG(u) \le \int_y^x d\log u = \log(x/y)$, and the upper estimate $\eta(x) - \eta(y) \le \log(x/y)$ is trivial. Since η is bounded, the lower estimate $\eta(x) - \eta(y) \ge -\log(x/y)$ holds trivially whenever $\log(x/y)$ is sufficiently large, say whenever $\log(x/y) \ge K$. Thus it suffices to find a lower estimate $\eta(x) - \eta(y) \ge -\log(x/y) - O(1/\log x)$ under the additional assumption $\log(x/y) \le K$. This can be done by using Selberg's inequality to find

$$
\begin{aligned}
\eta(x) - \eta(y) &= \frac{1}{\log x}\left[\eta(x)\log x - \eta(y)\log y - \eta(y)\log\frac{x}{y}\right] \\
&\ge \frac{1}{\log x}\left[-\int_0^\infty \eta\left(\frac{x}{u}\right) dP(u) + O(1)\right. \\
&\qquad \left. + \int_0^\infty \eta\left(\frac{y}{u}\right) dP(u) + O(1)\right] - O(\eta(y))\frac{K}{\log x} \\
&= \frac{1}{\log x}\left[-\int_0^\infty \left\{\eta\left(\frac{x}{u}\right) - \eta\left(\frac{y}{u}\right)\right\} dP(u)\right] - O\left(\frac{1}{\log x}\right) \\
&\ge -\frac{1}{\log x}\int_0^x \log\frac{x}{y}\, dP(u) - O\left(\frac{1}{\log x}\right) \\
&= -\frac{[\log(x/y)][G(x) + \eta(x)]}{\log x} - O\left(\frac{1}{\log x}\right) \\
&= -\log\frac{x}{y} - \left(\log\frac{x}{y}\right)\frac{-\gamma + \eta(x)}{\log x} - O\left(\frac{1}{\log x}\right) \\
&\ge -\log\frac{x}{y} - K\frac{O(1)}{\log x} - O\left(\frac{1}{\log x}\right) \\
&= -\log\frac{x}{y} - O\left(\frac{1}{\log x}\right).
\end{aligned}
$$

This completes the proof of (7).

The estimate $\sum (1/n)\eta(x/n) = \text{const} + O[(\log x)/x]$ is a good indication that η cannot approach any limit other than zero. A formulation of this estimate which is more convenient for present purposes is $\int_1^\infty (1/u)\eta(x/u)\,du = O(1)$, or what is the same,

(8) $$\int_0^x \eta(v)\, d\log v = O(1).$$

This shows that the average value of η on $[1, x]$ relative to the invariant measure $d\log v$ approaches zero; hence, because η changes slowly, η must be arbitrarily near zero infinitely often. The estimate (8) can be proved as follows:

$$
\begin{aligned}
\int_0^x \eta(v)\, d\log v &= \int_1^\infty \frac{1}{u}\eta\left(\frac{x}{u}\right) du \\
&= \sum_1^\infty \frac{1}{n}\eta\left(\frac{x}{n}\right) - \sum_1^\infty \frac{1}{n}\eta\left(\frac{x}{n}\right) + \int_1^\infty \frac{1}{u}\eta\left(\frac{x}{u}\right) du \\
&= O(1) - \frac{1}{2}\eta(x) - \int_1^\infty \bar{B}_1(u)\, d\left[\frac{1}{u}\eta\left(\frac{x}{u}\right)\right] \\
&= O(1) + \int_1^\infty \bar{B}_1(u)\eta\left(\frac{x}{u}\right)\frac{du}{u^2} - \int_1^\infty \frac{\bar{B}_1(u)}{u}\, d\eta\left(\frac{x}{u}\right)
\end{aligned}
$$

$$= O(1) + O(1) + \int_1^\infty \frac{\bar{B}_1(u)}{u}\, dG\left(\frac{x}{u}\right) - \int_1^\infty \frac{\bar{B}_1(u)}{u}\, dP\left(\frac{x}{u}\right)$$
$$= O(1) + \int_1^{x/a} \frac{\bar{B}_1(u)}{u}\, d\log\left(\frac{1}{u}\right) - \int_1^\infty \frac{\bar{B}_1(u)}{u} \cdot \frac{u}{x}\, d\psi\left(\frac{x}{u}\right)$$
$$= O(1) - \frac{1}{x}\int_1^\infty \bar{B}_1(u)\, d\psi\left(\frac{x}{u}\right)$$
$$= O(1) + O\left(\frac{\psi(x)}{x}\right) = O(1).$$

The prime number theorem can now be deduced from the following theorem.

Theorem Let B be an upper bound for $|\eta(x)|$ and let M be an upper bound for $|\int_1^x \eta(u)\, d\log u|$. Without loss of generality assume $M > B^2$. Then if $\beta > 0$ is any number less than or equal to B with the property that $|\eta(x)| \leq \beta$ for all sufficiently large x, the number

$$\beta' = \beta - (\beta^3/400M)$$

has the same property.

Proof Let β be given and let x_0 be such that $|\eta(x)| \leq \beta$ for $x \geq x_0$. Divide the interval $[x_0, \infty)$ into subintervals on which η changes by less than $\frac{1}{3}\beta$. Specifically, define λ by $\log\lambda = \frac{1}{4}\beta$ and set $x_j = \lambda^j x_0$. Then $x_j \leq y < x \leq x_{j+1}$ implies by (7) that $|\eta(x) - \eta(y)| \leq \log x - \log y + O(1/\log x) \leq \log x_{j+1} - \log x_j + O(1/\log x)_j \leq \log\lambda + O(1/\log x_0)$. By increasing x_0 if necessary this gives $|\eta(x) - \eta(y)| < \frac{1}{3}\beta$ as desired.

Call $|\eta|$ "small" on the interval $[x_j, x_{j+1}]$ if its value at either or both ends is less than $\frac{1}{2}\beta$ and otherwise call it "large" on the interval. Since η cannot change sign on an interval where $|\eta|$ is large, the average of $|\eta|$ can be estimated using (8). In fact, if $|\eta|$ is large on all intervals between x_j and x_{j+k}, then the average of $|\eta|$ is at most

$$\frac{1}{\log(x_{j+k}/x_j)}\int_{x_j}^{x_{j+k}} |\eta(v)|\, d\log v$$
$$= \frac{1}{k\log\lambda}\left|\int_{x_j}^{x_{j+k}} \eta(v)\, d\log v\right|$$
$$= \frac{4}{k\beta}\left|\int_1^{x_{j+k}} \eta(v)\, d\log v - \int_1^{x_j} \eta(v)\, d\log v\right| \leq \frac{8M}{k\beta}.$$

This can be made strictly less than β by making k large; for example, $(8M/k\beta) \leq \frac{5}{6}\beta$ when $k \geq 48M/5\beta^2$. On the other hand, if $|\eta|$ is small on at least one of the intervals between x_j and x_{j+k}, then the average of $|\eta|$ is at most $k^{-1}[(k-1)\beta + \frac{5}{6}\beta] = \beta - (\beta/6k)$ because $|\eta| \leq \beta$ throughout, and because among the k intervals of equal weight $\log(x_{n+1}/x_n) = \log\lambda$ there is at least one on which $|\eta| \leq \frac{5}{6}\beta$ throughout. Thus the average is strictly less than β in either case. Fix k as the smallest integer satisfying $k \geq 48M/5\beta^2$.

Then $k < (48M + 5\beta^2)/5\beta^2 \leq (48M + 5B^2)/5\beta^2 \leq 53M/5\beta^2$, $6k < 100M/\beta^2$, $\beta/6k > \beta^3/100M$, and finally

$$\beta - (\beta/6k) < \beta - (\beta^3/100M);$$

so the average of $|\eta|$ on $[x_j, x_{j+k}]$ (always with respect to the measure $d \log x$) is either less than this amount or less than $\frac{5}{6}\beta$. But since $\beta^2 \leq B^2 < M$ gives $\beta^2/100M < 1/100 < \frac{1}{6}$, this shows that

$$\frac{1}{\log(x_{j+k}/x_j)} \int_{x_j}^{x_{j+k}} |\eta(v)| \, d \log v \leq \beta - \frac{\beta^3}{100M}$$

for all $j = 0, 1, 2, \ldots$ when k is defined as above.

Consider now the average $(\log u)^{-1} \int_1^u |\eta(v)| \, d \log v$ for large u. Let the interval $[1, u]$ be divided into three parts, the interval $[1, x_0]$, the interval $[x_0, x_{nk}]$, and the interval $[x_{nk}, u]$, where n is the largest integer such that $x_{nk} \leq u$, with k as before. In the whole interval $[1, u]$ these three intervals count with weights $\log x_0/\log u$, $nk \log \lambda/\log u$, and $\log (u/x_{nk})/\log u < \log (x_{nk+k}/x_{nk})/\log u < k \log \lambda/\log u$, respectively. Thus the first and last intervals count with weights which approach zero as $u \to \infty$, and the averages on these intervals are constant on the first one and at most β on the last one. On the middle interval the average is at most $\beta - (\beta^3/100M)$. Therefore as $u \to \infty$, the average on the entire interval $[1, u]$ can be only slightly greater than $\beta - (\beta^3/100M)$, say

$$\frac{1}{\log u} \int_1^u |\eta(v)| \, d \log v \leq \beta - \frac{\beta^3}{200M}$$

for all sufficiently large u. But then the average of this amount over $1 \leq u \leq (x/a)$ relative to the measure $2 \log u \, d \log u$ for large x can be only slightly greater than $\beta - (\beta^2/200M)$. Thus the inequality (6) implies the desired conclusion.

Corollary $\eta(x) \to 0$ as $x \to \infty$.

Proof Start with $\beta = B$ and apply the theorem repeatedly. This gives a decreasing sequence $\beta_0 > \beta_1 > \beta_2 > \cdots$ of positive numbers such that for each β_n the inequality $|\eta(x)| \leq \beta_n$ holds for all sufficiently large x. Since $\beta_n \to 0$ as $n \to \infty$, this proves the corollary.†

This completes the elementary proof of the prime number theorem. It is natural to ask whether the stronger theorem $\eta(x) = O(\log^{-n} x)$ can also be proved by elementary methods. This was accomplished in the 1960s by both Wirsing [W5] and Bombieri [B9]. Thus the prime number theorem with the error estimate

$$\psi(x) = x + O(x/\log^n x)$$

can be proved by "elementary" methods.

† The β's have a limit, say β_∞, which satisfies $\beta_\infty = \beta_\infty - (\beta_\infty^3/400M)$, $\beta_\infty = 0$.

12.11 OTHER ZETA FUNCTIONS. WEIL'S THEOREM

In concentrating exclusively on the study of the zeta function and its relation to the prime number theorem, this book ignores one of the most fruitful areas of development of Riemann's work, namely, number theory. The use of functions like the zeta function in number theory was a major feature of the work of Dirichlet—both in his L-series and in his formula for the class number of a quadratic number field—many years before Riemann's paper appeared, and the use of such functions has been a prominent theme in number theory ever since. Riemann's contributions in this area were primarily function-theoretic, not number-theoretic, and consisted of focusing attention on the functions as functions of a *complex* variable, on the possibility of their satisfying a functional equation under $s \leftrightarrow 1 - s$, and on the importance of the location of their complex zeros. A few of the most important names in the subsequent study of these number-theoretic functions are those of Dedekind, Hilbert, Hecke, Artin, Weil, and Tate.

Ignorance prevents me from entering into a discussion of these functions and what is known about them. However, it seems that they provide some of the best reasons for believing that the Riemann hypothesis is true—for believing, in other words, that there is a profound and as yet uncomprehended number-theoretic phenomenon, one facet of which is that the roots ρ all lie on $\operatorname{Re} s = \frac{1}{2}$. In particular, there is a "zeta function" associated in a natural number-theoretic way to any function field over a finite field, and Weil [W2] has shown that *the analog of the Riemann hypothesis is true for such "zeta functions."*

APPENDIX

On the Number of Primes Less Than a Given Magnitude

by BERNHARD RIEMANN†

I believe I can best express my gratitude for the honor which the Academy has bestowed on me in naming me as one of its correspondents by immediately availing myself of the privilege this entails to communicate an investigation of the frequency of prime numbers, a subject which because of the interest shown in it by Gauss and Dirichlet over many years seems not wholly unworthy of such a communication.

In this investigation I take as my starting point the observation of Euler that the product

$$\Pi \frac{1}{1 - \frac{1}{p^s}} = \Sigma \frac{1}{n^s},$$

where p ranges over all prime numbers and n over all whole numbers. The function of a complex variable s which these two expressions define when they converge I denote by $\zeta(s)$. They converge only when the real part of s is greater than 1; however, it is easy to find an expression of the function which always is valid. By applying the equation

$$\int_0^\infty e^{-nx} x^{s-1}\, dx = \frac{\Pi(s-1)}{n^s},$$

one finds first

$$\Pi(s-1)\zeta(s) = \int_0^\infty \frac{x^{s-1}\, dx}{e^x - 1}.$$

If one considers the integral

$$\int \frac{(-x)^{s-1}\, dx}{e^x - 1}$$

†Translated from *Ueber die Anzahl der Primzahlen unter einer gegebenen Grösse* [R1, p. 145] by H. M. Edwards.

from $+\infty$ to $+\infty$ in the positive sense around the boundary of a domain which contains the value 0 but no other singularity of the integrand in its interior, then it is easily seen to be equal to

$$(e^{-\pi si} - e^{\pi si}) \int_0^\infty \frac{x^{s-1}\,dx}{e^x - 1},$$

provided that in the many-valued function $(-x)^{s-1} = e^{(s-1)\log(-x)}$ the logarithm of $-x$ is determined in such a way that it is real for negative values of x. Thus

$$2 \sin \pi s\, \Pi(s-1)\, \zeta(s) = i \int_\infty^\infty \frac{(-x)^{s-1}\,dx}{e^x - 1}$$

when the integral is defined as above.

This equation gives the value of the function $\zeta(s)$ for all complex s and shows that it is single-valued and finite for all values of s other than 1, and also that it vanishes when s is a negative even integer.

When the real part of s is negative, the integral can be taken, instead of in the positive sense around the boundary of the given domain, in the negative sense around the complement of this domain because in that case (when $\operatorname{Re} s < 0$) the integral over values with infinitely large modulus is infinitely small. But inside this complementary domain the only singularities of the integrand are at the integer multiples of $2\pi i$, and the integral is therefore equal to the sum of the integrals taken around these singularities in the negative sense. Since the integral around the value $n2\pi i$ is $(-n2\pi i)^{s-1}(-2\pi i)$, this gives

$$2 \sin \pi s\, \Pi(s-1)\, \zeta(s) = (2\pi)^s \sum n^{s-1}[(-i)^{s-1} + i^{s-1}],$$

and therefore a relation between $\zeta(s)$ and $\zeta(1-s)$ which, by making use of known properties of the function Π, can also be formulated as the statement that

$$\Pi\left(\frac{s}{2} - 1\right)\pi^{-s/2}\zeta(s)$$

remains unchanged when s is replaced by $1 - s$.

This property of the function motivated me to consider the integral $\Pi((s/2) - 1)$ instead of the integral $\Pi(s-1)$ in the general term of $\sum n^{-s}$, which leads to a very convenient expression of the function $\zeta(s)$. In fact

$$\frac{1}{n^s}\Pi\left(\frac{s}{2} - 1\right)\pi^{-s/2} = \int_0^\infty e^{-nn\pi x} x^{(s/2)-1}\,dx;$$

so when one sets

$$\sum_1^\infty e^{-nn\pi x} = \psi(x),$$

it follows that

$$\Pi\left(\frac{s}{2} - 1\right)\pi^{-s/2}\zeta(s) = \int_0^\infty \psi(x) x^{(s/2)-1}\,dx$$

or, because

$$2\psi(x) + 1 = x^{-1/2}\left[2\psi\left(\frac{1}{x}\right) + 1\right] \qquad \text{(Jacobi, Fund., p. 184),}$$

that

$$\Pi\left(\frac{s}{2} - 1\right)\pi^{-s/2}\zeta(s) = \int_1^\infty \psi(x)x^{(s/2)-1}\,dx + \int_0^1 \psi\left(\frac{1}{x}\right)x^{(s-3)/2}\,dx$$
$$+ \frac{1}{2}\int_0^1 (x^{(s-3)/2} - x^{(s/2)-1})\,dx$$
$$= \frac{1}{s(s-1)} + \int_1^\infty \psi(x)(x^{(s/2)-1} + x^{-(1+s)/2})\,dx.$$

I now set $s = \frac{1}{2} + ti$ and

$$\Pi\left(\frac{s}{2}\right)(s-1)\pi^{-s/2}\zeta(s) = \xi(t)$$

so that

$$\xi(t) = \tfrac{1}{2} - (tt + \tfrac{1}{4})\int_1^\infty \psi(x)x^{-3/4}\cos(\tfrac{1}{2}t\log x)\,dx$$

or also

$$\xi(t) = 4\int_1^\infty \frac{d[x^{3/2}\psi'(x)]}{dx}\,x^{-1/4}\cos\left(\frac{1}{2}t\log x\right)dx.$$

This function is finite for all finite values of t and can be developed as a power series in tt which converges very rapidly. Now since for values of s with real part greater than 1, $\log\zeta(s) = -\sum\log(1 - p^{-s})$ is finite and since the same is true of the other factors of $\xi(t)$, the function $\xi(t)$ can vanish only when the imaginary part of t lies between $\frac{1}{2}i$ and $-\frac{1}{2}i$. The number of roots of $\xi(t) = 0$ whose real parts lie between 0 and T is about

$$= \frac{T}{2\pi}\log\frac{T}{2\pi} - \frac{T}{2\pi}$$

because the integral $\int d\log\xi(t)$ taken in the positive sense around the domain consisting of all values whose imaginary parts lie between $\frac{1}{2}i$ and $-\frac{1}{2}i$ and whose real parts lie between 0 and T is (up to a fraction of the order of magnitude of $1/T$) equal to $[T\log(T/2\pi) - T]i$ and is, on the other hand, equal to the number of roots of $\xi(t) = 0$ in the domain multiplied by $2\pi i$. One finds in fact about this many real roots within these bounds and it is very likely that all of the roots are real. One would of course like to have a rigorous proof of this, but I have put aside the search for such a proof after some fleeting vain attempts because it is not necessary for the immediate objective of my investigation.

If one denotes by α the roots of the equation $\xi(\alpha) = 0$, then one can express $\log\xi(t)$ as

$$\sum\log\left(1 - \frac{tt}{\alpha\alpha}\right) + \log\xi(0)$$

because, since the density of roots of size t grows only like $\log(t/2\pi)$ as t grows, this expression converges and for infinite t is only infinite like $t \log t$; thus it differs from $\log \xi(t)$ by a function of tt which is continuous and finite for finite t and which, when divided by tt, is infinitely small for infinite t. This difference is therefore a constant, the value of which can be determined by setting $t = 0$.

With these preparatory facts, the number of primes less than x can now be determined.

Let $F(x)$, when x is not exactly equal to a prime, be equal to this number, but when x is a prime let it be greater by $\frac{1}{2}$ so that for an x where $F(x)$ jumps

$$F(x) = \frac{F(x+0) + F(x-0)}{2}.$$

If one sets

$$p^{-s} = s \int_p^\infty x^{-s-1}\,dx, \qquad p^{-2s} = s \int_{p^2}^\infty x^{-s-1}\,dx, \qquad \ldots$$

in the formula

$$\log \zeta(s) = -\sum \log(1 - p^{-s}) = \sum p^{-s} + \tfrac{1}{2}\sum p^{-2s} + \tfrac{1}{3}\sum p^{-3s} + \cdots,$$

one finds

$$\frac{\log \zeta(s)}{s} = \int_1^\infty f(x) x^{-s-1}\,dx$$

when one denotes

$$F(x) + \tfrac{1}{2}F(x^{1/2}) + \tfrac{1}{3}F(x^{1/3}) + \cdots$$

by $f(x)$.

This equation is valid for every complex value $a + bi$ of s provided $a > 1$. But when in such circumstances

$$g(s) = \int_0^\infty h(x) x^{-s}\,d\log x$$

is valid, the function h can be expressed in terms of g by means of Fourier's theorem. The equation splits when h is real and when $g(a + bi) = g_1(b) + ig_2(b)$ into the two equations

$$g_1(b) = \int_0^\infty h(x) x^{-a} \cos(b \log x)\,d\log x,$$

$$ig_2(b) = -i \int_0^\infty h(x) x^{-a} \sin(b \log x)\,d\log x.$$

When both equations are multiplied by $[\cos(b \log y) + i \sin(b \log y)]\,db$ and integrated from $-\infty$ to $+\infty$, one finds in both cases that the right side is $\pi h(y) y^{-\alpha}$ so that when they are added and multiplied by iy^a

$$2\pi i h(y) = \int_{a-\infty i}^{a+\infty i} g(s) y^s\,ds,$$

where the integration is to be carried out in such a way that the real part of s remains constant.†

The integral represents, for a value of y where the function $h(y)$ has a jump, the middle value between the two values of h on either side of the jump. The function f was defined in such a way that it too has this property, so one has in full generality

$$f(y) = \frac{1}{2\pi i} \int_{a-\infty i}^{a+\infty i} \frac{\log \zeta(s)}{s} y^s \, ds.$$

For $\log \zeta$ one can now substitute the expression

$$\frac{s}{2} \log \pi - \log(s-1) - \log \Pi\left(\frac{s}{2}\right) + \sum_{\alpha} \log\left[1 + \frac{(s-\frac{1}{2})^2}{\alpha\alpha}\right] + \log \xi(0)$$

found above; the integrals of the individual terms of this expression will not converge, however, when they are taken to infinity, so it is advantageous to reformulate the equation as

$$f(x) = -\frac{1}{2\pi i} \frac{1}{\log x} \int_{a-\infty i}^{a+\infty i} \frac{d \dfrac{\log \zeta(s)}{s}}{ds} x^s \, ds$$

by integration by parts.

Since

$$-\log \Pi\left(\frac{s}{2}\right) = \lim\left[\sum_{n=1}^{m} \log\left(1 + \frac{s}{2n}\right) - \frac{s}{2} \log m\right]$$

for $m = \infty$ and therefore,

$$-\frac{d \dfrac{1}{s} \log \Pi\left(\dfrac{s}{2}\right)}{ds} = \sum_{1}^{\infty} \frac{d \dfrac{1}{s} \log\left(1 + \dfrac{s}{2n}\right)}{ds},$$

all of the terms in the expression for $f(x)$ except for the term

$$\frac{1}{2\pi i} \frac{1}{\log x} \int_{a-\infty i}^{a+\infty i} \frac{1}{ss} \log \xi(0) x^s \, ds = \log \xi(0)$$

take the form

$$\pm \frac{1}{2\pi i} \frac{1}{\log x} \int_{a-\infty i}^{a+\infty i} \frac{d\left[\dfrac{1}{s} \log\left(1 - \dfrac{s}{\beta}\right)\right]}{ds} x^s \, ds.$$

But

$$\frac{d\left[\dfrac{1}{s} \log\left(1 - \dfrac{s}{\beta}\right)\right]}{d\beta} = \frac{1}{(\beta - s)\beta}$$

†This argument is not quite correct. See the relevant note in Riemann's collected works [R1] (translator's note).

and, when the real part of s is greater than the real part of β,

$$-\frac{1}{2\pi i}\int_{a-\infty i}^{a+\infty i}\frac{x^s\,ds}{(\beta - s)\beta} = \frac{x^\beta}{\beta} = \int_\infty^x x^{\beta-1}\,dx$$

or

$$= \int_0^x x^{\beta-1}\,dx$$

depending on whether† the real part of β is negative or positive. Thus

$$\frac{1}{2\pi i}\frac{1}{\log x}\int_{a-\infty i}^{a+\infty i}\frac{d\left[\frac{1}{s}\log\left(1-\frac{s}{\beta}\right)\right]}{ds}x^s\,ds$$

$$= -\frac{1}{2\pi i}\int_{a-\infty i}^{a+\infty i}\frac{1}{s}\log\left(1-\frac{s}{\beta}\right)x^s\,ds$$

$$= \int_\infty^x \frac{x^{\beta-1}}{\log x}\,dx + \text{const}$$

in the first case and

$$= \int_0^x \frac{x^{\beta-1}}{\log x}\,dx + \text{const}$$

in the second case.

In the first case the constant of integration can be determined by taking β to be negative and infinite. In the second case the integral from 0 to x takes on two values which differ by $2\pi i$ depending on whether the path of integration is in the upper halfplane or in the lower halfplane; if the path of integration is in the upper halfplane, the integral will be infinitely small when the coefficient of i in β is infinite and positive, and if the path is in the lower halfplane, the integral will be infinitely small when the coefficient of i in β is infinite and negative. This shows how to determine the values of $\log[1 - (s/\beta)]$ on the left side in such a way that the constants of integration drop out.

By setting these values in the expression for $f(x)$ one finds

$$f(x) = \mathrm{Li}(x) - \sum_\alpha [\mathrm{Li}(x^{(1/2)+\alpha i}) + \mathrm{Li}(x^{(1/2)-\alpha i})]$$

$$+ \int_x^\infty \frac{1}{x^2-1}\frac{dx}{x\log x} + \log\xi(0),$$

where‡ the sum $\sum_\alpha$ is over all positive roots (or all roots with positive real parts) of the equation $\xi(\alpha) = 0$, ordered according to their size. It is possible, by means of a more exact discussion of the function ξ, easily to show that with this ordering of the roots the sum of the series

$$\sum_\alpha [\mathrm{Li}(x^{(1/2)+\alpha i}) + \mathrm{Li}(x^{(1/2)-\alpha i})]\log x$$

†Note that this excludes the possibility Re $\beta = 0$ and therefore does not apply to roots, if any, on the imaginary axis (translator's note).

‡Concerning the erroneous value of $\log\xi(0)$ in this formula, see Chapter 1 (translator's note).

is the same as the limiting value of

$$\frac{1}{2\pi i}\int_{a-bi}^{a+bi}\frac{d\frac{1}{s}\sum\log\left[1+\frac{(s-\frac{1}{2})^2}{\alpha\alpha}\right]}{ds}x^s\,ds$$

as b grows without bound; by a different ordering, however, it can approach any arbitrary real value.

From $f(x)$ one can find $F(x)$ by inverting

$$f(x)=\sum\frac{1}{n}F(x^{1/n})$$

to find

$$F(x)=\sum(-1)^{\mu}\frac{1}{m}f(x^{1/m}),$$

where m ranges over all positive integers which are not divisible by any square other than 1 and where μ denotes the number of prime factors of m.

If $\sum_{\alpha}$ is restricted to a finite number of terms, then the derivative of the expression for $f(x)$ or, except for a part which decreases very rapidly as x increases,

$$\frac{1}{\log x}-2\sum_{\alpha}\frac{\cos(\alpha\log x)x^{-1/2}}{\log x}$$

gives an approximate expression for the density of primes + half the density of prime squares $+\frac{1}{3}$ the density of prime cubes, etc., of magnitude x.

Thus the known approximation $F(x)=\mathrm{Li}(x)$ is correct only to an order of magnitude of $x^{1/2}$ and gives a value which is somewhat too large, because the nonperiodic† terms in the expression of $F(x)$ are, except for quantities which remain bounded as x increases,

$$\mathrm{Li}(x)-\tfrac{1}{2}\mathrm{Li}(x^{1/2})-\tfrac{1}{3}\mathrm{Li}(x^{1/3})-\tfrac{1}{5}\mathrm{Li}(x^{1/5})$$
$$+\tfrac{1}{6}\mathrm{Li}(x^{1/6})-\tfrac{1}{7}\mathrm{Li}(x^{1/7})+\cdots.$$

In fact the comparison of $\mathrm{Li}(x)$ with the number of primes less than x which was undertaken by Gauss and Goldschmidt and which was pursued up to $x=$ three million shows that the number of primes is already less than $\mathrm{Li}(x)$ in the first hundred thousand and that the difference, with minor fluctuations, increases gradually as x increases. The thickening and thinning of primes which is represented by the periodic terms in the formula has also been observed in the counts of primes, without, however, any possibility of establishing a law for it having been noticed. It would be interesting in a future count to examine the influence of individual periodic terms in the formula for the density of primes. More regular than the behavior of $F(x)$ is the behavior of $f(x)$ which already in the first hundred is on average very nearly equal to $\mathrm{Li}(x)+\log\xi(0)$.

†Strictly speaking, the terms $\mathrm{Li}(x^{(1/2)+\alpha i})$ are not periodic but merely oscillatory (translator's note).

References

A1 Abel, N. H., Solution de quelques problemes a l'aide d'integrales définies. *Mag. Naturvidenskaberne* **2** (1823). (Also "Oeuvres," 1, pp. 11–27.)

A2 Abel, N. H., Letter to Holmboe dated 4 August 1823. *In* "Niels Henrik Abel, Memorial, Publié à L'Occasion du Centenaire de sa Naissance."

A3 Ahlfors, L. V., "Complex Analysis." McGraw-Hill, New York, 1953.

B1 Backlund, R., Sur les zéros de la fonction $\zeta(s)$ de Riemann. *C. R. Acad. Sci. Paris* **158**, 1979–1982 (1914).

B2 Backlund, R., Über die Nullstellen der Riemannschen Zetafunktion. Dissertation, Helsingfors, 1916.

B3 Backlund, R., Über die Nullstellen der Riemannschen Zetafunktion. *Acta Math.* **41**, 345–375 (1918).

B4 Backlund, R., Über die Beziehung zwischen Anwachsen und Nullstellen der Zetafunktion. *Ofversigt Finska Vetensk. Soc.* **61**, No. 9 (1918–1919).

B5 Bochner, S., "Vorlesungen über Fouriersche Integrale," Akademische Verlagsgesellschaft M. B. H., Leipzig, 1932. (Reprinted by Chelsea, Bronx, New York, 1948.)

B6 Bochner, S., Ein Satz von Landau und Ikehara. *Math. Z.* **37**, 1–9 (1933).

B6a Bohr, H., and Landau, E., Über das Verhalten von $\zeta(s)$ und $\zeta_K(s)$ in der Nähe der Geraden $\sigma = 1$. *Göttinger Nachr.* 303–330 (1910).

B7 Bohr, H., and Landau, E., Beiträge zur Theorie der Riemannschen Zetafunktion. *Math. Ann.* **74**, 3–30 (1913). (Also Bohr's "Collected Works," Vol. I, B11.)

B8 Bohr, H., and Landau, E., Ein Satz über Dirichletsche Reihen mit Anwendung . . . . *Rend. Circ. Mat. Palermo* **37**, 269–272 (1914). (Also Bohr's "Collected Works," Vol. I, B13.)

B9 Bombieri, E., Sulle formule di A. Selberg generalizzate per classi di funzioni aritmetiche e le applicazioni al del resto nel "Primzahlsatz." *Riv. Mat. Univ. Parma* [2] **3**, 393–440 (1962).

C1 Cantor, M., "Vorlesungen über Geschichte der Mathematik," Vol. 3, p. 663. Teubner, Leipzig, 1898.

C2 Chebyshev, P. L., Sur la fonction qui détermine la totalité des nombres premiers inferieurs à une limite donnée. *J. Math. Pures Appl.* [1] **17** (1852). (First published 1848. Available in both French and Russian Editions of Collected Works.)

C3 Chebyshev, P. L., Mémoire sur les nombres premiers. *J. Math. Pures Appl.* [1] **17** (1852). (First published 1850. Available in both French and Russian Editions of Collected Works.)

C4 Chebyshev, P., "Oeuvres." St. Petersburg. (Republished by Chelsea, Bronx, New York, 1962 (?).)

C5 Chebyshev, P., "Polnoe Sobranie Sochineniy." Press of the USSR Acad. of Sci., Moscow and Leningrad, 1947.

D1 Debye, P. J. W., Näherungsformeln fur die Zylinderfunktionen .... *Math. Ann.* **67**, 535–558 (1909). (Engl. Transl. in "The Collected Papers of Peter J. W. Debye." Wiley (Interscience), New York, 1954.)

D2 Denjoy, A., L'Hypothèse de Riemann sur la distribution des zéros de $\zeta(s)$, reliée à la théorie des probabilites. *C. R. Acad. Sci. Paris* **192**, 656–658 (1931).

D3 Dirichlet, P. G. Lejeune, Sur l'usage des series infinies dans la théorie des nombres. *J. Reine Angew. Math.* **18**, 257–274 (1838). (Also "Werke," Vol. I, pp. 359–374.)

E1 Edwards, H. M., "Advanced Calculus." Houghton, Boston, Massachusetts, 1969.

E2 Euclid, "Elements," Book 9, Prop. 20.

E3 Euler, L., De progressionibus transcendentibus .... *Comm. Acad. Sci. Petropolitanae* **5**, 36–57 (1730). (Also "Opera" (1), Vol. 14, pp. 1–24.)

E4 Euler, L., Variae observationes circa series infinitas. *Comm. Acad. Sci. Petropolitanae* **9**, 222–236 (1737). ("Opera" (1), Vol. 14, pp. 216–244.)

E5 Euler, L., "Introductio in Analysin Infinitorum," Chapter 15. Bousquet and Socios., Lausanne, 1748. ("Opera" (1), Vol. 8.)

E6 Euler, L., "Institutiones Calculi Differentialis," Pt. 2, Chapters 5 and 6. Acad. Imp. Sci. Petropolitanae, St. Petersburg, 1755. ("Opera" (1), Vol. 10.)

E7 Euler, L., Remarques sur un beau rapport entre les séries des puissances tant directes que réciproques. *Mem. Acad. Sci. Berlin* **17**, 83–106 (1761). ("Opera" (1), Vol. 15, pp. 70–90.)

E8 Euler, L., Survey of Euler's work on the factorial function (German). Introduction to Euler's "Opera" (1), Vol. 19, 1932.

F1 Franel, J., and Landau, E., Les suites de Farey et le problème des nombres premiers. *Göttinger Nachr.*, 198–206 (1924).

F2 Fuss, P.-H., "Correspondance Mathematique et Physique." Imp. Acad. Sci., St. Petersburg, 1843. (Reprinted, Johnson Reprint Corp., 1968.)

G1 Gauss, C. F., Circa seriem infinitam

$$1 + \frac{\alpha\beta}{1\cdot\gamma}x + \frac{\alpha(\alpha+1)\beta(\beta+1)}{1\cdot 2\cdot\gamma(\gamma+1)}x^2 + \cdots,$$

Comm. Soc. Reg. Sci. Gottingensis **2** (1813). (Also "Werke," Vol. III, pp. 125–160.)

G2 Gauss, C. F., Letter to Enke dated 24 December 1849. "Werke," Vol. II, pp. 444–447. Königlichen Gesellschaft der Wissenschaften zu Göttingen.

G3 Gauss, C. F., "Werke," Vol. X_1, p. 11. Königlichen Gesellschaft der Wissenschaften zu Göttingen.

G4 Genocchi, A., Formule per determinare quanti siano i numeri primi fino ad un dato limite (Resumé of Riemann's Paper). *Ann. Mat. Pura Appl.* [1] **3**, 52–59 (1860).

G5 Gram, J.-P., Sur les Zéros de la Fonction $\zeta(s)$ de Riemann. *Acta Math.* **27**, 289–304 (1903).

H1 Hadamard, J., Étude sur les Propriétés des Fonctions Entières et en Particulier d'une Fonction Considérée par Riemann. *J. Math. Pures Appl.* [4] **9**, 171–215 (1893). (Also in "Oeuvres.")

H2 Hadamard, J., Sur la distribution des zeros de la fonction $\zeta(s)$ et ses consequences arithmétiques. *Bull. Soc. Math. France* **24**, 199–220 (1896). (Also in "Oeuvres.")

H3 Hardy, G. H., Sur les Zéros de la Fonction $\zeta(s)$ de Riemann. *C. R. Acad. Sci. Paris* **158**, 1012–1014 (1914). (Also in "Collected Papers.")

H3a Hardy, G. H., Prime numbers, *Brit. Ass. Rep.*, 1915, 350–354. ["Collected Papers," Vol. 2, pp. 14–19.]

H4 Hardy, G. H., "Ramanujan." Cambridge Univ. Press, London and New York, 1940. (Reprinted by Chelsea, Bronx, New York.)

H5 Hardy, G. H., "Divergent Series." Oxford Univ. Press, London and New York, 1956. (First Edition, 1949).

H6 Hardy, G. H., and Littlewood, J. E., The zeros of Riemann's zeta-function on the critical line. *Math. Z.* **10**, 283–317 (1921). (Also in Hardy's "Collected Papers.")

H7 Hardy, G. H., and Wright, E. M., "An Introduction to the Theory of Numbers." Oxford Univ. Press, London and New York, 1938.

H8 Haselgrove, C. B., "Tables of the Riemann Zeta Function," Roy. Soc. Math. Tables, Vol. 6. Cambridge Univ. Press, London and New York, 1960.

H9 Hilbert, D. Problèmes futures des mathématiques. *C. R. 2nd Congr. Int. Math., Paris, 1902*, p. 85.

H10 Holmgren, E., On Primtalens Fördelning. *Öfversigt Kgl. Vetensk. Acad. Stockholm* **59**, 221–225 (1902–1903).

H11 Hutchinson, J. I., On the roots of the Riemann zeta-function. *Trans. Amer. Math. Soc.* **27**, 49–60 (1925).

I1 Ikehara, S., An extension of Landau's theorem in the analytic theory of numbers. *J. Math. and Phys.* **10**, 1–12 (1931).

I2 Ingham, A. E., "The Distribution of Prime Numbers," Cambridge Tracts in Math. and Math. Phys., No. 30. Cambridge Univ. Press, London and New York, 1932. (Republished by Stechert-Hafner, New York and London, 1964.)

J1 Jacobi, C. G. J., Suite des notices sur les fonctions elliptiques. *J. Reine Angew. Math.* **3**, 303–310 (1828). (Also "Werke," Vol. 1, pp. 255–263.)

J2 Jeffreys, H. J., and Jeffreys, B. S., "Methods of Mathematical Physics." Cambridge Univ. Press, London and New York, 1946.

J3 Jensen, J. L. V. W., Sur un nouvel et important théorèm de la théorie des fonctions. *Acta Math.* **22**, 359–364 (1899).

K1 Von Koch, H., Sur la distribution des nombres premiers. *Acta Math.* **24**, 159–182 (1901).

K2 Kuzmin, R., Contribution to the theory of a class of Dirichlet series (Russian). *Izv. Akad. Nauk SSSR, Ser. Math. Nat. Sci.* **7**, 115–124 (1930).

K3 Kuzmin, R., On the roots of Dirichlet series (Russian). *Izv. Akad. Nauk SSSR Ser. Math. Nat. Sci.* **7**, 1471–1491 (1934).

L1 Lakatos, I., Proofs and refutations (I–IV). *British J. Philos. Sci.* (1963).

L2 Landau, E., Nouvelle démonstration pour la formule de Riemann .... *Ann. Sci. École Norm. Sup.* [3] **25**, 399–442 (1908).

L3 Landau, E., "Handbuch der Lehre von der Verteilung der Primzahlen." Teubner, Leipzig, 1909. (Reprinted by Chelsea, Bronx, New York.)

L3a Landau, E., Zur Theorie der Riemannschen Zetafunktion, *Vierteljahrsschr. Naturf. Ges. Zurich.* **56**, 125–148 (1911).

L4 Legendre, A. M., "Théorie des Nombres," 4th ed., Vol. 2, 4th Pt., §VIII Paris, 1830. (Reprinted, Librairie Sci. Tech., A. Blanchard, Paris, 1955.)

L5 Lehman, R. S., Separation of zeros of the Riemann zeta function, I. *Math. Comp.* **20**, 542–550 (1966).

L6 Lehman, R. S., On the difference $\pi(x) - \text{Li}(x)$. *Acta Arith.* **11**, 397–410 (1966).

L7 Lehmer, D. H., On the roots of the Riemann zeta-function. *Acta Math.* **95**, 291–298 (1956).

L8 Lehmer, D. H., Extended computation of the Riemann zeta-function. *Mathematika* **3**, 102–108 (1956).

L9 Lehmer D. N., List of prime numbers from 1 to 10,006,721. Publ. No. 165. Carnegie Inst. of Washington, Washington, D.C. 1913. (Reprinted, Hafner, New York, 1956.)

L10 Levinson, N., A motivated account of an elementary proof of the prime number theorem. *Amer. Math. Monthly* **76**, 225–245 (1969).

L11 Lindelöf, E., Quelques remarques sur la croissance de la fonction $\zeta(s)$. *Bull. Sci. Math.* [2]. **32**$_1$, 341–356 (1908).

L12 Littlewood, J. E., Quelques conséquences de l'hypothèse que la fonction $\zeta(s)$ n'a pas de zéros dans le demi-plan Re $(s) > \frac{1}{2}$. *C. R. Acad. Sci. Paris* **154**, 263–266 (1912).

L13 Littlewood, J. E., Sur la distribution des nombres premiers. *C. R. Acad. Sci. Paris* **158**, 1869–1872 (1914).

L14 Littlewood, J. E., On the zeros of the Riemann zeta-function. *Proc. Cambridge Philos. Soc.* **22**, 295–318 (1924).

M1 von Mangoldt, H., Zu Riemann's Abhandlung 'Ueber die Anzahl der Primzahlen unter einer gegebenen Grösse'. *J. Reine Angew. Math.* **114**, 255–305 (1895).

M2 von Mangoldt, H., Beweis der Gleichung $\sum_{k=1}^{\infty} \mu(k)/k = 0$. *S.-B. Kgl. Preuss. Akad. Wiss. Berlin*, 835–852 (1897).

M3 von Mangoldt, H., Zur Verteilung der Nullstellen der Riemannschen Funktion $\xi(t)$. *Math. Ann.* **60**, 1–19 (1905).

M4 Mellin, H., Eine Formel für den Logarithmus transcendenter Funktionen von endlichem Geschlecht. *Acta Soc. Sci. Fenn.* **29** (1900).

M5 Mertens, F., Ein Beitrag zur Analytischen Zahlentheorie. *J. Reine Angew. Math.* **78**, 46–62 (1874).

M6 Mertens, F., Über eine Eigenschaft der Riemann'schen ζ-Function. *S.-B. Akad. Wiss. Wien Math.-Natur. Kl. Abt. 2A* **107**, 1429–1434 (1898).

P1 Polya, G., Über die Nullstellen gewisser ganzer Funktionen. *Math. Z.* **2**, 352–383 (1918).

P2 Polya, G., Über trigonometrische Integrale mit nur reellen Nullstellen. *J. Reine Angew. Math.* **158**, 6–18 (1927).

R1 Riemann, B., "Gesammelte Werke." Teubner, Leipzig, 1892. (Reprinted by Dover Books, New York, 1953.)

R1a Riemann, B., Unpublished Papers, Handschriftenabteilung Niedersächische Staats- und Universitätsbibliothek, Göttingen.

R2 Riemann, B., "Partielle Differentialgleichungen" (Lectures edited and prepared for publication by K. Hattendorf). Vieweg, Braunschweig, 1876.

R3 Rosser, J. B., Yohe, J. M., and Schoenfeld, L., Rigorous computation and the zeros of the Riemann zeta-function. *Cong. Proc. Int. Federation Information Process. 1968*, pp. 70–76, Spartan, Washington, D.C. and Macmillan, New York, 1969.

S1 Selberg, A., On the zeros of Riemann's zeta-function. *Skr. Norske Vid.-Akad. Oslo* No. 10 (1942).

S2 Selberg, A., The zeta-function and the Riemann hypothesis. *Proc. Skand. Math. Kongr., 10th*, pp. 187–200, Jul. Giellerups Forlag, Kobenhavn, 1947.

S3 Selberg, A., An elementary proof of the prime number theorem. *Ann. of Math.* [2] **50**, 305–313 (1949).

S4 Siegel, C. L., Über Riemanns Nachlaß zur analytischen Zahlentheorie. *Quellen Studien zur Geschichte der Math. Astron. und Phys. Abt. B: Studien* **2**, 45–80 (1932). (Also in "Gesammelte Abhandlungen," Vol. 1. Springer-Verlag, Berlin and New York, 1966.)

S5 Stieltjes, T. J., Sur le developpement de log $\Gamma(a)$. *J. Math. Pures Appl.* [4] **5**, 425–444 (1889). (Also in "Oeuvres.")

S6 Stieltjes, T. J., Letters in "Correspondance d'Hermite et de Stieltjes," 2 Vols. Gauthier-Villars, Paris, 1905. (Especially Letter No. 79 and No. 4 in the Appendix.)

S7 Stirling, J., "Methodus Differentialis Sive Tractatus de Summatione et Interpolatione Serierum Infinitarum." London, 1730.

T1 Tauber, A., Ein Satz aus der Theorie der unendlichen Reihen. *Monatsh. Math.* **8**, 273–277 (1897).

T2 Taylor, A. E., "Advanced Calculus." Ginn (Blaisdell), Boston, Massachusetts, 1955.

T3 Titchmarsh, E. C., "The Zeta-Function of Riemann," Cambridge Tracts in Math. and Math. Phys. No. 26. Cambridge Univ. Press, London and New York, 1930.

T4 Titchmarsh, E. C., On van der Corput's method and the zeta-function of Riemann (IV). *Quart. J. Math. Oxford Ser.* **5**, 98–105 (1934).

T5 Titchmarsh, E. C., The zeros of the Riemann zeta-function. *Proc. Roy. Soc. Ser. A* **151**, 234–255 (1935).

T6 Titchmarsh, E. C., The zeros of the Riemann zeta-function. *Proc. Roy. Soc. Ser. A* **157**, 261–263 (1936).

T7 Titchmarsh, E. C., "Introduction to the Theory of Fourier Integrals." Oxford Univ. Press, London and New York, 1937.

T8 Titchmarsh, E. C., "The Theory of the Riemann Zeta-Function." Oxford Univ. Press, London and New York, 1951.

T9 Turing, A. M., Some calculations of the Riemann zeta-function. *Proc. London Math. Soc.* [3] **3**, 99–117 (1953).

V1 de la Vallée Poussin, C.-J., Recherches analytiques sur la théorie des nombres (première partie). *Ann. Soc. Sci. Bruxelles* [I] $\mathbf{20}_2$, 183–256 (1896).

V2 de la Vallée Poussin, C.-J., Sur la fonction $\zeta(s)$ de Riemann et le nombre des nombres premiers inférieurs a une limite donnée. *Mém. Courronnés et Autres Mém. Publ. Acad. Roy. Sci., des Lettres Beaux-Arts Belg.* **59** (1899–1900).

W1 Walfisz, A., "Weylsche Exponentialsummen in der Neueren Zahlentheorie." VEB Deutcher Verlag der Wiss., Berlin, 1963.

W2 Weil, A., "Courbes Algébriques et Variétés Abéliennes." Hermann, Paris, 1948.

W3 Wiener, N., Tauberian theorems, *Ann. of Math.* **33**, 1–100 (1932).

W4 Weyl, H., Zur Abschätzung von $\zeta(1 + it)$. *Math. Z.* **10**, 88–101 (1921).

W5 Wirsing, E., Elementare Beweise des Primzahlsatzes mit Restglied I. *J. Reine Angew. Math* **211**, 205–214 (1962).

Index

A

ABEL
  integral formula for $\zeta(s)$, 9, 221
Abel's theorem, 278
Alpha notation for roots $\rho = \frac{1}{2} + i\alpha$, 36
Analytic continuation
  of a particular integral, 276–277
  Riemann's conception of, 9
  of zeta function, 11, 16n, 115
Approximate functional equation, *see* Hardy and Littlewood
Asymptotic expansions, 87, *see also* Euler–Maclaurin summation for evaluation of zeta, Logarithmic integral, Riemann–Siegel asymptotic formula, Stirling's series

B

BACKLUND, 97
  determination of $N(T)$ for $T = 200$, 128–129
  estimation of $N(T)$, 132–134
  theorem on Lindelöf hypothesis, 182, 188–190
Bernoulli numbers, 11, 103
  $B_{2n+1} = 0$, 11n, 14n, 103
Bernoulli polynomials $B_n(x)$, 100–103
  periodified $\bar{B}_n(x)$, 104
BOHR and LANDAU
  Riemann hypothesis implies S unbounded, 181, 201–202
  theorem on roots near Re $s = \frac{1}{2}$, 19, 193–195
BOMBIERI, 297

C

Cesaro averages, 279
CHEBYSHEV
  estimates of π from estimates of ψ, 68, 76–77
  estimation of $\pi(x)$, 3–4, 281–284
  mentioned by Riemann, 5
Chebyshev's identity, 281–284
Conditional convergence
  in Fourier analysis, 23
  of sums over ρ, 20–21, 30, 35, 49–50

D

DEBYE, 139
Definite integrals, *see* Riemann, evaluation of definite integrals
DENJOY, 268–269
DIRICHLET, 298, 299
  acquaintanceship with Chebyshev, 4
  use of Euler product formula, 7

E

ERDÖS, 282, 288
EUCLID, 1
EULER
  and factorial function, 7–8
  formula $\sum_{n=-\infty}^{\infty} x^n = 0$, 212
  and functional equation of zeta, 12
  ϕ-function, 250–251
  product formula, 1, 6–7, 22–23, 50n
  statement of $\sum \mu(n)/n = 0$, 92
  $\sum p^{-1}$ diverges, 1

Euler–Maclaurin summation, 97
for evaluation of zeta, 114–115
statement of method, 98–106
summary of method, 106
see also Stirling's series
Euler product formula, *see* Euler
Euler's constant, 67, 106n
Explicit formulas, *see* J function, psi function

F

Factorial function, 7–9, *see also* Stirling's series
Farey series, 263–264
related to Riemann hypothesis, 264–267
Fourier analysis, 23–25, 203–225
adjoint of an operator, 205
interpretation of Chebyshev's identity, 282–283
inversion formulas, 24, 27, 51, 54–56, 205, 213–215
transform of an (invariant) operator, 204
Fourier series of $\bar{B}_1(x)$, 196
Fourier transform, 209, 211, *see also* Fourier analysis, transform of an operator
FRANEL and LANDAU
Farey series and the Riemann hypothesis, 263–267

G

G function, 206
functional equation 209–210
Gamma function, *see* Factorial function
GAUSS, 299, 305
counts of primes, 2, 305
on density of primes, 2
notation Π for factorial, 8
GRAM
computation of 15 roots ρ, 96–97
Gram points, 125–126
Gram's law, 127, 171
exceptions, 126–127, 176n
statement of, 126

H

H function, 207
HADAMARD
proof of prime number theorem, 6, 38, 68
proof of product formula for ξ, 18, 21, 39
proof that $\zeta(1 + it) \neq 0$, 69–72
publication of three circles theorem, 187
HARDY, 136
infinitely many roots on Re $s = \frac{1}{2}$, 19, 226–229
HARDY and LITTLEWOOD
approximate functional equation, 201n, 229n
average of $|\zeta(s)|^2$ on Re = const, 195
estimate of $\zeta(\frac{1}{2} + it)$, 201
KT roots on Re $s = \frac{1}{2}$, 19, 226, 229–237
reformulation of Lindelöf hypothesis, 201
Tauberian theorems for Cesaro averages, 279
use of Tauberian theorem to prove prime number theorem, 280
HASELGROVE, 96, 121, 157, 161, 178
on exceptions to Gram's law, 126n
excerpts from tables, 122–123, 158
HILBERT, 6, 298
HUTCHINSON, 97
numerical analysis of roots ρ, 126–127 129–132

I

Ikehara's theorem, 281

J

J function, 22
Riemann's formula for, 33, 48, 61–65
in terms of $\pi(x)$, 33
JACOBI, 15
Jensen's theorem, 39–41

K

KUZMIN
proof of Riemann–Siegel integral formula, 273–278

L

LANDAU, 62, 62n, 136
  average of $|\zeta(s)|^2$ on Re = const, 195
  o, O notation, 200
  see also Bohr and Landau, Franel and Landau
LEGENDRE
  on density of primes, 3
  notation for gamma function, 8
Legendre relation for factorial function, 9
LEHMAN, 269
  verification of Riemann hypothesis to g_{250000}, 172
LEHMER, D. H.
  computations of roots ρ, 175–179
  verification of Riemann hypothesis to g_{25000}, 172
LEHMER, D. N.
  counts of primes, 3
  on Riemann's formula for $\pi(x)$, 35
Lehmer's phenomenon, 179
LEVINSON, 288
LINDELÖF
  estimates of growth of $\zeta(s)$, 182–186
Lindelöf hypothesis, 186, 177n, 188, 201
  as consequence of Riemann hypothesis, 188
Lindelöf's theorem, 184
  modified, 186
LITTLEWOOD
  improvement of $\beta < 1$, 200
  improvement of Bohr–Landau theorem, 195
  improvement of Tauber's theorem, 279
  $\int_0^T S(t)\,dt = O(\log T)$, 173
  $\pi(x) < \mathrm{Li}(x)$ fails, 269
  Riemann hypothesis and growth of *M*, 261
  use of three circles theorem, 187
  see also Hardy and Littlewood
Li(x), *see* Logarithmic integral
Logarithmic integral Li(x), 26
  asymptotic formula for, 86
  estimate of $\mathrm{Li}(x^\rho)$ as $x \to \infty$, 90
  value at x^β for complex β, 30

M

M function (sum of Möbius μ), 260
MACLAURIN, *see* Euler–Maclaurin summation
VON MANGOLDT
  estimate of $N(T)$, 173
  proof of formula for ψ, 50–61
  proof of Riemann's estimate of $N(T)$, 133
  proof of Riemann's formula for $J(x)$, 48, 61–65
  proof that $\sum \mu(n)/n = 0$, 92
  statement of formula for ψ, 49, 54, 66
MELLIN, 25n
  estimate of $\zeta(1 + it)$, 183
MERTENS
  proof that $\zeta(1 + it) \neq 0$, 79–80
Mertens's theorem, 6
Möbius inversion, 34, 217–218, 283, 285
Mu function of Lindelöf $\mu(\sigma)$, 186
Mu function of Möbius $\mu(n)$, 34, 91–92, 217

N

N function, 128
  Backlund's verification of Riemann's estimate, 132–134
  evaluated by Turing's method, 172–175
  Riemann's estimate, 18–19, 301
Nonsense, 212, 217

O

o, O notation, 200

P

Parseval's equation, 215–216
Pi function $\pi(x)$, 4, 33
  approximations to, 84–91
  in terms of $J(x)$, 34
  see also Prime number theorem
Poisson summation formula, 209–210
POLYA
  theorems on functions with zeros on Re $s = \frac{1}{2}$, 269–273

Prime number theorem, 4
  improved remainder, 84, 200
  proof, 68–77
Product formula for sine, 9, 18, 47, 224
Product formula for $\xi(s)$, 20–21
  proof, 46–47
Psi function $\psi(x)$ of Chebyshev, 49
  von Mangoldt's formula for, 49–61
  see also Chebyshev

R

Ramanujan's formula, 218–225
Rho, *see* Roots ρ
RIEMANN
  analytic functions treated globally, 20
  comments on $\pi(x) \sim \mathrm{Li}(x)$, 34–36, 269, 305
  computations of roots, 159–162
  error involving $\xi(0)$, 31
  estimate of $N(T)$, 18–19, 301
  evaluation of definite integrals, 12–13, 19, 26–33, 146–148
  "everywhere valid" formula for $\zeta(s)$, 9–11
  explicit formula for $J(x)$ $(= f(x))$, 33, 304
  introduction of function ξ, 16
  manuscript with statement of asymptotic formula, 156–157
  paper on $\pi(x)$, 1–38
  proofs of the functional equation, 12–16, 166–170, 274, 300–301
  questions unresolved by, 37–38
  skill as analyst, 136
  statement of Riemann hypothesis, 19, 30n, 301
  translation of paper on $\pi(x)$, 299–305
  use of "Fouriers theorem", 23–25, 302–303
  use of saddle point method, 139n
  view of analytic continuation, 9, 20
Riemann hypothesis, 6, 19
  implies Lindelöf hypothesis, 188
  in light of Riemann–Siegel formula, 164–166
  probabilistic interpretation, 268–269
  related to error in prime number theorem, 88–91
  related to Farey series, 263–267
  related to growth of $M(x)$, 260–263
  verified to $T = 200$, 129
  verified to $T = 300$, 129–132
  verified to $T = g_{1040}$, 171
  verified to $T = g_{25000}$, 172
  verified to $T = g_{250000}$, 172
  verified to $T = g_{3500000}$, 172, 179–180
Riemann–Siegel asymptotic formula, 136–164
  error estimates, 162–164
  manuscript, 156–157
  statement, 154
Riemann–Siegel integral formula, 137, 166–170
  proof by Kuzmin, 273–278
Roots ρ of $\xi(s) = 0$, 18–19
  computations of, 96, 157–162, 178
  crude estimate of density, 42–43
  von Mangoldt's estimate of density, 56–58
  real parts < 1, 70–72, 79–81, 200
  Riemann's estimate of density 21, 43, 302
  see also Conditional convergence, Bohr and Landau
Rosser's rule, 180–181
ROSSER, YOHE, and SHOENFELD
  error estimate for Riemann–Siegel formula, 163
  use of Turing's method, 175, 180
  verification of Riemann hypothesis to $T = g_{3500000}$, 179–180

S

S function $S(T)$, 173
  estimates of 174, 190–193, 201–202
Saddle point method, 139–140
SELBERG
  elementary proof of prime number theorem, 282, 288–297
  $KT \log T$ roots on $\mathrm{Re}\, s = \frac{1}{2}$, 19, 226, 237–259
Selberg's inequality, 284–288
Self-reciprocal operators, 210–211
SIEGEL
  discovery of Riemann–Siegel formula, 136

proof that Riemann–Siegel formula is asymptotic, 163
Slit plane, 111
Steepest descent, method of, 139–140
STIELTJES
alleged proof of Riemann hypothesis, 262–263
estimate of remainder in Stirling's series, 112
evaluation of the constant in Stirling's series, 113
STIRLING, 109
Stirling's formula, *see* Stirling's series
Stirling's series 106–114
for $\Pi'(s)/\Pi(s)$, 113
statement, 109
Stieltjes's estimate of the remainder, 112

T

TAUBER, 278
Tauberian theorems, 278–281
Theta function of Jacobi, 15, 170, 227
Theta function θ of Chebyshev, 76, 288
Theta function $\vartheta(t)$ in $\zeta(\frac{1}{2}+it) = Z(t) \exp i\,\vartheta(t)$, 119
asymptotic formula for, 120–121
Three circles theorem, 187
TITCHMARSH
average of $\zeta(\frac{1}{2}+ig_n)$ is 2, 229
error estimate for Riemann–Siegel formula, 162–163
as secondary source, 199, 226
verification of Riemann hypothesis to $T = g_{1040}$, 171
Turing's method of evaluating $N(T)$, 172–175

V

DE LA VALLÉE POUSSIN
estimate of error in prime number theorem, 78–95
proof of prime number theorem, 68
proof that $\sum \mu(n)/n = 0$, 91–95
Vinogradov's estimate of $\zeta(1+it)$, 200

W

WEIL, 298
Weyl's estimates of $\zeta(1+it)$, 200
Wiener's Tauberian theorem, 280–281
WIRSING, 288, 294, 297

X

Xi function $\xi(s)$, 16–18
rate of growth, 41
as transform of self-adjoint operator, 206–213

Z

Z function $Z(t)$, 119
asymptotic formula for, *see* Riemann–Siegel asymptotic formula
Zeta function
analytic continuation of, 9–11, 16n
evaluation by Euler–Maclaurin summation, 114–118
as Fourier transform of summation operator, 204
functional equation, 12–16, 222, 224–225, *see also* Euler, Riemann
growth related to zeros, 182, 188, 190, 193, 200
growth in strip $0 \leq \text{Re } s \leq 1$, 182–201
value of $\zeta'(0)/\zeta(0)$, 66–67, 134–135
values at integers, 11–12
values at negative integers, 216–217